THE ROYAL INSTITUTION
LIBRARY OF SCIENCE

(being the Friday Evening Discourses in Physical Sciences held at the Royal Institution: 1851–1939)

ASTRONOMY

Volume 1

THE ROYAL INSTITUTION
LIBRARY OF SCIENCE

(being the Friday Evening Discourses in Physical Sciences held at the Royal Institution: 1851–1939)

ASTRONOMY

Volume 1

EDITED BY

SIR BERNARD LOVELL

O.B.E., F.R.S., Ph.D.

Nuffield Radio Astronomy Laboratories,
Jodrell Bank, Macclesfield, Cheshire, England

ELSEVIER PUBLISHING COMPANY LTD

1970

ELSEVIER PUBLISHING COMPANY LIMITED
BARKING, ESSEX, ENGLAND

ELSEVIER PUBLISHING COMPANY
335 JAN VAN GALENSTRAAT, P.O. BOX 211, AMSTERDAM,
THE NETHERLANDS

AMERICAN ELSEVIER PUBLISHING COMPANY INC.
52 VANDERBILT AVENUE, NEW YORK N.Y. 10017

444 20102 5

LIBRARY OF CONGRESS CATALOG CARD NUMBER 75–132808

Printed photolitho in Great Britain by Page Bros. (Norwich) Ltd., Norwich and London

FOREWORD TO THE SERIES BY THE GENERAL EDITOR

Sir William Lawrence Bragg

The discourses at the Royal Institution are unique in character. On each Friday between October and June, a well-known authority is invited to give a general review of his subject, and he is asked not to assume any expert scientific knowledge on the part of his listeners. A tradition has grown up around the "Royal Institution Discourse" which is regarded as a special occasion calling for a talk distinguished by simplicity and clarity and, when appropriate, illustrated by interesting experiments and demonstrations. Most of the discourses are on scientific subjects, but the arts are also represented.

These talks present a broad survey of current scientific thought and achievement. They carry the authority of the famous men who have given them, and at the same time they are easy to understand. Their great interest provides the justification for their being published in the present form. No attempt has been made to select; all have been included because it is rare to find one lacking in interest today.

Since the number of discourses is large and they cover such a wide range, they have been grouped under main subject headings and the discourses under each of the following headings will be published as a separate series:

The Physical Sciences	(Physics and Chemistry)
Astronomy	(Astronomical Sciences)
The Earth Sciences	(Geology, Meteorology)
The Biological Sciences	(Botany, Zoology, Palaeontology)

After the above volumes have been published, it is anticipated that further volumes in the series will appear.

The series starts in the year 1851, because it was only then that regular publication of accounts or abstracts of the discourses began. The Royal Institution at this date had already been a major centre of research in Great Britain for fifty years, first under Humphry Davy and then under Michael Faraday and many famous men had worked there. A decision as to when to end the series must be arbitrary, but it was felt that the more recent discourses, while just as interesting, do not have the same attraction of catching the atmosphere of science in generations which have passed which is much of the charm of the earlier ones. This series was therefore brought to a close in 1939, just before World War II.

It is hoped that this series will be interesting to professional scientists, and especially to students of the History of Science in that it preserves so much fascinating material in a readily available form.

But it is also hoped that this collection will be welcome to all who are interested in science, since it was to them that the Discourses were originally addressed and for them that they were so carefully planned by the great masters in the past.

PREFACE

In the period covered by these Discourses, from 1851 to 1939, man's understanding of the universe was transformed. Today it is hard to believe that in 1909 Eddington's Discourse on 'Some recent results of astronomical research' was concerned largely with the discovery of a new satellite of the planet Jupiter and with the problems of comets. Soon, the epic researches with the 60-inch and 100-inch telescopes on Mt. Wilson were to reveal the vast extent of the Milky Way, the existence of the extra-galactic nebulae and the expansion of the universe. Then we find Eddington discoursing on the subject for which he is remembered in history—the 'Interiors of the Stars and the Expansion of the Universe.'

Today, when, with radio telescopes, it is possible to detect and measure the presence of water, the hydroxyl radical, formaldehyde, ammonia and carbon monoxide in interstellar space, it is salutary to be reminded through Huggins' Discourse of 1865 that a century ago the spectroscopy of stars was similarly a new subject. Only then was there observational proof that the common elements existing on earth and in the sun were common also to the stars.

When Nasmyth lectured in 1864 (with the Earl of Rosse as Chairman) he described the scene which would present itself to a spectator on the moon, concluding that 'When we seek to picture to ourselves the wild and unearthly scene that would thus be presented to our gaze, we must search for it in the recollection of some fearful dream'. Nasmyth's dream picture materialised 105 years later and it is indeed impressive to compare this with the astronauts' description of the scene.

This collection of the astronomical discourses makes fascinating reading. The historic figures of the past recreate themselves and their subject—Kapteyn, Milne, Eddington, Jeans and Slipher on the universe, Huggins on spectroscopy, A. S. Herschel on the great Leonid meteor showers, Lockyer and Schuster on solar eclipses, the Earl of Rosse on the moon's heat, and the succession of Astronomers' Royal from Airy to Spencer Jones all appear in these pages.

By our standards today the instrumental techniques available in that age were rudimentary; even so the observers achieved precision in their results. The discourses contain wonderful examples of intellectual penetration into the unknown from features of the universe so dimly observed. Above all, the discourses are masterpieces of exposition and for this alone they would be worthy of collection. In the event they are far more. They are invaluable historical documents providing a summary of the incredible progress in astronomical researches during those years and a glimpse into the workings of the mind of many of the legendary figures who were responsible.

BERNARD LOVELL

Jodrell Bank, June 1970

CONTENTS

1866

1867

1869

1870

1871

1872

1885

1886

1887

1889

1890

1891

CONTENTS FOR VOLUME 2

1933

1935

1936

Friday, May 2.

THE DUKE OF NORTHUMBERLAND, President, in the Chair.

THE ASTRONOMER ROYAL,

On the Total Solar Eclipse of 1851, *July* 28.

THE Lecturer remarked that the subject which he had suggested to the Managers of the Institution for the present Lecture might at first sight appear meagre and common-place, but that he believed it would be found to be one of the highest interest: first, because during a total eclipse we are permitted a hasty glance at some of the secrets of nature which cannot be seen on any other occasion: secondly, because the general phænomenon is perhaps the most awfully grand which man can witness. Many of his audience had probably seen large partial eclipses of the sun, and they might suppose that a total eclipse is merely an intensified form of a partial eclipse; but, having himself witnessed a total eclipse, he was able to assure them that no degree of partial eclipse up to the last moment of the sun's appearance gave the least idea of a total eclipse, as regarded either the generally terrific appearances, or the singular nature of some of the phænomena. Many years ago, in reading the admirable essay in the Philosophical Transactions by the late Mr. Baily on the eclipse (usually called that of Thales), the occurrence of which suspended a battle between the Lydians and the Medes, he had been struck by the cogency of Mr. Baily's arguments, which showed that only a *total* eclipse could be admitted as sufficient to produce the effect ascribed to it; and by the remark (cited by Mr. Baily) of Maclaurin and Lemonnier, that in an annular eclipse of the sun, even educated astronomers when viewing the sun (nearly covered by the moon,) with the naked eye could not tell that it was not full. The appearances, however, in a total eclipse, as he should afterwards mention, were so striking, that there could be no difficulty in believing the historian's account to be literally correct.

Proceeding first to explain the simple causes of a solar eclipse, the Lecturer remarked that the moon's distance from the earth is nearly one four-hundredth part of the sun's distance, and that the moon's diameter is very nearly one four-hundredth part of the sun's diameter, and that therefore, on the average, the sun's *apparent* diameter and the moon's *apparent* diameter are very nearly equal. But in consequence of the elliptic forms of their orbits, the sun's distance is liable to small variations, and the moon's distance to very considerable variations: when the moon is at the most distant part of her orbit, her *apparent* diameter is smaller than the sun's, and if she happens at that time to be between a spectator and the sun, she will be seen as a black disk covering the central part of the sun and leaving a ring of light all round: when the moon is at the nearest part of her orbit, her *apparent* diameter is larger than the sun's, and she will, to a spectator in the proper locality, completely cover the sun, and produce a Total Eclipse. But neither of these things can happen unless the plane of the moon's orbit be in such a position that the moon, when approaching the state of conjunction or new moon, is seen to pass not above the sun or below the sun but over the sun.

The Lecturer then called attention to the circumstance that four successive total eclipses occur in the month of July at intervals of nine years, namely 1833, July 17; 1842, July 8; 1851, July 28; and 1860, July 18. For the explanation of this curious circumstance it was necessary to show, first, how it happened that at intervals of nine years the moon's orbit was in such a position that, for a nearly definite apparent position of the sun, the moon's path would cross the sun's disk: secondly, how it happened that at intervals of nine years the moon was at nearly her smallest distance from the earth, so that her apparent diameter was larger than the sun's. In reference to the former, it was shown that the moon revolves in an orbit whose plane is inclined to the plane of the ecliptic (the *apparent* orbit of the sun round the earth), and that the inclination is nearly invariable, but that the position of the line in which the plane of the moon's orbit intersects that of the ecliptic is constantly changing, revolving steadily in the direction opposite to the moon's motion, and performing a complete revolution in something more than nineteen years. Therefore if one node or extremity of this line of intersection were directed nearly to the July sun in 1833, the opposite node would be directed nearly to the July sun in 1842, and so on for four successive periods of nine years; and eclipses would be possible in July at the end of each period. But to show that they might be total eclipses, it was necessary to remark that the moon revolves in an ellipse of which the earth occupies one focus (a point much nearer to one end than to the other) and that the position of this ellipse is constantly varying, its long axis turning round in the same direction as the moon's motion, and completing a revolution in nine years and a half. Therefore if in 1833 the shorter end of the ellipse were nearly turned to the July sun, in 1842 the axis of the ellipse would have completely revolved, so that the shorter end of the ellipse would again be nearly turned to the July sun: and thus the eclipse which occurred, if total in 1833, would, if central, be total (not annular) in 1842; and so on for four periods of nine years.

The Lecturer then called attention to the great difference in the directions of the shadow-paths across Europe, for the eclipses of 1842 and 1851 : (the former being from W. S. W. to E. N. E. nearly, the latter from N. W. to S. E. nearly). This arose in part from the circumstance that (as above explained) the former of these eclipses occurred when the node or end of the intersection-line of the planes of orbits, turned towards the July sun, was that at which the moon rises to the north of the ecliptic, the latter when it is that at which the moon is descending to the south of the ecliptic. But the principal cause of the difference is this ; that the former eclipse occurred early in the morning, the latter in the afternoon : on placing a terrestrial globe in the proper position for July, with its north pole inclined considerably towards the sun, it is seen that, even if the moon moved precisely in the ecliptic, the path of her shadow across Europe before Europe came to the meridian would trend from the south to the north ; but if Europe had passed the meridian it would trend from the north to the south.

Quitting the geometrical explanations, the Lecturer then proceeded to describe some peculiar phænomena which had been observed in eclipses, and first, one which had been observed most distinctly in annular eclipses, and which is known by the name of "Baily's beads and strings." When the preceding limb of the moon, traversing the sun's disk, approaches very near the sun's limb, or when the following limb of the moon is in the act of separating from the sun's limb to enter on the sun's disk, the two limbs are joined for a time — (no one has estimated the duration with accuracy) — by alternations of black and white points or strings. Phænomena, evidently of the same class, have been observed in the transits of Venus and Mercury over the sun's disk ; the black planet, when just lodged on the sun's disk, being pear-shaped, with its point attached to the black sky. The Lecturer was able to state, in his own experience at the Royal Observatory, that at the same transit of Mercury this phænomenon was seen with some telescopes and was not seen with others. In the annular eclipse of 1836 observed at Königsberg, where the moon's limb but just entered completely on the sun's, and where consequently it grazed along the sun's for many seconds of time, the phænomenon appeared to resolve itself simply into points of light seen between lunar mountains. The Lecturer expressed himself generally satisfied with Professor Powell's explanation, that the phænomenon originates in that inevitable fault of telescopes and of the nervous system of the eye which tends to extend the images of luminous objects (producing what is generally termed irradiation), and thus enlarges the sun's disk towards the sky, towards the moon or planet, and towards the bottoms of its hollows.

In describing the total eclipse of 1842 (which perhaps was better observed than any one preceding it) the Lecturer insisted on our obligation to M. Arago, who had prepared the preliminary notices, and had used his powerful personal influence in inducing persons to make observations at numerous stations in the south of France; and had afterwards collected and compared the observations. Besides these

French observations, and the observations made by astronomers officially located in the path of the shadow, we have the observations of M. Schumacher who went to Vienna, of MM. Otto Struve and Schidlowsky at Lipetsk, (the former of whom was sent expressly by the Russian government,) of Mr. Baily who went to Pavia, and of the Lecturer himself who went to the Superga (near Turin).

It appears that, with M. Arago's telescope, the whole circumference of the moon was visible when the moon had entered on only about two-thirds of the sun's diameter. Whatever may be the cause of this unusual appearance, it seems to require the use of a telescope with a small number of glasses in the highest state of polish.

As the totality approached, a strange fluctuation of light was seen by M. Arago and others upon the walls and the ground, so striking that in some places children ran after it and tried to catch it with their hands.

Of the awful effect of the totality, and of the suddenness with which it came on, it is difficult to give an idea. The Lecturer cited an expression from Dr. Stukely's account of the total eclipse of 1744, observed on a cloudy day, "that the darkness came dropping like a mantle:" and compared it with his own, in similar weather, "that the clouds seemed to be descending." But all agree in the description of livid countenances, indistinct and sometimes invisible horizon, and general horror of appearance. It is well that we are enabled, by means of instances collected by M. Arago, to show that these are not simply the inventions of active human imaginations. In one case, a half-starved dog, who was voraciously devouring some food, dropped it from his mouth when the darkness came on. In another, a swarm of ants, who were busily carrying their burdens, stopped when the darkness came on and remained motionless till the light reappeared. In another, a herd of oxen, as soon as the totality was formed, collected themselves into a circle and stood with their horns outwards. Some plants (as the convolvulus and silk-tree acacia) closed their leaves.

The darkness at Venice was so great that the smoke of the steamboats could not be seen. In several places, birds flew against houses, &c. Where the sky was clear, several stars were seen. In several places a reddish light was seen near the horizon. A heavy dew was formed at Perpignan.

The Lecturer cited an instance which had been related to him by M. Arago, in which the Captain of a French ship had beforehand arranged in the most careful way the observations to be made: but, when the darkness came on, discipline of every kind failed, every person's attention being irresistibly attracted to the striking appearances of the moment, and some of the most critical observations were thus lost.

The most remarkable phænomenon observed in all preceding total eclipses, and seen equally in this, is the ring of light surrounding the moon, called the *corona*. The Lecturer described the magical change, from the state of a very narrow lune of solar light (the contour of the moon being totally invisible,) to the state of an entire dark moon surrounded by a ring of faint light, as most curious and striking. The progress of the formation of the ring was seen by his companion, and by some other persons: it commenced on the side of the moon opposite to that at which the sun disappeared. In the general decay and disease which seemed to oppress all nature, the moon and corona appeared almost like a local disease in that part of the sky. In some places, the corona was seen as distinctly double; it would appear that the ring which the Lecturer saw (whose breadth, by estimate of repeated duplication, he found to be about one-eighth part of the moon's diameter, or four minutes of arc nearly) was the inner of the two rings seen by M. Arago and others. The texture of the corona appeared in some places as if fibrous, or composed of entangled thread; in some places, brushes or feathers of light proceeded from it. One photometric estimate of the quantity of light in the corona, cited by M. Arago, gave it equal to one-seventh part of full moonlight. From a chromatic analysis of its light by means of an ordinary prism, it appeared to be deficient in green rays.

The Lecturer characterised the inquiry into the origin and locality of this corona as one of the most interesting connected with the eclipse. It had been specially indicated by M. Arago (see the Annuaire du Bureau des Longitudes, 1842) as a very important subject of inquiry whether the corona is concentric with the moon or with the sun; but his recommendation had received very limited attention. The general tenor of the evidence went to prove that the corona belongs to the sun. This, however, was not the opinion of more ancient writers, who tacitly consider it as the atmosphere of the moon.

But the most remarkable of all the appearances were the red mountains or flames apparently projecting from the circumference of the moon into the inner ring of the corona, to the height of one minute of arc at the smallest estimation, or a much greater height by other estimations. It was afterwards discovered that these had been seen before by Vassenius, a Swedish astronomer, who observed the eclipse of 1733 at Göteborg, (a place very favourable for the approaching eclipse), and whose account is given in the Philosophical Transactions, vol. xxxviii. He terms them "subrubicundæ nonnullæ maculæ, extra peripheriam disci lunaris conspectæ, numero tres aut quatuor." This observation, however, was not known to any of the observers in 1842, and all were therefore taken by surprise. Drawings were exhibited of these red mountains as seen at Perpignan, Narbonne, Vienna, Pavia, Superga, and Lipetsk. It was shown that, by a trace still visible on the engraving, the drawing first made at Vienna had coincided very exactly with that made at Pavia; that the Narbonne observations would be very exactly reconciled with them by supposing the error (very likely to occur to unpractised astronomers) of taking the north limb to be the upper limb; that at Perpignan, Superga, Lipetsk, the lowest of the red prominences was not seen: and that at Superga and Lipetsk only was the middle one of the upper prominences seen, though in several places an irregular band of red light had been seen of which one salient point might be the prominence in question. In all the places where the order of formation had been observed, the same prominence (the left-hand upper prominence) was defined as the first seen. At Perpignan this was observed by M. Mauvais to show itself first as a small point and to project gradually as from behind the moon. The discordance in these representations did not appear to the Lecturer at all startling; it was not greater than the discordance in the accounts given by two good observers in different rooms of the same building at Padua.

The determination of the locality and nature of these red prominences is one of the most difficult of all connected with the eclipse. The first impression undoubtedly was that they are parts of the sun. If so, their height, at the lowest estimation, is about thirty thousand miles. The principal objection, however, to their solar location is the difference in their forms as seen at different places: thus at Perpignan they are represented as widest at the top: at all other places they are widest at the base. Moreover at some places, as Pavia and Vienna, where they were seen a long time, they underwent no change: whereas at Perpignan one at least was seen to slide out as from behind the moon. In all cases, however, much is to be allowed for the hurried nature of the observation.

The only theory which has been formally propounded as explaining them is that of M. Faye, who conceives them to be the result of a kind of mirage.

The Lecturer explained the nature of ordinary mirage (the kind of reflection produced by the hot air adhering to a heated surface of any solid) and described the distortion produced in the image of a star as seen in the Northumberland telescope of the Cambridge Observatory, when first mounted in a square pyramidal tube, whose angles were constructed more solidly than its sides, reducing the inner form to an octagon. When this tube had become warm before observation in the open air, the angle-blocks remained warm after the sides and the internal air had become cool, and a kind of mirage was produced which distorted the image of a star into four long rays like the sails of a windmill. M. Faye has particularly adverted to this instance, and conceived that in the circumstances of our atmosphere at the time of the eclipse, where the air on one side only of the path of light is somewhat heated by the sun, sufficient explanation might be found for the distortion of some inequalities of the moon. The Lecturer professed himself totally unable to follow this theory into details, remarking only that in the rapid passage of the moon's shadow he conceived it impossible to find air in the state required for the explanation.

The Lecturer then adverted to that part of his subject of which all that had been already said was only introductory, namely the approaching eclipse of July 28. After quoting an American newspaper, showing the great interest excited by this eclipse beyond the Atlantic as one of the strongest inducements for Americans to visit Europe in the coming summer, he invited attention to its course across Europe. Entering Norway near Bergen, the shadow crosses both coasts of Norway, both coasts of Sweden, and the eastern coast of the Baltic: then ranges through Poland and the south frontier of Russia across the sea of Azof through Georgia to the Caspian Sea. It passes Christiania, Göteborg, Carlscrona, Danzig, Königsberg, Warsaw, and Tiflis. A great part of this course, especially that from Bergen to Königsberg, is very accessible by sea, and Warsaw by land. The Lecturer trusted that many English travellers might be induced to observe this eclipse. If possible, stations should be chosen as well near the northern and southern boundaries of the shadow as near the centre. No particular skill in astronomical observation is required, the phænomena being rather of a more generally physical kind: and indeed, as far as the observations of the eclipse of 1842 showed, the travelling physicists had been more successful than the stationary astronomers. The apparatus required would depend on the special objects of the observer; a telescope and a watch might be considered indispensable in every case: for analysis of light, a common prism and a polariscope might be taken by some persons: photometry, actinometry, &c., might be interesting to others, and appropriate instruments would be required: other observers would be interested in meteorology. The apparatus which the Lecturer considered it most important to perfectionate now, for use during the eclipse, is photogenic apparatus; it would be impossible to set too high a value on a series of Daguerréotypes or Talbotypes of the sun and corona taken during the eclipse.

The Lecturer concluded by saying that a series of suggestions for the observation, accompanied by a map, had been prepared by a Committee of which he is a member, and were nearly ready to leave the printer's hands: and he undertook to transmit a copy of these suggestions to any person who would make application to him.

G. B. A.

Friday, February 4.

W. R. Grove, Esq. M.A. F.R.S. Vice-President, in the Chair.

G. B. Airy, Esq. F.R.S. Astronomer Royal.

On the results of recent calculations on the Eclipse of Thales and Eclipses connected with it.

The Lecturer commenced by remarking that he should not have thought the calculations connected with any other eclipse a subject worthy of his audience. The eclipse commonly called that of Thales is however one of extraordinary interest. It refers to a point of time which connects in a remarkable way the history of Asia Minor and the Greek colonies settled there with the history of the great Eastern empires. Its precise date has been for a long time a subject of discussion among the ablest astronomical computers and chronologers. It shews in a remarkable degree the power of astronomy; for it is no small thing that we are able to go back so many centuries and confidently to describe a phænomenon which then occurred, almost to its minutest features. But it shews also the weakness of astronomy. It requires the combination of theory and observation, with a full sense of the possible inaccuracies of both, and with an endeavour by the use of each to correct the failings of the other. It requires general criticism, history, tradition, and a careful examination of geographical and military circumstances. But when all these aids are properly brought to bear upon it, a conclusion is obtained upon which there appears to be no room for further doubt.

In the last century, the computations, or rather the assumptions, of distant eclipses, were extremely vague. The theory of the moon's motion, as applicable to distant eclipses, was imperfect; and it would almost seem that computers, under a sense of this imperfection, felt themselves free to interpret the calculations as loosely as they might find convenient. Eclipses were adopted by them, as corresponding to historical accounts, which did not represent the physical phænomena when visible: some were even taken which occurred before sunrise or after sunset at the places of observation.

The great step made in theory, in reference to these inquiries, was the discovery made by Laplace near the end of the last century, of the secular change in the moon's mean motion in longitude (accompanied by similar changes in the motion of the perigee and the node). In explanation of this, the Lecturer pointed out that the force which acts upon the moon tending to draw it towards the earth is not simply the attraction of the earth, but consists of that attraction diminished by a disturbing force which is produced by the sun's attraction. The sun sometimes attracts the moon towards the earth or the earth towards the moon, sometimes it produces the opposite effect; but on the whole it tends to pull the moon away from the earth. And this diminution of the earth's attraction is greater as the sun is nearer; and therefore, in an elliptic orbit such as the earth describes about the sun (or such as the sun *appears* to describe about the earth), the diminution of the earth's attraction is greater when the earth is nearest to the sun than when the earth is farthest from the sun. It might be supposed that one of these effects exceeds that which would happen when the earth is at its mean distance from the sun, as much as the other falls short of it; but in reality the excess is greater than the deficiency, and therefore the more excentric the earth's orbit is, the greater is this disturbing force. So long as these circumstances remain the same, the magnitude of the moon's orbit will not be sensibly altered. But the fact is, that, in consequence of the perturbations produced by the planets, though the earth's mean distance from the sun remains unaltered from age to age, yet the excentricity of its orbit is diminishing from age to age; the sun's disturbing force is therefore diminishing from age to age: and the real force which acts upon the moon as tending to draw it towards the earth is therefore increasing from age to age; and, from age to age, the moon approaches a little nearer to the earth and performs her revolutions a little quicker. This effect is extremely small. Between one lunation and the next (taken one with another) the moon's distance from the earth is diminished by about $\frac{1}{14}$ of an inch; it would seem at first that this could produce no discoverable effect in the moon's motion: but one of the most wonderful things in the application of the laws of mechanics generally, and the law of gravitation in particular (where the magnitude of the force varies with the variation of distance), is, that the effect of a variation of a small fraction of an inch is as certain, in proportion to its magnitude, as that of a thousand miles. Still the effect produced in the moon's apparent motion is very small: in a century it amounts to only ten seconds; an angle which, when expressed in the usual way by the breadth of a known object as seen at a known distance, is less than the angle subtended by the human hand as seen at the distance of a mile. Yet in the course of twenty-four centuries the effect of this becomes so important as, in the case of eclipses, completely to change the face of the heavens; an eclipse might happen in Asia or Africa which, but for this consideration, we might expect to occur at that time in America.

Shortly after the discovery of this secular change, the French lunar tables (Bürg's) were constructed, the first which introduced this element. The late Mr. Francis Baily soon made use of these in an investigation of the date of the eclipse of Thales, which deserves to be ranked among the most valuable that has been directed to that subject. The historical account of the eclipse is, that the Medes attacked the Lydians, and that a war continued several years, until at length, when the two armies were preparing for battle, the day suddenly became night (an event which Thales is said to have predicted), and both parties were so much alarmed that they made peace at once. Mr. Baily in the first place pointed out, from a collation of the best accounts of total and annular and other partial eclipses, that nothing but a total eclipse could produce such a striking effect and that a total eclipse could do it. Mr. Baily afterwards saw the total eclipse of 1842, but he saw it from the window of a house: the Lecturer, who had seen the total eclipses of 1842 and 1851, in each case from the top of a hill and in command of the open country, wished much that Mr. Baily could so have seen it, when he could not have failed to be reminded of his former assertions with regard to the eclipse of Thales: the phænomenon in fact is one of the most terrible that man can witness, and no degree of partial eclipses gives any idea of its horror. Mr. Baily then, using Bürg's tables, computed all the eclipses which could by possibility be visible in Asia Minor through a period of time exceeding that to which the eclipse of Thales is limited by chronological considerations, and found that only the eclipse of B. C. 610, September 30, could be total; and that the track of its shadow would pass across the mouth of the river Halys. He accordingly fixed upon that as the true date. But he then made a calculation which threw great doubt upon the result. Upon using the same tables to compute the eclipse of Agathocles (to be described shortly) he found that the track of the shadow would be nearly 200 miles in error; and, with a degree of good faith which was characteristic of him, he at once avowed his belief that if the elements of the tables were so altered as to make the eclipse of Agathocles possible, the eclipse of B. C. 610 could no longer be shewn to be total in or near Asia Minor. He expressed his confidence however that no other eclipse could, under any possible change of the tables, have been total in Asia Minor. Mr. Baily's conduct in this avowal was favorably contrasted with that of a German astronomer, Oltmanns; who, in one paper, using the same tables as Baily, fixed upon the same date as Baily for the eclipse of Thales; and in another paper, after alteration of the elements, shewed that the eclipse of Agathocles was possible; but, although he then alluded to his own calculations of the eclipse of Thales, had not the courage to announce that his former conclusions must now be considered to be unfounded.

The Lecturer then proceeded to explain how it happens that there exists such a connexion between two eclipses nearly 300 years apart, that the errors of calculation of one can have any influence upon the other. He explained that the moon's orbit is inclined to the sun's apparent orbit round the earth, but not always in the same direction, the line of nodes (or the intersection of the planes of the two orbits) revolving so as to complete a revolution in about 19⅓ years; and that an eclipse of the sun can happen only when the line of nodes is turned nearly towards the sun (as, in other cases, the shadow falls above or below the earth). If for a given day of the year, (when the sun is in one certain position), the moon is in that part of its orbit most nearly in the direction of the sun, the shadow of the moon will fall upon a certain point of the earth; but now if the place of the node be changed, the effect will be that of driving a wedge under the moon, and she will be thrown further north or south, and the shadow upon the earth will be thrown further north or south. Thus the place of the node will define the part of the earth on which the shadow will be thrown; and conversely, a knowledge of the part of the earth on which the shadow is thrown will give information on the place of the node. Thus the alteration of the lunar elements, which is necessary to throw the shadow further north in the eclipse of Agathocles, consists in an alteration of the place of the node (other elements being supposed moderately correct); and this requires an alteration in the annual motion of the node, reckoning backwards from the present time when the position of the node is well known; and applying the same annual correction by the rule of three backwards to the place of the node at an assumed time of the eclipse of Thales, the corrected place of the node at that time is found, and then the corrected track of the moon's shadow can be found.

Subsequently to the time of the calculations of Baily and Oltmanns, the improvements in astronomy have been very great. Many advances have been made in theory, and one of the secular changes (that of motion of perigee) has been greatly modified. The Greenwich Lunar Observations from 1750 to 1830 (which are the foundation of Lunar Astronomy) have been completely reduced, on one uniform plan. Improvements have been made in the details of construction, but still greater improvements in the principles, of astronomical instruments. Our knowledge, also, of the geography of the countries to which the eclipses before us have relation, is much more accurate and extensive than it was.

Still there remain causes of uncertainty in the results of any calculations made for such distant periods.

First, in the theory. No person who has not fairly entered into the details of the Lunar Theory can conceive the complexity of the algebraical expressions and the operations which occur in it. Besides the usual chances of error from mistake of figures and mistake of signs, there is the risk of mistake in the selection of some terms to the exclusion of others, and the possibility of positive error in the metaphysical reasoning which guides some of the operations. And we are driven at last to admit that what is sometimes called mathematical evidence is after all but moral evidence. And thus it has happened that the conclusions of different theorists on some very important points are by no means accordant.

Secondly, in the observations from which are determined the elements that are to be combined with theory. Upon the same principle by which it was shewn that the track of shadow in one eclipse depends upon the track of shadow in another eclipse, it will be easily seen that the track of shadow in a distant eclipse will depend upon the observed elevation of the moon in the beginning of the modern period of comparatively accurate astronomy; (for that elevation determines the place of the node; and an error in the elevation produces an error in the computed place of the node for that time; and this exhibits an error in the annual motion of the node; and that error carried through the long period to a distant eclipse produces a very great error in the place of the node there, and consequently in the track of the shadow). If a ladder of centuries be constructed, each stave corresponding to a century, the extent of tolerably accurate and well-reduced observations of the moon (1750 to 1830) is represented by only $\frac{4}{5}$ of an interval of staves. Thus it appears that an error of two seconds in Bradley's observations (the angle which a finger-ring subtends at the distance of a mile, and which is smaller than can be perceived by the unassisted eye) would destroy our conclusions with regard to the distant eclipses in question. The fault in the principle of the Greenwich instrument used for observing the elevation of the moon (namely a quadrant, the use of which was for many years the bane of astronomy), and the slovenly way of using it in Bradley's time (no attention being given to the taking the elevation of the moon at the precise instant of her passing the meridian, though her elevation then changes rapidly) might well allow of this error. The Lecturer stated that both in the careful examination of the principles on which instruments are constructed, and in the rigorous attention to the proper rules for their use, it might be hoped that great improvement would be found in modern times.

In consequence of these causes of uncertainty, it becomes very desirable, in the investigation of the eclipse of Thales, to correct the elements of the moon's motions by some other well determined eclipse. Omitting the eclipses since the year 1200 A.D., and two in the second century B.C. which are somewhat discordant, there are two eclipses of peculiar value. One is the eclipse at the battle of Stiklastad at which Olaf king of Norway was killed, A.D 1030, August 31; in which the precise spot is known, and the precise position of the moon is known (the breadth of the shadow being very small, inasmuch as, when the eclipse commenced on the earth, it was annular). The only objection is, that if there is any uncertainty in the secular change of mean motions, the adjustment of the mean motions to represent the eclipse of Stiklastad will still leave a large uncertainty on an eclipse about 1600 years before it. Using the illustration of the ladder of centuries, it is like fixing the ladder at the bottom and at a point at one third of its height, which fastening, if the ladder is bent in some uncertain degree, still leaves great uncertainty in the place of its top. The other eclipse is that of Agathocles, B.C. 310, August 15, which will leave little uncertainty of that kind, if we can but determine its exact place upon the earth.

Agathocles, the Lecturer stated, being blockaded by the Carthaginians in Syracuse, placed men on board a fleet, ready to escape on the first opportunity; the approach of a provision-fleet drew off the Carthaginian ships, and Agathocles burst out of the harbour, and was pursued by the Carthaginians, but escaped. The next morning there was an eclipse of the sun which was evidently total. After six days he landed in the Carthaginian territories at a place called the Quarries, and, traversing their provinces, reduced the citizens of Carthage to the utmost difficulty, (in their terror they sacrificed 500 children to their god Kronos). The Lecturer acknowledged his obligations to Capt. W. H. Smyth, R.N. who had called his attention to the enormous quarries at Alhowareah (Aquilaria) under Cape Bon, from which Utica and Carthage were built; which place appears to have been used by the Romans as a landing-place from Sicily; and which the Lecturer adopted without doubt as the landing-place of Agathocles. He then stated that from J. W. Bosanquet Esq., he had received the suggestion that Agathocles might have passed the Strait of Messina; and that gentleman had pointed out the passages in the historical accounts which indicated the belief of the sailors that they were going either to Italy or to Sardinia. The Lecturer stated that, on minute examination, he had found that only the city of Gela remained in alliance with Syracuse, and the provision-fleet must have come from Gela, and must have approached Syracuse from the south, and from this it followed that Agathocles must have escaped to the north. This brings the probable position of Agathocles at the time of the eclipse near to Messina; if it were still supposed (as had been formerly supposed) that he sailed to the south, his position would probably have been near to Cape Passaro. The Lecturer explained the small corrections which must be made in the Greenwich determination of the place of the moon's node to satisfy these two conditions: and these were then taken as bases for the investigations connected with the eclipse of Thales.

The armies which were confronted at the time of the eclipse of Thales were evidently large armies (from the circumstances that they were commanded by the kings in person, who were ready to make a treaty on the spot, and that their principal allies, Syennesis and Labynetus, were present). And the principal question to be answered is, where such armies were likely to march. The Lecturer called attention to the general form of Mount Taurus and Anti-taurus (as one part is sometimes called), ranging from the mountains in the south of Asia Minor, in a general north-eastern direction, till they joined the mountains about Trebizond and Erzerum; and stated that, according to the best information that he could obtain, (in which he had been materially assisted by W. J. Hamilton, Esq. and M. Pierre Tchihacheff) the following were the principal roads through them. On the north coast there is one, of which the difficulties were well known from the retreat of the ten thousand Greeks. From Erzerum there are two roads towards Siwah (Sebaste) and Kaisarieh (the Cappadocian Cæsarea), rugged, and passing through barren countries. There is one road from Kaisarieh falling on a branch of the Euphrates, which flows by Malatieh (Melitene); a rugged road parallel to it from Guroun; and finally, the road which is the best of all, descending from the southern mountains into the plain of Tarsus and Adana, then skirting the sea by Issus to Antioch. The Lecturer stated that on examining history he found no instance of an easterly or westerly march through the northern mountains: he had found one march of an army (under the Byzantine emperor Heraclius) from Trebizond to the south, which army however returned by Issus: one march by Melitene, where the last great battle of Chosroes Nushirvan with the Byzantine armies was fought: but, from the time of the younger Cyrus and Alexander, the marches by Issus are very numerous. Some of these lines of march are evidently very much curved out of their straight direction in order to take advantage of the pass of Issus: thus Alexander marched thither from Angora (Ancyra): Valerian entered by Issus to attack Sapor: Sapor, when in Armenia, and on his way to attack Cæsarea, marched by Issus: Julian in return invaded Persia by the same road. From these circumstances it appeared most probable that the Medes entered by Issus to attack the Lydians, and that the battle-field would probably be included in the polygon whose angles are Issus, Melitene, Ancyra, Sardes, and Iconium.

The Lecturer then shewed what would be the track of the shadow in the eclipse of B. C. 585, May 28, either on the supposition that the place of the Moon's node was that given by the Greenwich observations, or on the supposition that the motion of the node was so corrected as to make the shadow in the eclipse of Agathocles pass centrally over the assumed southern position of Agathocles, or over the assumed northern position of Agathocles. The uncorrected Greenwich track, and the track over Asia Minor corresponding to a central eclipse on the southern position of Agathocles, though not inadmissible, are too far south to be accepted as probable; but the track over Asia Minor, corresponding to the elements which give a central or nearly central eclipse for the northern position of Agathocles (near the strait of Messina), would certainly pass over any probable position of the battle-field.

The conclusion as to the general fitness of the eclipse of B. C. 585 for representing the circumstances of the eclipse of Thales, by inference from modern elements of calculation, was first published by Mr. Hind in the Athenæum.

The Lecturer then stated that he had examined in greater or less detail every eclipse from B. C. 630 to B. C. 580, and that no other eclipse could pass over Asia Minor. That selected by Messrs. Baily and Oltmanns, it was now shewn, passed to the north of the sea of Azof.

In concluding this astronomical discussion, the Lecturer expressed his opinion that the date B. C. 585 was now established for the eclipse of Thales beyond the possibility of a doubt.

The Lecturer then alluded to the tradition preserved by Sir John Malcolm from the poetical History of Persia, that Kai Kaoos (whom Sir John Malcolm considers to be the same as Cyaxares or Astyages, or possibly to represent both), having marched on a military expedition into Mazenderam, himself and his army were struck with sudden blindness; and that this had been foretold by a magician. Sir John Malcolm considered, and it appears most probable, that this is the record of a total eclipse of the sun; but no total eclipse near this time passed over Mazenderam. The Lecturer conceived therefore that it might refer to the eclipse of Thales, though with a strange perversion of the name of the province. Such perversions however occur in the Persian poetical history with regard to other names, which there is reason to believe are correctly given by the Greeks. The name Xerxes, for instance, has been found by Colonel Rawlinson in the Behistun inscriptions under the form *Khshayarsha,* of which the Greek Xerxes was probably a fair oral representation; whereas the name preserved in the poetical history is Isfundear. These confusions however are incidental to poetical history: thus if the Henriade of Voltaire should remain as the only history of the times to which it relates, the name of the king who preceded Henri IV. would go down as Valois, instead of Henri III. [G. B. A.]

Friday, March 5.

SIR HENRY HOLLAND, BART. M.D. F.R.S. Vice-President, in the Chair.

PROFESSOR C. PIAZZI SMYTH,

ASTRONOMER-ROYAL FOR SCOTLAND.

Account of the Astronomical Experiment of 1856, *on the Peak of Teneriffe.*

THE object proposed in the astronomical experiment carried out on the Peak of Teneriffe, the year before last, was to ascertain how much telescopic observation can be improved, by eliminating the lower third or fourth part of the atmosphere; in other words, by elevating instruments and observers some 10,000 feet above the sea level.

By such a proceeding, not only was it hoped to rise above the greater part of the clouds, mist, haze, and other aerial impurities, so patent to the eyes of all men,—but also to get rid of certain other sources of optical disturbance, which only manifest themselves to star-gazers, employing telescopes with high magnifying powers. To such, those disturbances are rarely or never absent, even on the finest nights; and they act more and more prejudicially, the larger and more perfect the instruments that may be employed. From this cause it is, that the full magnifying power which a modern telescope is calculated to bear, can very seldom be applied; minute celestial phenomena remain undiscovered; and our opticians must lose heart in carrying on the improvement of object-glass manufacture, if their best performances are to be for ever condemned to have their finer qualities neutralized by the badness of the air in which they are tested.

Here is evidently an evil of no ordinary magnitude, and it has troubled astronomers long. Clearly foreseen and carefully weighed by Sir Isaac Newton, that great philosopher described the nature of the difficulty, and the only means of remedy in 1730, in words as felicitous as comprehensive. "Telescopes," says he, "cannot be so formed as to take away that confusion of rays, which arises from the tremors of the atmosphere. The only remedy is a most serene and quiet air, such as may perhaps be found on the tops of the highest mountains, above the grosser clouds."

This appears a very simple and probable piece of speculation, yet somehow it dropped out of notice, and never had the seal of practical trial applied to its theoretical prediction, until the late First Lord of the Admiralty, Sir Charles Wood, duly advised by the Astronomer-Royal, Mr. Airy, commissioned me to make the attempt in the summer of 1856, by carrying a large telescope to a considerable height up the flanks of the Peak of Teneriffe.

That mountain was chosen as the most elevated one within reach of a summer expedition, and at the same time of practicable ascent with large instruments. It is situated, moreover, in the middle of the N.E. trade wind region, where the weather is not only more regular than in any other part of the world, but where, *mutatis mutandis*, some pretty certain data as to the climate of the upper atmospheric strata, had been procured from another and grander scientific work, recently performed, also under the Lords of the Admiralty, viz., the remeasurement of La Caille's Arc of the Meridian at the Cape of Good Hope, by their Lordships' southern astronomer, Mr. Maclear.

Further particulars of a practical nature, relative to the character of the ground, as well as the temper and quality of the inhabitants, having been procured from Robert Stephenson, M.P., who had lately visited the island. and whose early experiences on South American cordilleras, had long since led him to look with favour on Newton's mountain method of improving astronomical observation,—the preparations for this novel sort of expedition went on quickly during May and June.

At this stage, so much kindly interest in the attempt was shown in the limited circle of working astronomers, that, beginning with Mr. Airy, I was favoured by them in the course of a few weeks with the loan of many valuable instruments; and finally, Mr. Stephenson, who had already paved the way to the expedition being called into being, tendered the magnificent contribution of no less than the use of his yacht *Titania*, and her able crew. With them, accordingly, we set sail on June 22nd from Cowes; I say we, for I was accompanied by my wife, the best assistant that either an astronomer or any other man can possibly have.

There was still just a trifle of uncertainty spread over our prospects, for some very opposite opinions were in the field, apparently supported by observed facts; and a few voices even loudly proclaimed, that high mountain tops, all the world over, are invariably loaded with clouds and mist and sleet, and tormented for ever with impetuous storms. Yet strong in our own belief, on we went in the swift *Titania*, and arriving in Teneriffe on July the 8th, were at once made free of the island by the liberal and hospitable Spanish authorities; beginning our first ascent six days afterwards, with a long line of mules laden with instruments and baggage.

The morning was desperately cloudy, quite a desponding sort of day; but the angle of ascent in the road was happily most moderate and uniform, so on and up we rode with the greatest facility. A sympiesometer, by Adie of Edinburgh, gave the heights without dismounting. At 3000 feet of altitude, still pacing up a constant slope, the level of the clouds was reached, —those clouds which had made the sky look so unhappy when we were starting from the port of Orotava. A whiff or two of damp mist flew about us for a while, and then we suddenly emerged into clear hot sunshine. From that moment, and hour after hour, as the decreasing column of the sympiesometer chronicled the height ascended, and as we continued toiling up the long slope of the mountain, the sun shone vehemently down upon us from a sky of the purest blue; and never did the clouds below attempt to leave their constant level of 3000 or 4000 feet above the sea. In this brilliant illumination, in the rarefied and arid air, amidst volcanic rocks of grotesque and imposing forms,—in fact, in this most moonlike region we travelled on, until by evening we had reached the top of Mount Guajara, on the southern side of the elevation crater; and thus within 24 days of leaving England, had the satisfaction of bivouacking on the top of a mountain 8900 feet high, and only 28° from the equator; in a calm air, too, with a temperature of 65°, and under a sky undimmed by a single cloud, and gloriously resplendent with stars.

Was not that at once a realisation of Newton's prophetic description, "a serene and quiet air on the top of the highest mountains above the grosser clouds"?—for all this time the lowlands beneath us, and the sea far and wide, were covered in by a broad expanse of mist, whose rollers were driving along under the influence of a violent N.E. wind.

That great plain of vapour floating in mid-air at a height of 4000 feet, was a separater of many things. Beneath, were a moist atmosphere, fruits, and gardens, and the abodes of men; above, an air inconceivably dry, in which the bare bones of the great mountain lay oxidising in all variety of brilliant colours, in the light of the sun by day, and stars innumerable at night.

Below that constant curtain of cloud, were towns and villages, —prisons, theatres and churches many,—above it, save a few goatherds wandering over the heights with their flocks of Guanche breed, were no traces of human life but in our little astronomical encampment.

Then how truly serene and quiet, and transparent, too, was the air above our 8900 foot elevation; for, on erecting our telescopes, not only was each star, whether high or low, seen with an exquisite little disc and nearly perfect rings, but the space-penetrating power was extended with the same instrument and same eye from the 10th magnitude, at the sea level, to the 14th, on Guajara.

Similar results in their ultimate bearing, followed other observations, as those of solar and lunar radiation. But time fails me to tell of two months' mountain experience of days, always better for astronomy than in the towns below, and sometimes supremely adapted therefor; and of how, accompanied by two of the sturdy seamen of the *Titania*, we tried our telescopes on the flanks of the Peak itself, at a height of 10,700 feet, ascertained at once the practicability and advisableness of greater heights still, and climbed the culminating point of the mountain 12,198 feet high.

To describe these operations in full, there is now neither time, nor perhaps necessity, as the original observations, with all the numerical and instrumental particulars, have been communicated to the Admiralty, and by them were transmitted to the Royal Society, for publication, in June last; while as to the more popular part of our daily experiences and little personal adventures, should any one care to read them, are they not contained in a little book, recently published by Mr. Lovell Reeve, and illustrated with genuine photo-stereographs?

Such plates being actual reproductions of nature by herself, I may, perhaps, be allowed to call some attention to them, not indeed altogether through means of the book, but by exhibiting, with the assistance of Prof. Tyndall and Mr. De la Rue, magnified optical pictures of some glass copies from the original negatives.

These will be better understood, if attention be turned for a few moments to this large model of the Peak of Teneriffe, and a tract of country about it, sixteen miles square. It has been prepared for this special occasion by the enthusiastic kindness of my friend Mr. James Nasmyth, C.E., long experienced in watching lunar craters in telescopes of his own making; and professionally intimate with metals, fluid and solid, with all the volcanics, indeed as well as the mechanics of the workshop. When he heard of a terrestrial crater, the great crater of Teneriffe, *eight miles* in diameter, he could not restrain his admiration and his zeal; so setting to work with all the map and measurement particulars which I could furnish, he produced the present model, as accurate as the existing state of our knowledge admits.

By a general vertical illumination the colours may be most distinctly seen. The green indicates vegetation, mainly confined to the lower 4000 feet of altitude, or to the region below the clouds. Above them are seen chiefly the colours of the lava rocks; the oldest, light yellow; the most recent, black; and the intermediate, red.

[The first collection of pictures shown illustrated the scenery of the green region on the northern coast; the second had its locale at a height of 8000 feet on Mount Guajara, or the southern wall of the great elevation crater, submarine at the time of its formation. And the third was confined to the eruption crater, or central cone, constituting the so-called Peak of Teneriffe; and exhibits the phenomena of subaerial volcanic action, at elevations extending from 9000 to 12,200 feet.

Having exhibited the prevailing colours by a vertical light, a ray of electric light was next thrown upon the model, at a low angle, so as to bring out the forms, and especially the angle of slope, of the various cones and craters.]

Studied in this manner, the model will yield so much information, that I will not venture to detain the audience longer, save with a very few words on the social bearing of this astronomical experiment on the Peak of Teneriffe. The claims of science to respect amongst men, for its services in promoting the union of nations and the brotherhood of mankind, have been often dwelt on. Of this admirable and humanizing tendency, is not our experiment on Teneriffe an example, within its little range? See an observer sent out by the English Government, received in a fortified town of the Spaniards, not only without distrust, but as frankly as if one of themselves. And did they suffer by it? We took no notes of their forts and guns, and military array, we applied ourselves to our scientific business alone; and if we have brought away anything more from Teneriffe than what I have already had the honour of describing to you, it is, respect and admiration for the Spanish character; and grand ideas of the results to astronomy as well as some other sciences, if this first experiment, this mere trial of a new method, be annually repeated, and energetically followed out.

[C. P. S.]

Friday, May 20, 1859.

HENRY BENCE JONES, M.D. F.R.S. Vice-President, in the Chair.

JOHN HALL GLADSTONE, Ph.D. F.R.S. M.R.I.

On the Colours of Shooting Stars and Meteors.

ALL are familiar with the smaller kinds of shooting stars, and most have observed those of a larger size which shoot across the sky like a rocket, and burst perhaps in a shower of sparks; many persons also have been witnesses of the grander displays called fire-balls, or bolides, and some few have seen those bright clouds that have occasionally appeared and rained down stones upon the earth. It is not certain that all these are connected phenomena, or that there is a solid nucleus to every shooting star; yet it is impossible to draw any exact line of distinction, and there is every gradation between the most striking and the most simple of these appearances. The investigations of scientific men have made us acquainted with many facts relating to these bodies: thus, their direction is never perpendicular to the earth, but more frequently almost horizontal, and though they fly from every quarter of the heavens, the majority come from that part towards which the earth is at the time moving; their velocity averages about 20 miles per second; their height above the earth is, of course, very various, yet the more brilliant fire-balls seem to begin their luminous course at somewhere about 40 miles above us; their size is probably small in all instances, although, from irradiation, they frequently appear to present a considerable diameter; they occur often in showers; and these showers have been observed to have an annual periodicity. At the present time these star-showers occur generally about August 10 and November 13, but at the end of the 11th century the most remarkable period was April 4; and those wonderful people the Chinese, who have kept records of showers of meteors since March 23, B.C. 687 (when Manasseh was ruling over Judah, and European history scarcely existed) tell of other periods, pre-eminent among which is July 22.

The meteorolites which fall from the sky are of two sorts; the stony, consisting of silicate of magnesia, with more or less admixture of lime, potash, or soda combined with silicic acid; and the metallic, consisting of iron, which always contains a small quantity of nickel, with phosphorus and sulphur, and often contains in addition cobalt, and zinc, tin, lead, manganese, or chromium, with carbon or chlorine. Other elements have also been mentioned as found in certain meteoric stones. Three specimens were exhibited; a broken piece of silicate interspersed with metal, which fell at Triguerre, in France, and a huge mass of supposed meteoric iron, the property of Prof. Tennant; and a fragment of a piece of iron found in Mexico, now in the Royal Institution, and which, from its chemical composition, is believed to be meteoric.

The cosmical theory is the only one capable of explaining the known facts of these meteorites, though that is not without its difficulties. It supposes that in the interplanetary spaces, at least near the earth's orbit, there are a vast number of minute solid bodies revolving round the sun, either singly or in streams, and that our globe in its passage comes into collision with some of these, or periodically cuts the orbit of these streams of planetary dust. The small pieces of solid matter are supposed to become incandescent or ignited by their rapid friction against the air.

As to the colour of meteors, we have much information given in the lists of the Chinese, in those of the Rev. Baden Powell, published in the Reports of the British Association, and in those of M. Coulvier Gravier. M. Poey of Havannah has taken the trouble of arranging all these observations according to the colour, and the month of appearance: and the totals of his tables form the basis of the accompanying table, in which however a little liberty has been taken with the classification, all the recorded colours being referred to the six principal divisions of the prismatic spectrum, and these combined with white, and white itself. The Chinese colour observations are rather under, and the English rather above a thousand, but for the sake of comparison they have been reduced to that proportion.

	Chinese.	English.	French.
Red	5·1	12·2	4
White-Red	0·5	4·9	6
Orange	56·8	10·5	4
Yellow	0·6	14·2	7
White-Yellow . . .	0·5	1·8	1
Green	0·0	0·6	0
White-Green . . .	0·0	0·6	1
Blue	0·8	30·8	0
White-Blue . . .	32·7	5·4	41
Purple	1·0	0·5	0
White	2·0	18·5	3
Total . . .	100·0	100·0	67

The very apparent dissimilarities in these three lists are capable of more or less explanation. The Chinese mention orange (or rather yellowish red and reddish yellow) meteors in great numbers, but these seem to be balanced in a great measure by the numerous observations of red, pale red, and yellow, as well as orange in the English and French lists. It should be remembered that there may be every gradation from red through orange to yellow, and it may be fairly open to doubt whether the inhabitants of the Celestial empire gazing at the stars one or two thousand years ago distinguished colours just as their French translator would do. Again, the Chinese and French give white-blue in great numbers; but this is evidently the same as the English blue. The English lists also make numerous mention of white meteors, because in Prof. Powell's Catalogue the fact of a meteor being white or colourless is usually noted, which is not generally the case with either the French or the Chinese observations. The number of French meteors classified as white-blue, is swollen by many described by M. Coulvier Gravier as "white becoming bluish in the horizon."

The points of similarity in the three lists are, the small number of green meteors—what there are occurring generally among those fireballs that change colour; the small number of purple; the absence of brown; and the fact that the large majority of meteors exhibit some distinctive colour. They may be generally divided into two groups, the one blue, the other orange, inclining more or less either to red or yellow.

If these appearances are really produced by the passage of pieces of stone or iron through the earth's atmosphere during its annual course round the central orb, it is very possible that the stream of little bodies that intersects our orbit at one time of the year may differ in composition from those that cut our path at another period. It occurred to the speaker that this might be evidenced by a difference of colour during their combustion, and that the monthly tables of M. Poey afforded the means of determining whether such was really the case. On examining the Chinese record it was found that the prevailing colour of a great shower of falling stars is very rarely given; the colour observations are almost confined to large single meteors; and little can be observed beyond the fact that the blue meteors are more numerous in comparison with the orange during the months of August, September, October, and November, than during the rest of the year. M. Poey has also made the remarkable observation, that the Chinese meteors "show a remarkable constancy of tints during a long period of years, when an equally constant but different scale of colour prevails, and this for several successive periods;" a fact that may possibly be due to the changes in the periodical showers already adverted to. If, however, we turn to the monthly tables of the English observations, we are at once struck with the marked difference in the relative proportions of the different colours. Thus, confining our attention to the months of August and November, when the great showers occur, we observe a difference that cannot be attributed to mere accident. In the following table red and white-red have been added together, yellow and white-yellow, blue and white-blue.

	August.	November.
Red	49	24
Orange	8	23
Yellow	44	16
Blue	164	30
	265	93

On glancing at these numbers, we cannot fail to remark, the great deficiency of orange meteors in August, and their comparative abundance in November, while conversely blue meteors occur in great numbers during August and are comparatively rare at the later period. The yellow appear in about average quantity in each month as compared with the whole year, but the red exceed the average somewhat in November. Hence it may be deduced, that at the autumnal period the meteorites generally burn with a red or orange glow, while those which cross our orbit about the 10th of August display in combustion a blue colour, and this is in perfect accordance with what is observed on a closer inspection of Prof. Powell's lists. The speaker stated that last August he had had the good fortune to witness these streaming blue meteors passing from east to west, and leaving a phosphorescent train in their wake.

Another fact of interest connected with this subject, is the change that is frequently remarked in the colour of a meteor during its passage. Thus the French observations make repeated mention of falling stars that changed from white when overhead, to bluish or reddish in the horizon; from white they have been also seen to change to orange-yellow and blue-green, to reddish and bluish with a reddish train; and from yellowish white to orange-yellow and greenish-white, the meteor being broken into several fragments, two of which passed from white to the colour of red-hot iron. The changes from orange-yellow to green, and from yellow and red to greenish-yellow have also been recorded by M. Coulvier Gravier. The Chinese tell of the converse change from red to white, and British observations record the passage from blue to red; from blue to green, and finally red; from green to crimson; and from green to orange and red. The trains left by meteors during their flight, are usually of the same colour as the meteors themselves, but not always so; thus red meteors have sometimes left a blue luminosity, and conversely red sparks have frequently fallen from meteors of another colour. The Chinese record such instances in such terms as "at the moment when the globe of fire fell, a flame appeared, and a score of little red stars jumped out of it."

On turning over the British Association Catalogues we observe many indications of these changes of colour; thus, in an account of a meteor seen at Poona on Sept. 7, 1847, we read:—"Before the first bursting the meteor was of exceeding brightness, of an intense blue colour, and at the instant of explosion it changed into red." The following is nearer home. A fire-ball is thus described by an observer at Lambeth :—"As it brightened it displayed the most lovely colours, which could be distinctly traced to the radial colours produced by the sun; at one period green, violet (deep), pale red, &c., and their effects through the thin stratum of clouds which were in its path were most gorgeous." In a most graphic description, given by a lady, of a meteor that appeared over Hampstead, we find the remarkable statement:—"It shot forth several fiery coruscations, and while we were gazing at it, broke into an *intensely* radiant cloud. . . . It cast a most brilliant light on the houses there, brighter than moonlight, and unlike any light I ever saw. It appeared of a blue tint on the bricks, but there was no *blue* light in the cloud itself."

In discussing these reputed facts it is necessary to take into account certain illusions to which observers are subject. Thus, at the outset, there is the diversity of names given to the same colour by different persons. No two individuals, however perfect their perception of colour, would perhaps agree in their mode of naming the colours of all the stationary objects around them, how very likely then would they be to disagree in naming the colour of a light which appeared suddenly and unexpectedly in the sky, and as suddenly disappeared! Many instances of this discrepancy might be cited from the accounts of observed meteors; but the most curious instance with which the speaker was acquainted had occurred in the descriptions of the beautiful meteor that travelled over England on the 12th of last September, being visible in the evening before even the daylight had disappeared. Of the many eye-witnesses who described the phenomenon in the *Times* newspaper, five mentioned its apparent colour; of whom F.A.B. states it to have been "green at first;" N.R. "green, surrounded by white;" W. Rowlett, "white;" and W. "vivid, whitish blue;" while B.H. asserts that it was "primrose." Here, however, the discrepancy is so great as to lead us to the conviction that the meteor of Sept. 12, 1858, was really one of those that change colour during their passage through our atmosphere, and thus present different appearances to observers in different places.

It is quite possible that a meteor may emit rays which in the aggregate would produce one colour, and yet may affect the observer with a sensation of a different colour. This may arise either from absorption, intensity, or contrast.

In illustration of the effect of absorption Dr. Gladstone exhibited the prismatic spectrum by means of the electric lamp, and showed how certain glasses produced a similar absorption of the rays to that which takes place in the common phenomena of the red sun, or orange moon. The effect of dispersion was rendered visible by the non-transmission of the extreme blue and violet rays through water into which a little milk had been poured. This "sky-blue" mixture, produced by a substance itself colourless, represented the light clouds and vapours which must frequently affect the apparent colour of meteors, and suggested a simple explanation of the fact that of the variable meteors observed in the misty skies of England, so many terminate in red. Smoke has much the same effect on the spectrum as milk and water.

In illustration of the effect of intensity in causing lights actually coloured to produce very nearly the sensation of white, the electric light, from charcoal points, was exhibited under red and blue glasses, when it appeared dazzling and almost white; though no white light was really transmitted, and such objects as common paper, when illuminated by it, reflected the coloured radiance. This will explain the phenomenon so frequently observed by M. Coulvier Gravier, of a white meteor becoming bluish or reddish as it approached the horizon; and also the paradox of an "*intensely* radiant cloud" white in itself, but throwing a *blue* light on the walls of houses. In these cases it seems highly probable that the incandescent meteors were really emitting coloured light, but that this colour did not become apparent till the light was reduced either by distance, or by reflection from other objects. In a similar manner, the fact of a yellowish meteor becoming more or less green as it passes away from the spectator, so frequently noticed in the French observations, may be explained by the well-known changes in the chromatic impressions produced by yellow rays according to their intensity.

In reference to the effect of contrast it was remarked, that every lady is aware of the alteration that may be produced in the apparent tint of any article of dress by the juxtaposition of some other bright colour; and indeed it may be laid down as a general law that the apparent colour of every object is affected to a greater or less degree by the colours of all other objects seen at the same time. This remark holds good equally of self-luminous bodies, as, for instance, the flame of a gas-lamp, which assumes a rather bluish tint when the intensely yellow soda flame is brought beside it. This also must be a source of error in the observations of meteors.

After making due allowance for these points of difficulty and probable fallacy, we may approach the question—How far are these chromatic phenomena in accordance with the cosmical theory? Were pieces of iron to be shot through our atmosphere at the rate of twenty miles per second, there is good ground for believing that the friction would make them red-hot, if not incandescent. An iron wire, heated in the galvanic circuit, was observed by the speaker to emit at first principally orange and green rays, but as the heat increases the true red rays are emitted in an increasing degree, till "bright redness" is attained; and when combustion begins blue rays are also given forth, the general impression being then that of a reddish whiteness. The combustion and scintillation both of ordinary and of meteoric iron were shown in several ways. The metallic masses which fall through the air are never composed solely of iron, and it is difficult to say which metal or which other constituent would be the first ignited. Nickel in combustion displays a larger amount of green rays; sulphur, as is well known, burns with a blue, and phosphorus with a white flame. Two pieces of iron pyrites, attached to the wires of a powerful galvanic battery, when brought momentarily into contact, were ignited with a very luminous flame, which exhibited the characters of both burning iron and sulphur; and on one occasion, when the experiment was tried, the lambent blue flame of the latter element was visible for some time after the circuit was completely broken, and the ferruginous scintillations had ceased. The other metals occasionally found in meteorites, such as cobalt, zinc, or lead, will of course burn with their distinct flames; and the silicates, though incapable of combining with oxygen, may give out an intense light when strongly heated. This was demonstrated by placing the fragment of the meteoric stone that fell at Triguerre in the oxy-hydrogen blowpipe, when it began to fuse, and became brilliantly incandescent. In all these observations on the colours displayed by them there is nothing antagonistic to the idea that these luminous meteors are produced, as some have certainly been, by the combustion of such solid masses of metal and stone as occasionally strike the earth; but we are not yet in a condition to deduce its composition from the colour of any particular meteor.

[J. H. G.]

WEEKLY EVENING MEETING,

Friday, May 3, 1861.

THE DUKE OF NORTHUMBERLAND, K.G. F.R.S. President, in the Chair.

PROFESSOR FARADAY, D.C.L. F.R.S.

On Mr. Warren De la Rue's Photographic Eclipse Results.

THE speaker commenced by drawing attention to the sun as the great source of light and heat to the planets of our system; and to the phenomena which occur from time to time when the earth and the moon are brought by their orbital revolutions nearly or absolutely in the same plane. The sun, casting shadows of the moon and of the earth in an opposite direction to their illumined sides, there would always be produced a total eclipse of the sun, or the moon, when these bodies were situated in the same line on the same side of the sun, if the distances of the earth or moon invariably admitted of the one falling within the shadow of the other. In consequence, however, of the elliptical form of the orbits of the earth and moon, the distances of these planets from each other and the sun are constantly varying, and sometimes the shadow of the earth does not reach the moon, or that of the moon does not reach the earth. We might consequently have, in the case of the sun, either a partial eclipse when the sun, moon, and earth were not exactly in the same plane, or an annular or a total eclipse when they were so situated. A total eclipse might be only just total, or be of a shorter or longer duration as the apparent diameter of the moon exceeded by little or much the apparent diameter of the sun; no eclipse of the sun is so great, however, as to shut off the light of the sun from the whole hemisphere of the earth; on the contrary, the shadow of the moon can never cover more than a very small extent of the earth's surface. On the 18th of July of last year, it happened that under unusually favourable circumstances there occurred a total solar eclipse; the sun was nearly at his greatest possible distance from us, and therefore had almost a minimum apparent diameter, and the moon was nearly at her least possible distance, and therefore had a nearly maximum apparent diameter, so that the breadth and duration of the total eclipse were nearly at a maximum; moreover, the shadow of the moon passed over a country easily accessible to European astronomers. The speaker, after pointing out the course of the eclipse, from sun rising in Greenland, across the Atlantic, across Spain, the Mediterranean, and Africa, stated that for a belt of 60 miles broad, the duration of the eclipse in Spain was fully three minutes, and about three minutes and a half in the central line of that belt. Such favourable circumstances were not likely to occur again within the probable lifetime or opportunities of the observers, who went out to Spain to observe the eclipse. Hence the desirability of placing before scientific men the means used by various persons to record and measure the appearances visible at total eclipses, with a view to facilitate the labours of future observers.

Curious and paradoxical as it might at first appear, it is necessary to shut out the sun in order to see it; for example, said the speaker, look on this electric light, could any one conjecture from its aspect what it is? One sees a brilliant centre surrounded by rays, but one cannot see the two charcoal points which are producing it; and if it were magnified and projected against a screen by means of lenses, although the ignited points would be rendered evident, still there are matters about them which would remain invisible. So it is with the sun; one can so reduce his light, by means of dark glasses, or other contrivances, as to enable us to scrutinize by means of telescopes his photosphere; or we might project his image on to a screen, and thus examine it: but we should not see the sun, that is the whole sun. His mottled surface, his bright markings, his dark spots would undoubtedly be thus shown; but we should fail to discern those curious appendages which were first observed a few years back on the occasion of a total eclipse. These so-called red flames, about the nature of which many conjectures have been made, are, as it now appears, true belongings of the sun, and are not subjective phenomena, produced, as some philosophers suppose, by a deflection or diffraction of the sun's light in passing through the valleys on the moon's profile.

The object of the speaker being, as before stated, to bring under the notice of the Members of the Institution, Mr. De la Rue's photographic results: he now described the Kew Photoheliograph. A photograph of the instrument and temporary observatory, taken in Spain, being projected on the screen by the electric lamp. The heliograph consists of a telescope, the tube of which is square in section, and which can be made to follow the sun accurately by means of clockwork. The optical part consists of an object glass (actinically corrected), and a secondary combination of lenses, situated near the focus of the object glass, for the purpose of magnifying the sun's image to four inches in diameter. The image so magnified together with position wires are depicted on the collodion plate placed in the telescope. In taking ordinary sun pictures, or pictures of the partial phases of the eclipse, the light is allowed to act on the sensitive plate by the passage of a narrow slit in a brass plate drawn with great rapidity across the secondary magnifying lens. By this contrivance, the sun's image is allowed to act for a very small fraction of a second of time. Thirty-one pictures of the various partial phases of the eclipse were obtained in this way by Mr. De la Rue and his assistants. Several of the most interesting were projected on to the screen by means of the electric lamp. The two totality pictures were, however, obtained in another way; the brass plate with the slit was drawn aside, and the picture of the protuberances allowed to fall for a whole minute on the collodion plate.

The first picture was procured exactly from the commencement of the totality and during the minute precisely succeeding it. The second picture from about a minute preceding the reappearance of the sun until just before he reappeared.

These pictures were shown on the screen by means of the electric lamp, and it was seen that the luminous prominences extended for a long distance beyond the moon's dark limb. In the first picture, some prominences were completely detached, and were at some considerable distance from the moon's limb ; these, in the second picture, were reached by the moon, which during the interval had been travelling across the sun's disc. Fresh prominences had come into view in the second picture on the western limb, while some of the prominences on the eastern limb had been shut off by the moon's motion.

The speaker then referred to diagrams, which had been enlarged from actual photographs, etched upon glass by hydrofluoric acid, and graduated in accordance with the data furnished by the images of position wires on the photographs. These diagrams showed that a luminous prominence situated at a *right angle* to the path of the moon's motion across the solar disc had hung back in reference to the moon's centre an angular distance of about $5\frac{1}{2}°$, while prominences situated *in the direction* of the moon's path had not shifted angularly, but were covered and uncovered to an extent of about 93″ during the period of totality. Such results were in accordance with the hypothesis that the prominences belong to the sun, and opposed to the theory that they are subjective phenomena produced by the deflection of the sun's light.

In order to render evident the relative positions of the whole of the protuberances visible during the eclipse, attention was drawn to a diagram enlarged from an etched photograph of the sun, on which were etched also the protuberances visible in the first and second totality —photographs which coincided exactly when superposed in respect of those parts visible in both. It was seen on the diagram, that in consequence of Rivabellosa not being exactly in the centre of the shadow path, the moon's centre was depressed below the sun's centre, and thus at the nearest approach of the two centres they were distant about 14″. This had the effect of rendering visible a little more of the prominences on the northern limb, and of shutting off a portion on the southern limb of the sun.

The speaker drew attention to the heights of many of the prominences ; to the circumstance of their brilliancy in some cases being greatest in those parts nearest the sun, while in others the brightest part was on that edge most distant from the sun. One prominence, upwards of 70,000 miles distant from the sun's limb, was particularly pointed out—this had not been seen by human eyes, but there was its image fixed and recorded by the heliograph in both the pictures. Photography could therefore render evident to us phenomena of the sun which the human eye could not discern ; and here we had another of the many proofs of the importance of varying our means of observation. This prominence was not the less real because we could not see it; it existed and emitted a radiant force; invisible it was true, but still nevertheless a force, and even possibly a greater chemical force than that of the visible prominences. In order to render this apparent, a spectrum was produced by means of the electric light and projected on to a collodion plate recently made sensitive, and placed in full view of the audience; during the thirty seconds of exposure, marks were made by scratching through the collodion film to indicate the position of red, yellow, green, and violet bands; on developing the picture and projecting the image on to the screen, it was seen, by reference to the scratches, that up to the violet band very little effect had been produced, but that the invisible rays beyond the violet had produced a very intense image, to an extent equal to the breadth of the visible spectrum, consisting of several well-marked bands of varying intensity.

The speaker now drew attention to the corona: on Mr. De la Rue's photographs the corona to some extent was visible, but recourse was had to a photograph of a drawing showing the whole phenomena, which was also projected against the screen. It was pointed out that observations in Spain had proved that the corona polarized light, and as light coming direct from a luminous body is not polarized, but that after reflection it is so; the fact of polarization tended to show that the corona must be a consequence of an atmosphere around the sun reflecting the sun's light.

The speaker concluded by drawing attention to some phenomena connected with the sun's spots, their rotation, the sudden bursting out of a brilliant light observed by two astronomers distant from each other; and also to a curious foliated appearance in the sun's spots, observed by Mr. Nasmyth. What are these vast masses which reach to such enormous distances beyond the sun, as we see him under ordinary circumstances? One, we perceive, extends nearly as far from the sun as three times the entire circumference of the earth, and another is suspended away from the sun's limb about once that distance! Modern science places at our disposal methods of determining the nature of some of the constituents of the sun's atmosphere, with a degree of certainty equal, perhaps, to any of our laboratory methods, could we bring away a sample and analyse it by chemical means. Employing the principles and methods of Bunsen and Kirchhoff, which Professor Roscoe recently brought under our notice, might we not, suggested the speaker, on the occasion of another solar eclipse give some answer to this question, and add another proof of the reality of these prominences by revealing something as to the nature of their constituent particles.

Lastly, there was exhibited a photograph of the moon, obtained by Mr. De la Rue at his own observatory, for the double object of showing the great beauty and perfection of the pictures, and in illustration of the light and shadow being different in proportion from the visible image, in consequence of the actinic force of the light reflected by different parts of the moon not according precisely with its luminosity.

[M. F.]

Friday, May 20, 1864.

The EARL OF ROSSE, F.R.S. M.R.I., in the Chair.

JAMES NASMYTH, ESQ. M.R.I.

On the Physical Aspects of the Moon's Surface.

As the Moon's hemisphere, which is ever turned towards us, has its features illuminated in opposite directions during her monthly passage in her orbit around the earth, every part of it is exposed in turn to the rays of the sun, which fall on the details of its features in constantly varying inclinations; and it is from this circumstance that we have such favourable opportunities afforded to us of obtaining a very correct knowledge of the configuration of the details in question, as well as of their height or depression above or below the mean level of the Moon's general surface. Thus it is that we are enabled most carefully to scrutinize her remarkable surface; and should we have drawn any hasty inferences from one set of observations, the opportunity is usually presented to us in the course of a fortnight, or at farthest a month, to correct them if erroneous, or to verify them if accurate, and to pursue further investigations that may be suggested by reflection on what we had last observed.

In these respects telescopic visits to the surface of the Moon yield more correct and reliable results than would many a visit to portions of our world where the scenery to be surveyed is not, perhaps, conveniently accessible: and even when it is reached, the traveller may be surrounded by circumstances which very seriously interfere with his personal comfort, or disturb that tranquillity which is so requisite a condition for close and accurate observation, and thus lead him to hasty conclusions, which he has no future opportunity to rectify. In strong contrast with such circumstances is the position of the astronomer, comfortably placed beside his telescope, in the silence and tranquillity of a fine clear night, with all distracting objects excluded from his view. The whole of his attention is thus brought to focus, as it were, on the point under investigation there and then presented to his scrutiny, and ready to yield perfectly truthful replies to his questions; nothing being requisite for a correct interpretation of facts, other than a quick eye backed by a sound and unbiassed judgment.

It is from circumstances such as these that we have acquired, by a long course of assiduous observation and reflection, an amount of intimate acquaintance with the physical structure of the Moon's exterior, in many important respects far more accurate than is our knowledge of that portion of the earth.

In order rightly to interpret the details of the Moon's surface, as revealed to us by the aid of the telescope, we ought, in the first place, to bear in mind the true nature of volcanic action, namely, that while it has reference to the existence of intense temperature and molten matter, it does not derive its origin from *combustion*, considered as such in a strictly chemical sense, but proceeds from an incandescent condition, induced in matter by the action of that great cosmical law which caused an intense heat to result from the gravitation of particles of matter towards a common centre. These particles, originally existing in a diffused condition, were, by the action of gravitation, made to coalesce, and so to form a planet. Volcanic action, then, has in all probability for its source the heat consequent upon the collapse of such diffused matter, resulting in that molten condition through which there is strong reason to believe all planetary bodies to have passed in their primitive state, and of which condition the geological history of our earth furnishes abundant evidence. Thus the molten lava which we see issuing from an active volcano on the earth, is really and truly a residual portion of that molten matter of which the entire globe once consisted.

In reference to the nature and origin of that eruptive force which had again and again, in the early periods of the Moon's history, caused the remaining molten matter of her interior to be ejected from beneath her solidified crust, and so to assume nearly every variety of volcanic formation in its most characteristic aspect, the key to these may be found in the action of that law which pervades almost all matter in a molten condition, namely, that "molten matter occupies less bulk, weight for weight, than the same material when it has ceased from the molten state;" or, in other words, "that matter in a molten state is specifically more dense than the same material in a solidified condition." Thus it is that in passing from the molten to the solid state the normal law is resumed, and expansion of bulk either just immediately precedes or accompanies solidification. It is, therefore, in this expansion in the bulk of the solidifying matter, beneath the Moon's crust, that we are to look for the true cause of that eruptive or ejective action which has resulted in the displacement, *surfaceward*, of the fluid portion of the Moon's internal substance; a displacement which has manifested itself in nearly every variety of volcanic formation, such as circular craters with their central cones or mountains of exudation, cracked districts, &c.; all these variations of well-recognized volcanic phenomena being intermingled and overlaid one upon the other in the most striking and wonderful manner.

It may, however, be very reasonably and naturally asked, "What evidence have I that the features I refer to have any relation to volcanic action at all?" In reply to such a question I would direct the inquirer's attention to one single feature which, I conceive, demonstrates more completely than any other the fact of volcanic action having (at however remote a period) existed in full activity in the Moon. The special feature to which I would refer is the central cone that may be observed within those "Ring-formed mountains," as they have been termed. "The central cone" is a well-known and distinctive feature in terrestrial volcanoes. It is the residue of the last expiring efforts of a once energetic eruptive volcanic action, which had thrown the ejected matter to such a considerable distance round about the volcanic vent, that in its descent it had accumulated around in the form of a ring-shaped mountain or crater; whilst on the subsidence of this volcanic energy, the ejected matter was deposited in the immediate vicinity of the vent or volcanic orifice, and thus arose the "central cone."

Anyone who is familiar with terrestrial volcanic craters must, at the first glance at those which are scattered in such infinite numbers over the Moon's surface, detect this well-known analogous feature, the central cone, and at once reasonably infer that these similar forms arose from a common cause, that cause being no other than volcanic action, accompanied by all its most marked characteristics.

Fig. 1.*

Fig. 1 represents a fair average type of the structure of a Lunar Volcanic Crater with its central cone A.

Fig. 2.

Fig. 2. The same in section.

* The woodcuts have been kindly lent by Mr. Churchill, publisher of the 'Quarterly Journal of Science.' They illustrated Mr. J. Nasmyth's article in the July number of that work.

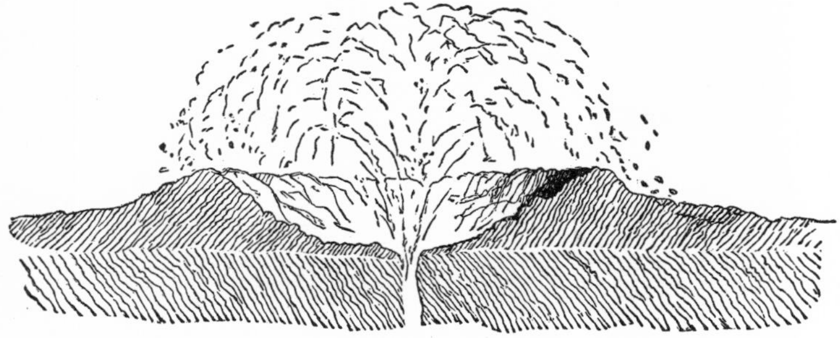

Fig. 3.

Fig. 3 is the section of a Lunar Crater, showing how by the eruption, and subsequent deposition of the ejected matter, the circular outer wall or crater had been formed.

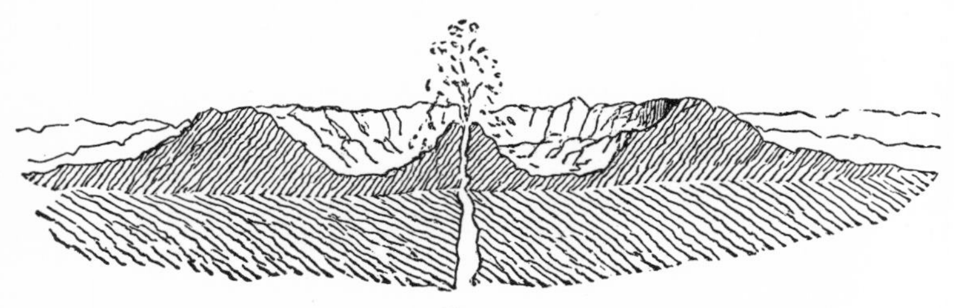

Fig. 4.

Fig. 4. The section of the same, exhibiting the manner in which the central cone had resulted from the expiring efforts of the eruptive action.

In examining the Moon's surface, we cannot but be impressed with the vast dimensions of many of the volcanic craters with which her surface is studded. Craters of thirty miles and upwards in diameter are by no means uncommon, and the first impression on the mind in reference to such magnitudes is one of astonishment, that so small a planet as the Moon (whose magnitude is only about $\frac{1}{49}$th that of the earth) should exhibit evidence of volcanic violence so far greater than any that we have on the earth. This apparent paradox will, however, disappear when we come to consider that in consequence of the Moon being so much less than the earth, the force of gravity on its exterior is not above $\frac{1}{6}$th of that on the earth, and that the weight of the lunar materials on its surface is reduced in the latter proportion, while, on the other hand, by reason of the small magnitude of the Moon and its proportionately much larger surface in ratio to its magnitude, the rate at which it parted with its original cosmical heat must have been vastly more rapid than in the case of the earth. Now, as the disruptive and eruptive action and energy are in proportion to the greater rate of cooling, those forces must have been much greater in the first instance; and, operating as they did on matter so much reduced in weight as it must be on the surface of the Moon, we thus find in combination two conditions most favourable to the display of volcanic force in the highest degree of violence. Moreover, as the ejected material in its passage from the centre of discharge had not to encounter any atmospheric resistance, it was left to continue the primary impulse of the ejection in the most free and uninterrupted manner, and thus to deposit itself at distances from the volcanic vent so much greater than those of which we have any example in the earth, as to result in the formation of the craters of vast magnitude so frequently encountered in a survey of the Moon's surface. In like manner we find the ejected matter piled up to heights such as create the utmost astonishment. Lunar Mountains of 10,000 feet high are of frequent occurrence, while there are several of much greater altitude, some reaching the vast height of 28,000 feet, and that almost at one bound, as they start up directly from the plane over which they are seen to cast their long black, steeple-like shadows for many a mile; whilst at other times they intercept the rays of the sun upon their highest peaks many hours before their bases emerge from the profound darkness of the long lunar night.

Among the many terribly sublime scenes with which the Moon's surface must abound, none can be grander than that which would present itself to the spectator, were he placed inside of one of these vast volcanic craters (Tycho, for instance), surrounded on every side by the most terrific evidences of volcanic force in its wildest features.

In such a position he would have before him, starting up from the vast plane below, a mighty obelisk-shaped mountain of some 9,000 feet in height, casting its intense black shadow over the plateau; and partly up its slope he would see an amphitheatrical range of mountains beyond, which, in spite of their being about forty miles distant, would appear almost in his immediate proximity (owing to the absence of that "aerial perspective," which in terrestrial scenery imparts a softened aspect to the distant object), so near, indeed, as to reveal every cleft and chasm to the naked eye! This strange commingling of near and distant objects, the inevitable visual consequence of the absence of atmosphere or water, must impart to lunar scenery a terrible aspect; a stern wildness, which may aptly be termed unearthly. And when we seek to picture to ourselves, in addition to the lineaments and conditions of the lunar landscape, the awful effect of an absolutely black firmament, in which every star, visible above the horizon, would shine with a steady brilliancy (all causes of scintillation or twinkling being absent, as these effects are due to the presence of variously heated strata, or currents in our atmosphere), or of the vivid and glaring sunlight, with which we have nothing to compare in our subdued solar illumination, made more striking by the contrast of an intensely black sky; if, we say, we would picture to ourselves the wild and unearthly scene that would thus be presented to our gaze, we must search for it in the recollection of some fearful dream.

That such a state of things does exist in the Moon we have no reason whatever to doubt, if we may be permitted to judge from inferences reasonably and legitimately deduced from the phenomena on its surface revealed by the telescope; neither can there be a question as to the presence there of the same brilliant tints and hues which accompany volcanic phenomena in terrestrial craters, and which must lend additional effect to the aspect of lunar scenery. Nor must we omit, whilst touching upon the scene that would meet the eye of one placed on the Moon's surface, the wonderful appearance that would be presented by our globe, viewed from the side of the Moon which faces earthward. Possessing sixteen times the superficial area, or four times the diameter, which the Moon exhibits to us, situated high up in the lunar heavens, passing through all the phases of a mighty moon, its external aspect ever changing rapidly as it revolves upon its axis in the brief space of four-and-twenty hours, what a glorious orb it would appear! Whilst its atmospheric phenomena, due to its alternating seasons, and the varying states of the weather, would afford a constant source of interest. But, alas! there can be none to witness all these glories, for if ever man was justified in forming a conclusion which possesses the elements of certainty, it is that there can be no organized form of life, animal or vegetable, of which we have any cognizance, that would be able to exist upon the Moon.

Every condition essential to vitality, with which we are conversant, appears to be wanting. No air, no water, but a glaring sun, which pours its fierce burning rays without any modifying influence for fourteen days unceasingly upon the surface, until the resulting temperature may be estimated to have reached fully 212°; and no sooner has that set on any portion of the lunar periphery than a withering cold supervenes; the "cold of space" itself, which must cause the temperature to sink, in all probability, to 250° below zero. What plant, what animal could possibly survive such alternations of heat and cold recurring every fourteen days, or the accompanying climatic conditions?

But let us not suppose, because the Moon is thus unfitted for animal or vegetable existence as known to us, that it is necessarily a useless waste of extinct volcanoes. Apart from its value as "a lamp to the earth," it has a noble task to perform in preventing the stagnation that would otherwise take place in our ocean, which would, without its influence, be one vast stagnant pool, but is now maintained in constant, healthy activity, through the agency of the tides that sweep our shores every four-and-twenty hours, bearing away with them to sea, all that decaying refuse which would otherwise accumulate at the mouths of rivers, there to corrupt, and spread death and pestilence around. This evil, then, the Moon arrests effectively, and with the tides for a mighty broom, it daily sweeps and purifies our coasts of all that might be dangerous or offensive.

But there is still another duty that she fulfils—namely, in performing the work of a "tug" in bringing vessels up our tidal rivers. Dwellers in seaports, or those who reside in towns situated up our tidal streams, have excellent opportunities of observing and appreciating her value in her towing capacity; and, indeed, it may with truth be said that no small portion of the corn with which we are nourished, and of the coal that glows in our firesides, is brought almost up to our very doors by the direct agency of the Moon.

[J. N.]

Friday, March 17, 1865.

Major-General Edward Sabine, R.A. Pres. R.S. Vice-President, in the Chair.

Balfour Stewart, Esq. F.R.S.

On the latest Discoveries concerning the Sun's Surface.

It is well known that a distant body does not impress the eye with the idea of relief.

Thus distant mountains do not *stand out* like the nearer parts of a landscape, and the sun and moon appear only as flat discs.

But although neither to direct nor telescopic vision do these heavenly bodies appear spherical, yet it is possible to produce stereoscopic impressions by combining together two pictures of these bodies taken at different times.

This has been done by Mr. Warren De la Rue with great success, and this gentleman has produced numerous stereoscopic representations of our satellite, in which the general sphericity as well as the unevenness of surface come well out. He has also produced a more limited number of stereoscopic impressions of the sun; but our knowledge of the solar surface is only to a certain extent derived from these impressions, while it is to a greater extent derived from the careful study of ordinary sun pictures.

The bounding luminous surface of the sun is termed his photosphere, and our subject may be divided with reference to the surface into three parts: the first comprising the region above the photosphere; the second, the photosphere itself; and the third, the region beneath the photosphere.

To commence with the region above the photosphere, it may be easily shown that this region contains a very dense atmosphere.

In the first place, according to a well-known law the dark lines in the solar spectrum denote the presence of certain substances in a state of vapour, and yet in a comparatively cold state, above the photosphere of the sun. These substances comprise among others sodium, magnesium, iron, and nickel.

Again, the rim of the sun when viewed by the eye, but more conspicuously in photographs, appears less luminous than the central portions; and this would also appear to indicate the presence of an absorbing atmosphere of lower temperature than the photosphere, so that a ray of light from the border having to pass through a considerable thickness of this atmosphere would become diminished in brightness. (This was illustrated by a beautiful photograph taken by Mr. De la Rue.)

In the third place, the red flames and part of the corona which surround the sun during a total eclipse, reveal to us the presence of an extensive solar atmosphere.

It appeared to the Astronomer Royal and some others as probable that these bodies belong to the sun; but their connection with our luminary was put beyond doubt by Mr. De la Rue, who by means of the Kew heliograph was enabled to take photographic pictures of the sun at the total eclipse which happened in Spain in July, 1860.

(The photographs were exhibited; and it was seen that as the moon proceeded over the sun's disc, the red flames and part of the corona discovered themselves at that side which she had left, and were covered up by her disc at that side towards which she was approaching, thus showing that they belong to the sun.)

Another proof in favour of the idea that the red flames belong to the sun is derived from the nature of the light which they emit. This has great photographic power compared to its luminosity, so much so that one prominence was photographed by Mr. De la Rue, which was invisible to the naked eye. Now such rays can, as far we know, belong only to intensely heated gas, and such gas can only belong to the sun.

Next, with regard to the photosphere, or luminous envelope of the sun, this surface when viewed through powerful telescopes appears granulated or mottled. (Reference was here made to a diagram lent by the Rev. J. Howlett, and to a photograph by Mr. De la Rue.) But besides this, there is reason to believe that great defining as well as magnifying power discloses the fact that the whole photosphere of the sun is made up of detached bodies, interlacing one another and preserving a great amount of regularity, both in form and size. Mr. Jas. Nasmyth was the first to proclaim this curious fact—he called these bodies willow leaves; Mr. Stone has called them rice-grains; Padre Secchi, *coups de pinceau.* (Some photographs from drawings lent by Mr. Nasmyth were then exhibited.)

The faculæ belong to this part of the subject: they are detached portions of the photosphere which are much brighter than the region around them; but this difference in brightness is chiefly apparent near the sun's limb. The reason of this is believed to be, that they are portions of the sun's photosphere thrown up into the higher regions of the atmosphere, by which means they are enabled to escape a great part of the absorbing effect of this atmosphere which is particularly strong near the border; while near the centre the absorption is not great, so that they do not gain much by escaping it.

The idea that faculæ are elevated has been confirmed by a stereoscopic impression of a sun-spot and some faculæ taken by Mr. De la Rue, in which the spot appears to be a hollow while the faculæ appear as elevated ridges.

It should also be remarked, that faculæ retain the same appearance often for a considerable time, sometimes even for days together, so that they are clearly not composed of heavy matter, but are rather of the nature of a cloud.

The phenomena of the third region, or that beneath the photosphere, may be embraced in one word, "sun-spots." These consist of an umbra, or central darkness, surrounded by a less dark penumbra. Mr. Dawes has discovered in some spots even a deeper darkness in the centre of the umbra.

Now if that theory of spots be correct which supposes that they are cavities of which the umbra forms the bottom, while the penumbra constitutes the sloping sides, then the umbra ought apparently to encroach on that side of the penumbra which is nearest the visual centre of the disc.

Prof. Alex. Wilson, of the University of Glasgow, was the first to remark that spots really behaved in this manner, and his observations have been abundantly confirmed by the Kew photographs. (One of these was exhibited.)

It therefore follows, that the umbra of a spot is at a lower level than the penumbra; and since luminous ridges, and sometimes detached portions of luminous matter, cross over spots, it must be concluded that the whole phenomenon is below the surface.

In the next place, spots are the means of exhibiting to us the rotation of our luminary. If we turn to the south and view the sun spots always cross the disc from east to west, that is to say from left to right. (The apparent path of a spot at different seasons was here traced on a diagram, constructed and lent by Rev. F. Howlett; and it was remarked that when allowance is made for the inclination of the earth's axis, the path is really the same at different seasons.)

Besides the apparent motion due to rotation of the sun, spots have also a proper motion of their own, which was discovered by Mr. Carrington. This motion is also from left to right; those near the solar equator moving fastest. Mr. Carrington also remarked that spots confine themselves to the equatorial regions of the sun.

Hofrath Schwabe, of Dessau, has remarked that spots have a period of maximum and minimum nearly every ten years, and General Sabine has found that the year of maximum sun-spots is at the same time that of greatest disturbance of the earth's magnetism.

Finally, the behaviour of sun-spots appears to some extent to be influenced by the planet Venus, in such a manner that when a spot comes round by rotation to the ecliptical neighbourhood of this planet, it has a tendency to dissolve; and on the other hand, as the sun's surface recedes from this planet, it has a tendency to break out into spots.

[B. S.]

Friday, May 19, 1865.

SIR HENRY HOLLAND, Bart. M.D. D.C.L. F.R.S. President, in the Chair.

WILLIAM HUGGINS, Esq. F.R.A.S.

On the Physical and Chemical Constitution of the Fixed Stars and Nebulæ.

THE speaker commenced with a few preliminary remarks on the peculiar relation in which the heavenly bodies stand to man. It is alone from these lights shining upon us from distant space that we can obtain any knowledge of the parts of the universe which are without the earth. The experimentalist who seeks to bring to light the truths which lie hidden beneath terrestrial phenomena, can subject the objects of his research to each other's influence and to the various forms of force; but the astronomer is left to the indications afforded by the sense of sight alone, for the interpretation of the heavens. The distances, magnitudes, and relative motions of the sun, moon, and planets have been obtained with great exactness, but in an important respect our knowledge of the heavenly bodies was at fault. Until quite recently we possessed no knowledge *from observation* of the structure and chemical constitution of the sun, the fixed stars, and the nebulæ. Fortunately, for the increase of our knowledge, the cause of failure lay not in that the light of these bodies is wanting in sufficient indications of their nature, but in that the unaided eye has no power to perceive the indications by which the light of each of the heavenly bodies is specially distinguished.

Newton opened the way to a knowledge of these unperceived qualities of light. He was the first to show that by the refractive power of a prism of glass, the component rays of a beam of light could be separated and presented to the eye as a band of blending colours. Wollaston and Fraunhofer discovered that the colours of the spectrum of solar light are not continuous, but are interrupted by transverse linear spaces where the light is wanting. In these lines of darkness of the solar and stellar spectra, the chemical nature of the sun and stars stands written in cipher, but for half a century the lines of Fraunhofer remained uninterpreted.

In 1859 Kirchhoff announced the law by which these dark lines can be deciphered. He immediately applied his method of interpretation to the light of the sun, and discovered in that body the presence of several of the elementary forms of matter which enter into the composition of the earth.

The speaker stated that it was his intention on this occasion to bring before the members of the Institution the results of the extension of this method of analysis by the prism to the heavenly bodies other than the sun. These researches have been carried on in his observatory during the last three years; and in respect of the greater part of these observations, *viz.* those on the moon, the planets, and the fixed stars, he has had the great pleasure of working conjointly with his distinguished friend, Prof. W. A. Miller.

The speaker then referred briefly to the principles of spectrum analysis, upon which their interpretation of the phenomena observed in the spectra of the heavenly bodies was based, stating that spectra may be arranged under three orders, *viz.* :—

1. *A continuous spectrum unbroken by bright or dark lines*, which indicates that the light has not undergone any modification on its way to us. Also that its source is an opaque body, almost certainly in the solid or liquid state. Such a spectrum gives no information of the chemical nature of the substance from which the light emanates.

2. *A spectrum of bright lines separated by dark spaces;* this informs us that the source of the light is matter in the gaseous state. By a comparison of the bright lines of such a spectrum with the lines of terrestrial flames we may discover whether any of these terrestrial substances exist in the distant and unknown source of light. The spectra of many of the nebulæ are of this order.

3. *A continuous spectrum interrupted by dark lines;* this shows that the light has passed through vapours which have deprived it of certain refrangibilities by a power of selective absorption.

Since Kirchhoff has shown that these dark lines agree exactly in position with the bright lines which the vapours would emit if in a luminous state, a comparison of these dark lines with the bright lines of terrestrial vapours will indicate whether any of these are present in the vapours through which the light has passed. The spectra of the fixed stars are of this order. In the case of these bodies, the vapours that produce the dark lines immediately surround them, and are those of the substances of their surfaces. The chemical constitution of the stellar atmospheres will correspond, at least in part, with that of the stars themselves.

The speaker then described the special form of apparatus by which he and Prof. Miller have compared the spectra of the stars *by a method of simultaneous observation*, with the spectra of many of the terrestrial elements.*

This apparatus is adapted to the eye-end of an achromatic telescope of eight inches aperture. The telescope is mounted equatorially, and follows the star to which it is directed by means of an accurately adjusted clock-motion.

The point of light which a star forms in the focus of the object-glass is lengthened in one direction only by a cylindrical lens. The short line of light falls upon a narrow slit, and the diverging rays are rendered parallel by an achromatic lens. They are then refracted by two prisms of dense flint glass of 60°. The spectrum is viewed with a small achromatic telescope which is carried by a micrometer screw. By means of this the observer can measure with great precision the position of the stellar lines relatively to those of the solar spectrum.

The spectra for comparison were obtained from the spark of an induction coil taken between electrodes of various metals. Sometimes a wire of platinum surrounded with cotton and moistened with a solution of the substance required was employed. The light from the spark is reflected by a small movable mirror upon a reflecting prism covering one half of the slit. By this arrangement the spectrum of the star and the spectrum of the metal compared with it, are seen in juxtaposition, and the coincidence or the relative position of a dark line in the stellar spectrum with a bright line in the metallic spectrum can be determined with very great precision. These comparisons are observations of great delicacy, and can be satisfactorily made on the finest nights only.

Results of the Observations on the Moon and Planets.

MOON.—Limited portions of the moon's surface were examined under varied conditions of illumination. As yet no strongly marked modification of the solar light has been detected, which would indicate a lunar atmosphere of considerable extent. The mode of disappearance of the spectrum of a star when occulted by the moon, is negative as to the existence of an atmosphere about the moon.

JUPITER.—Several lines in the spectrum of Jupiter indicate a powerful absorption by the atmosphere of this planet. These were compared with the lines of our atmosphere. The atmosphere of this

---

* It is upon this method of *direct comparison* that the trustworthiness of the results which they have obtained chiefly depends. In this respect too, their observations stand alone. In 1815 Fraunhofer recognized several of the solar lines in the spectra of the Moon, Venus, Mars, and four of the fixed stars. In 1862 Donati published diagrams of three or four lines in fifteen stars. Recently Secchi, Rutherfurd, and the Astronomer Royal have given diagrams of the positions, obtained by measurement only, of a few strong lines in several stars.

planet contains some of the gases or vapours present in our atmosphere, but it is not identical with it in constitution.

SATURN.—The observations of this planet are less certain because of the feebleness of its light. Some of the lines produced by its atmosphere appear to be identical with those seen in the spectrum of Jupiter.

MARS.—The lines characterizing the atmospheres of Jupiter and Saturn are not present in the spectrum of Mars. Groups of lines appear in the blue portion of the spectrum, and these by causing the predominance of the red rays, may be the cause of the red colour which distinguishes the light of this planet.

VENUS.—All the stronger lines of the solar spectrum were seen in the brilliant light of Venus, but no additional lines indicating an absorptive action of the planet's atmosphere.

In the case of most of the planets the solar light is probably reflected not from the planetary surface, but from clouds at some elevation above it ; under such circumstances the light would not be subjected to the absorbent action of the lower and denser portions of the planet's atmosphere, which are precisely those of our atmosphere, which are most effectual in producing the so-called atmospheric lines.

Results of the Observations on the Fixed Stars.

Since these bodies are self-luminous, we may hope to gain by prismatic analysis more information of their nature, than it is possible to do of the planets which all shine by reflecting the sun's light.

What are the stars? Endeavour with the most powerful telescopes to approach them, still they assume no apparent size ; they remain under the highest magnifying powers what they appear to the unaided eye, diskless, brilliant *points*.

Until quite recently, our knowledge of the stars might be summed up thus :—that they shine ; that they are immensely distant ; that the motions of some of them show them to be composed of matter endowed with a power of mutual attraction.

Photographs of drawings of the spectra of several stars were projected on a screen by means of the electric lamp. In these were seen the coincidences and relative positions of the dark lines of the stars with the bright lines of the elements which had been compared with them.

The results on the light of the stars, *Aldebaran* and *α Orionis* (*Betelgeux*) are given in the following table :—

Elements compared with ALDEBARAN.

COINCIDENT.

1. Hydrogen	with	lines C and F
2. Sodium	,,	double line D
3. Magnesium	,,	triple line *b*
4. Calcium	,,	four lines
5. Iron	,,	four lines and E
6. Bismuth	,,	four lines
7. Tellurium	,,	four lines
8. Antimony	,,	three lines
9. Mercury	,,	four lines

NOT COINCIDENT.

Nitrogen	three lines
Cobalt	two lines
Tin	five lines
Lead	two lines
Cadmium	three lines
Barium	two lines
Lithium	one line

70 lines measured.

Elements compared with α ORIONIS (BETELGEUX).

COINCIDENT.

1. Sodium	with	double line D
2. Magnesium	,,	triple line *b*
3. Calcium	,,	four lines
4. Iron	,,	three lines and E
5. Bismuth	,,	four lines
6. Thallium ?		

NOT COINCIDENT.

Hydrogen	C and F
Nitrogen	three lines
Tin	five lines
Lead	two lines
Gold ?	
Cadmium	three lines
Silver	two lines
Mercury	four lines
Barium	two lines
Lithium	one line

80 lines measured.

The 70 or 80 lines measured represent some of the stronger only of the numerous lines which are seen in the spectra of these stars. Some of these are probably due to the vapours of other terrestrial elements which have not been compared with these stars. It would be assumption to suppose that the 65 so-called elements constitute in its entirety the primary material of the universe. Doubtless in the spectra of the stars the chemist is introduced to many new elements—would that it were possible for him to recognize and to isolate them !

It is a very suggestive fact that the lines of hydrogen corresponding with C and F of the solar spectrum are wanting in the spectra of α Orionis and β Pegasi, and in these two stars only, out of more than fifty stars examined,

β Pegasi contains	sodium, magnesium, perhaps barium.
Sirius ,,	sodium, magnesium, iron, hydrogen.
α Lyræ (Vega)	sodium magnesium, iron.
Pollux ,,	sodium, magnesium, iron.*

No stars sufficiently bright to give a spectrum have been observed to be without lines. Star differs from star alone in the grouping and arrangement of the numerous fine lines by which their spectra are crossed.

* The spectra of the following stars have also been examined :—Castor ; ϵ, ζ, and η Ursæ Majoris ; α and ϵ Pegasi ; α, β, γ Andromedæ ; Rigel, η Orionis ; α Trianguli ; γ and ϵ Cygni ; α, β, γ, ϵ, and η Cassiopeiæ ; γ Geminorum ; β Canis Minoris ; β Canis Majoris ; Spica, γ, δ, and ϵ Virginis ; α Aquilæ ; Cor Caroli ; β Aurigæ ; Regulus, β, γ, δ, ϵ, ζ, and η Leonis.

The dark lines of absorption are strongest in the spectra of stars, the light of which is tinted with yellow or red. In *white* stars the lines, though equally numerous, are very fine and faint, with the exception of the lines of hydrogen which are relatively very strong. This suggests a peculiar condition of the investing atmospheres of white stars; a modification depending possibly upon their high temperature.

A comparison of the spectra of stars which differ in the colour of their light suggested the opinion that the colours of the stars have their origin in the chemical constitution of their atmospheres.

Since the source of the light of the stars is intensely heated solid (possibly liquid) matter, the light at the time of emission would be white alike in all the stars.

The colours of the spectrum in which the lines of absorption are most closely grouped, would be subdued in power relatively to the other colours, which would then predominate in the light of the star. This opinion is supported by the different arrangements of the lines in the spectra of the orange and blue components of the double star β Cygni.

From the additional knowledge which these spectrum observations give us, we are entitled to conclude that in plan of structure the stars closely resemble the sun. The source of their light is probably solid matter,* in a highly incandescent state. Around this photosphere, there exists an atmosphere of vapours of such of the elements of the stars as are volatile at their high temperature.

A community of matter appears to exist throughout the visible universe, for the stars contain many of the elements which exist in the sun and earth.

It is remarkable that the elements most widely diffused through the host of stars are some of those most closely connected with the living organisms of our globe, including hydrogen, sodium, magnesium, and iron. May it not be that, at least, the brighter stars are like our sun, the upholding and energizing centres of systems of worlds, adapted to be the abode of living beings?

Results of the Observations on the Nebulæ.

Besides the stars, the heavens are mottled over with feebly shining cloudlike patches and spots, often presenting strange and fantastic forms. Between 5000 and 6000 of these so-called *Nebulæ* are known. What is the nature of these strange objects? Dense swarms of suns melted into one mass by their enormous distance? Chaotic masses of the primordial material of the universe? The telescope alone

---

* The phenomena of the solar spots appear to show that the solid matter of the sun's photosphere exists in a finely-divided state, and may be compared to the carbon in an ordinary gas-flame, or to the cloud of phosphoric acid when phosphorus is burnt in oxygen. The sun's surface consists of this matter separated into masses, which are neither uniform in size nor regular in form; they may be compared to clouds.

would fail to give answers to these questions, and the analysis by the prism of objects so feebly luminous appeared hopeless.

In August last, the speaker directed his telescope, armed with the spectrum apparatus, to a small but comparatively bright nebula, 37 H. iv. His surprise was great to observe, that in place of a band of coloured light, such as the spectrum of a star would appear, the light of this object remained concentrated in three bright, bluish-green lines, separated by dark intervals. This order of spectrum showed the source of the light was luminous *gas*. The brightest of the three lines has a position in the spectrum about midway between *b* and F. More refrangible than this line, and separated from it by a dark interval, a fainter line occurs. The third and faintest line coincides with F, and with a line of hydrogen. The brightest line agrees in position with the brightest of the lines of nitrogen. The line intermediate in refrangibility does not correspond with any of the elements compared with it.

NEBULÆ,

The Spectrum of which indicates Gaseity.		*The Spectrum of which is Continuous.*
7 H. iv.	Three bright lines.	92 M.
66Σ		50 H. iv.
73 H. iv.		31 M. Great Neb. in Andromeda.
51 H. iv.		32 M.
I H. iv.		55 Andromeda.
The great Neb. in Orion		26 H. iv.
18 H. iv.	A fourth, faint line also.	15 M.
Annular Neb. in Lyra	The brightest line only.	2 M.
Dumb-bell Nebula		

A careful examination of different portions of the Dumbbell nebula, and of the great nebula in Orion, showed that both these nebulæ are uniform in constitution throughout, the light from one part differs from that of another in intensity alone.

The nebulæ 37 H. iv. and 73 H. iv., in addition to the bright lines, give a faint continuous spectrum; this was proved to be due to the light of the *nucleus*.

These observations appear to authorize the following opinions of the nature and structure of those of the nebulæ which give a spectrum of bright lines.

1. The light from these nebulæ emanates from intensely heated matter existing in *the state of gas*. This conclusion is corroborated by the great feebleness which distinguishes the light from the nebulæ. A circular portion of the sun's disk subtending 1′ would give a light equal to 780 full moons, yet many of the nebulæ, though they subtend a much larger angle, are invisible to the naked eye.* Upon the earth,

---

* See 'Outlines of Astronomy,' by Sir John F. W. Herschel, p. 616. 7th edition.

luminous gas emits a light which is very inferior in splendour to incandescent solid matter.

2. If these enormous masses of gas are luminous throughout, the light from the portions of gas beyond the surface visible to us, would be in a great measure extinguished by the absorption of the gas through which it would have to pass. These gaseous nebulæ would, therefore, present to us little more than a *luminous surface.* This consideration may assist in explaining the strange apparent forms of some of the nebulæ.

3. It is probable that two of the constituents of these nebulæ are the elements, hydrogen and nitrogen, unless the absence of the other lines of the spectrum of nitrogen indicates a form of matter more elementary than nitrogen. The third gaseous substance is at present unrecognized.

4. The uniformity and extreme simplicity of the spectra of all these nebulæ oppose the opinion that this gaseous matter represents the "nebulous fluid" suggested by Sir William Herschel, out of which stars are elaborated by a process of subsidence and condensation. In such a primordial fluid all the elements entering into the composition of the stars should be found. If these existed in these nebulæ, the spectra of their light would be as crowded with bright lines as the stellar spectra are with dark lines.

The supposition can scarcely be entertained that the three bright lines indicate a more primary and simple condition of matter; for then, if the process of elaboration into stars be now taking place, we should expect to find in some of the nebulæ, or in some parts of them, a more advanced state towards the formation of the separate elements of which we now know the stars to consist. Such an advance would be indicated by an increased number of bright lines. It is difficult to suppose that the excessively high temperature of the nebulæ keeps in check affinities by which, if unrestrained, the formation of the elements would take place; for in some of the nebulæ a nucleus exists, which from its continuous spectrum, its greater brightness, and apparent separation from the surrounding gas, we must regard as containing solid or liquid matter. At a temperature at which matter can become liquid or solid (though from peculiar conditions that temperature may be a very exalted one) we cannot suppose the formation of the chemical elements to be restrained by excessive heat.

5. A progressive formation of some character is suggested by the presence of more condensed portions, and in some nebulæ, of a nucleus. Nebulæ which give a continuous spectrum and yet show but little indication of resolvability, such as the great nebula in Andromeda, are not necessarily clusters of stars. They may be gaseous nebulæ, which by the loss of heat or the influence of other forces have become crowded with portions of matter in a more condensed and opaque condition.

6. If the observations of Lord Rosse, Professor Bond, and others are accepted in favour of the partial resolution of the annular nebula in Lyra, and the great nebula in Orion into discrete bright points,

these nebulæ must be regarded not as simple masses of gas, but as systems formed by the aggregation of gaseous masses. Is it possible that the permanence of general form of these nebulæ may be maintained by the motions of these separate masses?

7. The opinion of the enormous distance of the nebulæ from our system, since it has been founded upon the supposed extent of remoteness at which stars of considerable brightness would cease to be separately visible in our telescopes, has no longer any foundation on which to rest in reference at least to those of the nebulæ which give a spectrum of bright lines. It may be that some of these are not more distant from us than the brighter stars.

8. As far as the speaker's observations extend, they appear to be in favour of the opinion that these nebulæ are gaseous systems possessing a structure and a purpose in relation to the universe altogether distinct from the great cosmical masses to which the sun and the fixed stars belong. What is this special purpose? Many fascinating theories present themselves in connection with the great problems of the conservation of the energy of the universe, and of the source and maintenance of solar and stellar heat. In the opinion of the speaker science will be more advanced by the slow and laborious accumulation of facts, than by the easier feat of throwing off brilliant speculations.

[W. H.]

Friday, May 18, 1866.

SIR HENRY HOLLAND, BART., M.D. D.C.L. F.R.S. President, in the Chair.

The REV. CHARLES PRITCHARD, F.R.S.

PRESIDENT OF THE ROYAL ASTRONOMICAL SOCIETY.

On the Telescope: its Modern Form, and the Difficulties of its Construction.

IN the museum at Naples, among other articles exhumed from the volcanic mud of Herculaneum, will be found the contents of a lapidary's shop; the visitor may there see the half-finished gems, and the tools with which they were being engraved at the moment of the workmen's fright. By the side of them is a piece of glass, rudely shaped into a convex form; and this appears to be the first record of the existence of a lens. It had evidently been used for the purpose of magnifying the microscopic cuttings on the gems necessary to produce the intended effect.

Thirteen hundred years after this, spectacle-glasses had become somewhat common in Europe, and it was by a combination of these that one Hans Lippershey, in the year 1608, at Middleburg, in the Netherlands, invented a telescope, by means of which distant objects "were brought nearer." Several of these instruments, very early in the year 1609, were provided in a binocular form, for and at the expense of the States-General of the Netherlands; but there is no record of their having been directed to any astronomical object. In May, 1609, Galileo, at Venice, heard of this new invention, without, so far as is known, any precise intimation of the method by which the effect was produced. He hastened home to Padua, and on the day after his arrival he produced the form of telescope which still bears his name. His first instrument magnified but three times. Very shortly afterwards he constructed a second telescope, possessing a power of about six linear. With this he discovered the satellites of Jupiter, in January, 1610. He also observed spots in the sun, and mountains in the moon. Very soon after these discoveries, Galileo succeeded in constructing a telescope which magnified about thirty-three times linear; and with this he discovered the long-suspected phases of the planet Venus, thereby completing the proof which was still wanting of the truth of the Copernican system. There can be little doubt that the intellectual convictions which necessarily followed upon this discovery, emancipated mankind from the thraldom of the dogmas of the Aristotelians and the Schoolmen, while, at the same time, they consigned Galileo to the persecutions of the Inquisition, and deprived him ultimately of his personal liberty. It is remarkable that, so far as the means then in existence permitted, Galileo carried his particular form of telescope to its furthest practicable limit of perfection.

For some forty years the telescope remained in the form in which Galileo left it. In 1656, Huyghens, at the Hague, substituted two convex lenses in contact for the eye-glass instead of the single concave lens employed by Galileo. This contrivance materially increased the field of view, and with such a telescope, now extended to the lengths of twelve feet and of twenty-three feet, Huyghens discovered Saturn's ring, supposed by Galileo to consist of two small spheres, one on either side of the planet. He also discovered one of the satellites of Saturn, and he contrived that admirable form of eye-piece called the Huyghenian, which, with no material improvement, is still in constant use at the present day. The field of view was again so greatly increased by this invention that Huyghens was enabled to make use of telescopes of the length of even 120 to 160 feet.

The enormous length which was at this time necessary to assign to a telescope in order to produce what is now considered a small amount of amplifying power, put a natural and very confined limit to the practical application of the instrument. The causes which necessitated this great length of the instrument, combined at the same time with very narrow limits to the size of the object-glass, were principally two, independently of the impossibility of procuring homogeneous glass of large dimensions.

First, the effect of a lens upon a pencil of light incident upon it, is to spread it out into a coloured circle, of which the diameter is, with ordinary glass, about one-fiftieth part of the diameter of the object-glass. This defect—if the consequence of a natural law can be properly termed a defect—evidently set a natural limit not only to the size of the object-glass, but to the amount of magnifying power applicable to the eye-piece, thus necessitating the twofold inconvenience of contracting the size of the object-glass, and of producing power in the telescope by the increase of its length.

Newton, by unfortunately making use of a certain species of Venetian glass of low specific gravity, and very much resembling water in its optical properties, came to the conclusion that it was not possible to obtain optical or magnifying power without, at the same time, producing the dispersion of colour, which, as we have seen, is inconsistent with clear definition in telescopes of manageable dimensions.

In 1758, John Dollond, by experimenting with glass of a different character to that employed by Newton, discovered the source of that great philosopher's mistake. He found that a lens of the ordinary heavy metallic glass of his day, called flint-glass, dispersed the colour of a pencil of light about as much as a lens of crown or plate-glass, possessing double the general deflective or magnifying power of the former. Hence, by combining a convex lens of plate-glass with a concave lens of flint-glass, but possessing only half the power (in this case a diminishing power), he obtained a combination which was capable of forming a nearly colourless image.

We say *nearly* colourless, because the powerful plate lens acts, from the nature of the material, more powerfully in refracting the middle or green portion of the spectrum than can be recovered by the contrary action of the correcting or flint concave lens. Hence, there remains behind an uncorrected or residuary spectrum of about $\frac{1}{60}$th of the breadth of the original spectrum produced by the convex plate lens—that is to say, a pencil of white light is now dispersed over a circle whose diameter is about $\frac{1}{3000}$th part of that of the object-glass. In the best modern telescopes this defect is left to its fate. Mr. Cooke is at this moment engaged in the construction of an object-glass twenty-five inches in diameter, by far the largest ever yet attempted. The diameter of the circle and chromatic diffusion in this magnificent object-glass, when completed, cannot be less than about $\frac{1}{120}$th of an inch: unless, therefore, some secondary combination is introduced, this circumstance will unavoidably prevent the employment of any powerful eye-piece.

The necessity of employing small object-glasses was thus satisfactorily removed by the discovery of Dollond in 1758. Nevertheless, there still remains another serious cause of imperfection in the compound object-glass. Generally speaking, any lens, of which the surfaces are spherical, is much more powerful towards the margin than are the parts of it near to the centre. A pencil of light, incident on the whole aperture of a lens will, in general, be diffused over a circle whose diameter bears a very appreciable ratio to the thickness of the lens. Happily, however, the actions of a convex and of a concave lens are, in this respect, in opposite directions, and hence they have, when combined, a tendency to correct or compensate the spherical aberrations of each other. Still more happily, the amount of this spherical aberration depends very materially on the *relative* curvatures of the two surfaces of the lens. Without altering the focal length of a lens, it is quite possible very seriously to alter the amount of the spherical aberration. For instance, in lenses of any material, the aberration of a plano-convex lens is four times that of a lens of equal power, where the curvature of the side facing the incident light is three times that of the other surface. This remarkable effect arises from the circumstance that, although upon the whole, the same total deviation of the light is produced in the two lenses, the distribution of the amount of deviation to be produced by each surface in its turn is very different. By the application of these principles it has become comparatively easy to produce a combination free both from primary chromatic, and from spherical aberration. The correction of the colour by a concave flint lens depends, speaking roughly and in *general*, on its focal length, and not upon the relative curvatures of its surfaces; consequently these relative curvatures can be altered until those are found which balance the spherical aberration of the convex plate-glass lens. The investigation of the best methods of obtaining these appropriate relative curvatures is attended with extreme labour and much difficulty, and has occupied the thoughts of a long succession of accomplished mathematicians. The main source of difficulty has been supposed to arise from the necessity of taking the *thicknesses* of the lenses into the account. Mr. Pritchard, however, has recently shown that the thicknesses of the two lenses have a tendency to compensate one another in the amounts of spherical aberration which they respectively either introduce or remove, and he has demonstrated that tables constructed on the principle of neglecting the thicknesses, are practically applicable to all such cases as *ordinarily* arise in the construction of the aplanatic object-glasses of modern telescopes.

We may consider then that the once difficult question of the removal of spherical aberration in an object-glass as now practically solved, and that too in a manner which requires but little further trouble than the inspection of a set of tables. There does, however, still exist a practical difficulty, where theoretically there was, or even now is, supposed to be none. It is almost universally asserted in treatises upon the subject, that in order to produce an achromatic combination nothing further is required than to make the focal lengths of the two lenses proportional to the dispersive powers of the materials of which they consist. Practically, and even theoretically, this is not the case; but, on the contrary, the proper ratio of the focal lengths of the two lenses is perceptibly influenced by the *forms* or curvatures of the lenses themselves. It is herein that the eye and the skill of the optician are required; and (perhaps unexpectedly) it is in this direction that we are to look for one of the weakest and most troublesome elements in the construction of the object-glasses of telescopes. Such, however, has been the improvement in the manufacture of glass in England, and such is the ability of modern English artists, that there is good reason for believing they have now regained the ancient supremacy, which in this respect existed in the time of Dollond.

[C. P.]

Friday, May 25, 1866.

Sir HENRY HOLLAND, Bart. M.D. D.C.L. F.R.S. President, in the Chair.

A. S. HERSCHEL, Esq. B.A.

On the Shooting-stars of the years 1865–6, *and on the Probability of the Cosmical Theory of their Origin.*

ATTENTION was recently directed by Professor Newton,* of Yale College, U.S., to the probability, on well-considered grounds, that in the current year, 1866, a prodigious flight of meteors, the most imposing of its kind, and visible over a large area of the earth's surface, will make its appearance—perhaps for the last time in the present century—either on the morning of the 13th, or on the 14th of November. The rare opportunity thus afforded of deciding some important questions in the theory of shooting-stars makes it a matter of special interest for persons skilled in such accurate observations, to watch for its return on each of the mornings named (wherever practicable, between one and two o'clock†), to obtain the necessary data. The phenomenon at its maximum was seen by Humboldt, at Cumana, on the morning of the 12th of November, 1799; and again by Dr. Denison Olmsted, in its greatest brilliancy, at Newhaven, U.S., on the morning of the 13th of November, 1833. Olmsted was the first to sum up, in the following general language, the chief characteristics of the display :—

1. The number, *especially of bright meteors*, is much larger than usual.

2. An uncommonly large proportion *leave luminous trains.*

3. They proceed, with few exceptions, from a common centre in some part of the constellation *Leo.*

4. They are seen from midnight to sunrise, *and in greatest abundance between three and four* A.M.

---

* 'American Journal of Science,' 2nd series, vol. xxxvii. p. 377; and vol. xxxviii. p. 53.

† The object in restricting the watch to a particular hour is, that as many meteors as possible *may be simultaneously observed at distant places.*

These characteristics were regarded by Olmsted as sufficient to identify for the future the return of the November shower. Observers who endeavour to verify the truth of this description for themselves will be enabled by so doing, to furnish invaluable materials for future investigations. The same features are, moreover, of peculiar interest, since they have been found to characterize, with certain differences, a number of other meteoric showers.

The periodicity of the well-known shower of meteors on the 10th of August was discovered independently by Quetelet, in the year 1836, and by Herrick (not knowing what Quetelet had predicted) on the reappearance of the same meteors in the year 1837.* At their next return, on the 10th of August, 1839, it was shown by Littrow† that a star-shower, whose occurrence in A.D. 1451, on the 5th of August, N.S., is cited by Biot from the 'Chinese Annals,'—compared with another that appeared, according to the same authority, in A.D. 830, on the 26th July, N.S.,—gives 365 d. 6 h. 12 m. as the average length of the interval between any two successive returns of the August meteors; and again, that the same star-shower compared with that of the 10th of August, 1839, gives 365 d. 6 h. 8 m. as the average length of the same period,—the mean of the two determinations being 365 d. 6 h. 10 m., which is twenty minutes longer than the civil or tropical year; *but differs less than one minute from the sideral year*. Besides the displays of the August meteors which happened about the year A.D. 830, other returns of the August meteors, of the nature of star-showers, are recorded in the 'Chinese Annals'‡ to have happened at the end of July, N.S., about the year A.D. 933; and another is cited by Herrick which happened on the 2nd of August, N.S., in the year 1243. The meteoric year of August, therefore, coincides with the sidereal year; and a period of 103 years brings all the star-showers of A.D. 830, 933, 1243, and 1451 into conjunction with the remarkable star-shower of the 10th of August, 1863. A star-shower is cited, moreover, by Biot, which happened at the end of July, N.S., in the year A.D. 833. A return of this star-shower may not unreasonably be expected to occur on the night of the 10th of August, 1866.

Between the 13th of October and the 13th of November, during the years from A.D. 903 to 1833, not less than thirteen great star-showers have been recorded. They are separated from each other by the third part of a century, or by some multiple of this period, and are periodical re-appearances of one grand meteoric shower,

---

* 'American Journal of Science,' 1st series, vol. xxxiii. pp. 176, 354; and vol. xxxvii. p. 338.

† 'Astr. Nachr.,' vol. xviii. p. 374.

‡ Regarding the visibility of the August meteors in China, Captain Sherrard Osborn, C.B. R.N. &c., relates that they were seen by the watch on board the 'Furious,' near Yedo in Japan, on the night of the 9th—10th of August, 1858, shooting by hundreds from the north-east to the west overhead.—*Cruise in Japanese Waters*, p. 102.

viz. that seen by Humboldt in 1799, and by Olmsted in 1833—the star-shower expected to return in the present year, and known by the name of the 'Great November Shower.' Its encounter with the earth takes place one day later in the year at each of its principal returns. According to the exact calculations of Professor Newton, the next passage of the earth through the centre of the meteoric group will take place two hours after sunrise at Greenwich, on the morning of the 14th of November, 1866.* As, however, the phenomenon occurred upon the morning of the 13th, in each of the years 1864 and 1865 (and not so brightly upon the 14th), its moment of greatest brightness may, possibly, fall one day earlier than the predicted time, and a watch on both mornings, of the 13th and 14th, is accordingly recommended to be kept. The meteors being seen from midnight to sunrise, the hour from one to two o'clock is the best that can be chosen, as being the most convenient for making simultaneous observations.

On the morning of the 13th of November, 1865, the meteoric shower in England was observed at the Greenwich and Cambridge Observatories, at Hawkhurst, and at other places. The hourly number of meteors is stated by Mr. Glaisher and by Prof. Challis to have exceeded all before recorded at either of those Observatories. More than 250 meteors (279) were recorded at Greenwich from midnight until five o'clock, when for a space of nearly a quarter of an hour the paths of the meteors among the stars, &c., were not noted, but their number was simply counted. The result was, that at this time meteors of the first class were appearing at the rate of 250 per hour. At least a thousand meteors must have been visible at Greenwich from one to five o'clock on the morning of the 13th. They were only seen *after* midnight, until near sunrise, and they were most abundant during the hour from one to two o'clock. The maximum display of the November meteors expected in the year 1866, is still several hundred times greater than that observed at Greenwich on the 13th of November, 1865. *Two hundred and forty thousand* meteors are computed by Arago to have been visible above the horizon of Boston on the morning of the 13th of November, 1833.† Hundreds of the meteors seen on that occasion appeared brighter than the planets; and some of them were fireballs of the largest class.

Nearly two-thirds (172) of the meteors seen at Greenwich left luminous trains, visible for several seconds after the disappearance of the meteors. Their unusual number, and the appearance of the luminous trains which they left behind them, agree with Olmsted's description of the meteors in 1833, and leave no doubt of the return of the November shower. Its reappearance in the present year, there

---

* 'American Journal of Science,' 2nd series, vol. xxxviii. p. 60 *et seq.*

† Mr. Greg informs the writer that, as seen by Mr. Baxendell, F.R.A.S., on the west coast of Mexico, "The number of meteors seen at once, often equalled the apparent number of the fixed stars seen at a glance!"

is every probability, will entitle its next return to the name of the 'Great November Shower' of 1866.

Amongst the list of shooting-stars seen at Hawkhurst last November, seventeen were identical with meteors observed at Greenwich. Fifteen other meteors of the list were identical with meteors seen at Cambridge. The heights, lengths of path, &c., of ten of these accordant meteors were calculated, and this is also the number of accordances calculated by Dr. Heis of Münster. The average heights obtained at both places are entered in the following table for comparison:—

	Mean Heights of Ten November Meteors observed at	
	1. Hawkhurst. Brit. Stat. Miles.	2. Münster. Brit. Stat. Miles.
Mean height at first appearance	74	77
Mean ditto at disappearance	54	44
Mean ditto at centres of visible paths . . .	64	61

The average height of the centres *is a little greater* than the commonly received height of shooting-stars. Professor Newton has shown that the ordinary height of shooting-stars at the middle of their apparent paths is *not quite sixty miles above the surface of the earth.**

The divergence of the November meteors in the year 1865 from a common centre in the constellation *Leo*, was noted both in Europe and America, and the following positions were obtained:—

Place, and Observer's Name.	Position of Radiant Point, Nov. 13th. R.A.	N. Decl.
Greenwich (Mr. Glaisher)	160°	30°
Hawkhurst (Mr. Herschel)†	150	20
Münster (Dr. Heis)	148	24
Newhaven, U. S. (Prof. Newton) . . .	148	23
Philadelphia, U. S. (Mr. Marsh) . . .	148	24

The average of these positions is some degrees from the bright star γ *Leonis*, of which the position is in R.A. 153°, N. Decl. 21°; nevertheless the close agreement of the last three places of the list merits particular attention.

The following radiant points of special meteor-showers were recorded at Hawkhurst during the last few years, by the aid of star maps expressly suited to the purpose.—Views of the original observations drawn on punctured maps showing the radiant points, were exhibited, by Dr. Tyndall's assistance, on the white screen.

---

* 'American Journal of Science,' 2nd series, vol. xxxix. p. 194.

† 'Astronomical Register,' No. 37, p. 18.—The radiant point of thirty-five meteors observed during the hours *from one to three o'clock*, was in right ascension 148°, north declination 23°,—exactly the position assigned by Professor Newton. (A. S. H.)

METEORIC PERIOD.*	Date of Observation.	Position of the Radiant Point. R.A.	N. Decl.
Jan. 2–3	1864	234°	51°
Apr. 9–11	„	192	4
„ 19–21	„	278	35
Aug. 9–11	1863	44	56
Oct. 18–20	1864–5	90	16
Nov. 12–14	1865	150	20†
Dec. 11–13	1863–4	101	34

A number of observations of luminous meteors contained in the Catalogue of the British Association, from the year 1845 until the present time, are capable of being classified in a similar manner, and upwards of fifty radiant points of shower-meteors throughout the year are recognized as coinciding in their epochs and positions with the epochs and positions of similar radiant points observed by Dr. Heis. Charts of these meteors are now in process of completion, and a first step will be gained thereby, towards establishing a regular shower-meteor theory. When the principal meteoric showers, and their connected radiant points, and epochs are pointed out, what are commonly called "sporadic" shooting-stars will become extremely scarce.

A few instances lately brought to light will show that aërolites and fireballs are not only independent of geographical position, but that they are also fixed in their dates of appearance and directions. Two stonefalls took place on the 25th of August, 1865; one at Aumale, in Algeria, the other at Shergotty, in India. Two detonating meteors of the largest class were seen in December, 1865; one on the 7th, at the mouth of the Loire, the other on the 9th, at Charleston, U.S. Three detonating meteors, on the east coast of England alone, during the last five years (1861–5), occurred between the 19th and the 21st of November. Two of the latter meteors, whose paths could be traced, proceeded from a common centre in some part of the constellation *Cetus*. A bright fireball was seen at Hawkhurst on the morning of the 9th of December, and another at Manchester on the morning of the 13th of December, 1864, whose paths (continued *backwards*) proceeded directly from the radiant point of the period mentioned last in the foregoing list. A notable peculiarity in respect of general geographical distribution, periodical returns, and fixed directions would evidently connect aërolites and fireballs in true astronomical relation with ordinary shooting-stars. Nevertheless it appears that out of 72 aërolites whose hour of fall is certainly known, by far the greater number (58) occurred *after midday*, during the hours from noon to nine P.M.‡ Shooting-stars, on the contrary, reach

---

* These meteoric periods are advanced by Baumhauer, as dates on which fireballs are most common. 'Poggendorff's Annals,' vol. lxvi. p. 471. 1845.

† *See* note (†), p. 648.

‡ Only thirteen fell in the forenoon, and fifty-eight during the hours from noon to nine P.M. 'B. A. Report—1860,' p. 26.

their maximum at an opposite hour of the day, being found to be most abundant *after midnight;* or twelve hours later. An astronomical difference accordingly exists between aërolites and shooting-stars, to which it is not impossible that a physical difference of a kind not yet established should correspond. It is noticed, for example, that on the 10th of August, and the 13th November, dates on which shooting-stars and fireballs are more abundant than on any other nights in the year, *but one stone has fallen* on each date. The average height of seventy-eight meteors observed in America on the 13th of November, 1863, exceeded the usual height of meteors by fifteen or twenty miles.* On these grounds Professor Newton supposes that the November shower-meteors are composed of more easily destructible, or of more inflammable materials than aërolitic bodies. An examination of the light of shooting-stars (and particularly of the November meteors) by means of the spectroscope, would at once reveal the fact if these meteors and their luminous trains are composed of substances in a gaseous state; *the presumption is that they are solid*, although probably in a state of fine division. Whether shooting-stars consist of particles in a state of dust, or vapour, which the spectroscope can alone reveal, the return of the Great November Shower is a favourable opportunity for advancing our knowledge by its use, and a discovery is in store for refracting prisms, which should not (if possible) be allowed to pass away unheeded.

The spectrum of an entire star-shower was shown, by Dr. Tyndall's assistance, upon the screen with a straight-vision prism having a direct field of view of at least 20° of the sky.† The meteors were illuminated first by the carbon points of the electric lamp, and afterwards by the vapour of silver in the electric arc, showing the different appearances in the prismatic field of view, of a solid, or a gaseous source of light. Some tinder was ignited by a fire-syringe; and experiments with Gassiot's Geissler tubes were shown to illustrate the difference that exists, between the diffuse glow of an electric discharge, resembling the aurora in the highest and rarest strata of the atmosphere, and the massive light of meteors, resembling an ordinary spark.

[A. S. H.]

* 'American Journal of Science,' 2nd series, vol. xl. p. 252.

† A pair of such prisms, mounted upon a binocular frame, is recommended as the best instrument for meteospectroscopic observations. Mr. Browning constructs a Binocular Pattern of Dense Glass Prisms shown at the lecture, under the name of a "Meteospectroscope."

Friday, April 12, 1867.

SIR HENRY HOLLAND, Bart. M.D. D.C.L. F.R.S. President, in the Chair.

BALFOUR STEWART, Esq. LL.D. F.R.S.

DIRECTOR OF THE KEW OBSERVATORY.

On the Sun as a Variable Star.

THE man of science who would extend his inquiries into the remoter regions of the universe is beset from the first, with the following difficulty.

How is he to know that the laws with which he is here familiar hold in those distant regions?

Now, without attempting to discuss the origin of our beliefs, it may be inferred that the same principle which induces us to think that the laws of light which were proved to be true to-day will hold true to-morrow, induces us likewise to believe that what is proved to be true here will hold elsewhere.

But while we are all, without exception, led by an innate belief in unity of design to attribute the same fundamental laws to different regions of the universe, it is only lately that we have been informed of a similarity or rather identity which we had no previous reason to expect.

Spectrum analysis, that very powerful and searching method of investigation, while it informs us of great varieties in the molecular constitution of different regions of the universe, seems to proclaim the fact that the elemental forms of matter are greatly the same throughout, and that familiar substances, such as sodium, magnesium, and iron, with which we are here so well acquainted, form also the staples of other worlds.

A study of binary stars has likewise shown that the law of gravitation is not the peculiar attribute of our solar system, and we may entertain the hope that as our knowledge extends the appearance of resemblance between ourselves and distant regions will extend along with it.

In fine, there can be little doubt that we are to-day in a position to argue with more confidence regarding these regions than we were fifty years ago, and that we may assume that the laws of heat and light are the same throughout the universe. If this principle be allowed, the discussion of the varying brightness of certain stars, and it may be of our own sun, is one which we can approach with great hopefulness.

Indeed, we are now by means of this principle enabled to limit the number of the immediate causes of this change of light, and to say the change must be due to one or other of these causes. Now, when these various causes are examined one by one there is a tendency in the mind to reject certain of these and to prefer others, apart altogether from the results of observation; and this is one of those tendencies which, if it is indulged in with distrust, ought not to be wholly discarded.

But, while attempting to exhibit this evening this method of selection, and to hit upon the most probable cause of variable luminosity, it becomes us all, as disciples of the school of Bacon, to see what observation has to say to our selection; does this ultimate court of appeal confirm our conclusions or does it not?

[The phenomena presented by variable stars were then described; and it was next argued, from the ten-yearly period of sun-spots that our own luminary is a variable star. The photographs of sun-spots were also exhibited on a screen by aid of the electric light.]

Let us now proceed to the subject of the diminution of light which characterizes variable stars, and according to the principles already stated, see what such a diminution most probably implies. Two cases may be discussed under this head, according as the temperature may be supposed to remain constant or to change. First,—If the temperature remains constant, then a change of light implies a change of the surface or of the state of the heated body. This was illustrated by two simple experiments. In the first of these a ball of iron marked with chalk, and a piece of porcelain of a black and white pattern, both heated to redness, were viewed in the dark, and the white of both was found to be less luminous than the black, thereby proving that change of the reflecting power of the surface produces change of luminosity, even although the temperature be the same. In the second experiment it was shown, by means of the oxyhydrogen flame viewed by itself and then directed against a piece of lime, that a solid substance gives out much more heat and light than a gas at the same temperature, and thus that a change of state may be expected to produce a change of luminosity.

If, therefore, a solid or liquid substance change suddenly from a black to a white or reflecting surface, or if it become gaseous, it will decrease in luminosity.

It was then argued that we cannot readily suppose the variable luminosity of certain stars to be caused by a periodical change of surface from white to black and *vice versâ*.

On the other hand, if we assume (as is most probable) that the photosphere of a star consists of incandescent, detached particles similar to a cloud, we may well imagine this cloud-surface to contract or expand in the atmosphere of the star, so as to present a variable area to a distant observer. Mr. De la Rue and some others, including the speaker, entertain a suspicion that something of this kind may possibly take place with regard to our sun, although to a very small extent; but this is not proved. In the meantime, suffice it to say, such an explanation will not account for a variability analogous to sun-spots; and to account for these and similar phenomena, we must look for some other explanation. We may therefore take it for granted that a change of surface or of state, without a change of temperature, will not account for the phenomena of variability, in as far as these are similar to sun-spots.

A change of luminosity may, however, take place in another way, even while the temperature of the hot body remains constant. It may be caused by the interposition of a cold absorbing screen between the source of light and the observer.

Under this head may be ranked one hypothesis regarding stellar variability, which supposed the decrease of luminosity to be caused by a dark body of great size coming between the star and our earth. The probability of such a body being in an exact line between the star and the earth is, however, very small; and as our system is supposed to move in space, we should soon escape its interference, unless its size was supposed to be enormously large. This is not likely; and if stellar variability is caused by such a screen, the screen must be supposed to lie close to the star—in fact, to be connected with the star itself, and to form part of the atmosphere of the star.

The second case of varying luminosity is that in which the temperature may be supposed to change; and here evidently a decrease of luminosity implies a decrease of temperature.

If we imagine that stellar variability is caused by a decrease of temperature, we are led to contemplate two possible causes of this decrease.

In the first place, there may be supposed to be some chemical or molecular change periodically recurring, which produces a marked decrease of temperature. Evaporation on an extensive scale might account for it; but the same objection applies here as before, when we considered change of surface.

We have not the shadow of a proof that such processes are periodically going on in the sun or any star; and we do not get rid of our burden by this means, but merely, as it were, shift it from one shoulder to another.

Now, if we do not readily admit that the supposed fall of temperature is produced by some such process, we can only account for it by the redistribution of some previously existing comparatively cold matter; and this comparatively cold matter must be either on one side, under, or above (by under is meant nearer the centre).

It has been supposed by some that this cold matter might be to one side; that, in fact, a star might have one hemisphere cold and the other hot: so that, if it revolved round an axis, the effect produced on a distant spectator would be that of a varying brightness. The difficulty in this idea is two-fold. In the first place, we cannot easily conceive such an extremely artificial distribution of heat; and even if this could be conceived, we could not imagine that it would be a permanent one. This, therefore, is an unlikely hypothesis.

Next, the idea has been entertained that the comparatively cold matter whose re-distribution we are at present supposing, exists below, or nearer the centre of a sun or star than its luminous envelope; and the fact that sun-spots appear to be depressions, has given countenance to this idea; but if we imagine that the sun or star has been for a very long period of time surrounded by this luminous envelope, we shall have great difficulty in imagining the interior to be colder than it. As far as we can judge from terrestrial experiments, a body surrounded with a heated envelope, such as that of the sun, will ultimately have in all its parts the temperature of this envelope. We cannot, therefore, readily assent to this hypothesis, although it might add the sun and stars to the list of habitable worlds.

Now, if the colder matter, whose re-distribution we are supposing, come neither from one side nor from below, it must come from above; that is to say, above the luminous envelope of the sun and stars we must have colder matter.

Thus we see that if we consider the decrease of luminosity to be due to the presence, between the source of light and the observer, of a comparatively cold, absorbing body, we are driven to an atmosphere; and if we consider the decrease of luminosity to be due to a decrease of temperature, we are still driven to look to an atmosphere as the immediate cause.

We must now bring our results to the test of observation, and ask: What reason have we, in the first place, for assuming the existence of solar and stellar atmospheres? and in the second, What reason have we for supposing the decrease of luminosity to be immediately caused by an increased action of this atmosphere? Now, in the sun we have various proofs of the existence of a comparatively cold, absorbing atmosphere above the luminous envelope.

(1.) The existence of dark lines in the spectra of the sun and certain stars denotes the existence in these bodies of a comparatively cold, absorbing atmosphere above the luminous envelope.

(2.) The existence of such an atmosphere surrounding our own luminary is indicated by the fact that the sun's limb is less bright than his centre; this effect being, no doubt, caused by an absorbing atmosphere, and being greatest round the edge, for the same reason that any similar effect of our own atmosphere is greatest near the horizon.

(3.) Finally, the presence of a solar atmosphere extending as far as 72,000 miles above the bright surface, is indicated by the presence of those red flames which occur during a total eclipse of the sun. These red flames have been proved by Mr. De la Rue to belong to our luminary, while, from the nature of the light which they emit, we may infer (although this is not yet proved) that the heated matter is gaseous.

We now come to consider what observational evidence there is that the changes of luminosity in the sun's disc are due to the effects cooling, absorbing, or both together, of a greater or less stratum of atmosphere.

In the first place (under this head), there is abundant evidence that the luminous surface of the sun does not consist of one uniform mass of incandescent, solid, or liquid matter, it is neither land nor ocean, but it is a cloud. The proof of this assertion is derived from the behaviour of certain bright patches, or faculæ, as they are called, which appear near the sun's border, and generally accompany a spot or dark patch. The brightness, when near the border of these faculæ, denotes that they have escaped, in a great measure, the absorbing influence of the solar atmosphere, which influence is very strong near the border; in other words, they are elevated above a great portion of this atmosphere, and as they remain suspended for some time, they cannot be heavy matter. Indeed, masses of luminous matter have been known to sail across a spot evidently above it, and leaving it afterwards quite undisturbed.

Now, if we imagine the bright surface of the sun to be a cloud surface, or surface of condensation of small particles in contradistinction to a solid or liquid heavy surface, we at once, by such an hypothesis, greatly increase the freedom of action. Such a boundary might easily be depressed by the accumulation of an enormous downrush of cold atmosphere, or it might be raised above its ordinary level, and, generally speaking, would be more impressible than a continuous solid or liquid surface.

Again, if we view a spot as the centre of a disturbance of some kind, it is, in the first place, worthy of remark that the faculæ or bright portions which accompany a spot for the most part fall behind, as far as rotation is concerned, a fact which has been shown in the solar researches of Messrs. De la Rue, Stewart, and Loewy. Now, this would always take place in the case of a body carried from a lower region to a higher one, or from a region possessing less to one possessing greater absolute velocity of rotation. We are, therefore, induced to suppose that faculæ are masses which have been carried upwards from the area of disturbance, and have thus fallen behind.

So much for the up-rush of matter or the ascending current; and now, it may be asked, have we any evidence that a spot is a descending current?

We have evidence of a precisely similar character to that we have for faculæ, and we are entitled to conclude, from the observations of sun-spots made by Carrington, that spots, instead of falling behind, as far as rotation is concerned, move forward as if they had come from a higher region.

Also Lockyer, one of our sun-observers in this country, noticed a piece of matter in the very act of moving down. It was first of all as bright as the faculæ, then it became like the ordinary surface, then it grew dark like the spot itself, still retaining its identity of form.

It would thus appear that the comparatively cold, absorbing atmosphere is accumulated above the area occupied by a spot, while the faculæ are so high up as to escape its influence; and finally, we are led to conclude that all the variations in brightness that appear on the surface of our luminary are due to the presence, to a greater or less extent, of a comparatively cold, absorbing atmosphere.

We thus perceive that the phenomena of variability, as far as these are analogous to sun-spots, are due, most probably, to a greater or less amount of a comparatively cold, absorbing atmosphere.

A down-rush and a corresponding up-rush would thus appear to be the immediate cause of these spots; yet why, it may be asked, have these phenomena a periodicity? Why is there a ten-yearly period of sun-spots besides other probable periods?

At the same time, the following question arises, Why are sun-spots confined to the equatorial regions of the sun, which are also those regions which border upon the ecliptic or plane of the planets' motion?

Arguing possibly from this fact, the illustrious Galileo seems to have imagined a connection between sun-spots and planetary configurations; but he did not publish his ideas, probably from want of evidence.

In order to obtain as much information as possible on this point, Messrs. De la Rue, Stewart, and Loewy have measured the areas of all the spots in Carrington's original pictures, extending from the beginning of 1854 to the end of 1860, and the result deduced from these measurements is favourable to the idea of a connection between the behaviour of sun-spots and planetary configurations.

These results were obtained by noticing that at one period of time all the spots, as a rule, increase towards the centre of the sun's visible disc as they pass over it by rotation, while at another period they all decrease from their first appearance, or perhaps increase from their first appearance. Considering the earth merely as the point of view from which these various phenomena are observed, we have to ask, What is the cause of this peculiar behaviour of sun-spots?

It evidently must refer to something without the sun, and that something is not so very difficult to find. When these phenomena denoting peculiarities in the behaviour of sun-spots are attentively studied, it is seen that every twenty months the same behaviour occurs again. Couple with this the fact that every twenty months the planet Venus returns to the same position with reference to the earth, and we can scarcely help attributing some predominating influence over the behaviour of sun-spots to this planet. A closer analysis of the phenomena observed shows us that this is the case; and that as any portion of the sun's surface retreats by rotation from the neighbourhood of Venus, the spots on that portion have a tendency to increase, attaining a maximum at the point furthest from Venus. Jupiter has also much influence. Now, is it not a very extraordinary circumstance that two planets which are never so near the sun as they are near the earth, should appear to cause phenomena of the vast magnitude of solar spots?

It naturally occurs to us that the sun must be in a most sensitive molecular state, in consequence of which that wonderful mass experiences great changes from very small outward influences. (Experiments were here made, showing examples of this state.) Professor Tait and the speaker have conjectured that the properties of a body, especially with reference to heat, light, and electricity, may be influenced by the neighbourhood of a large body. An influence of this kind would naturally be most powerful upon a body such as the sun, which is of a very high temperature, just as a poker thrust into a hot furnace will cause a greater disturbance of heat than if thrust into a chamber very little hotter than itself.

We have, moreover, very good grounds for supposing the sun to be in a very sensitive molecular state. We may infer from certain experiments, especially those of Cagniard De la Tour, that at a very high temperature and under a very great pressure, the latent heat of vaporization is very small, so that a comparatively small increase of heat will cause a considerable mass of liquid to assume the gaseous form, and *vice versâ*.

We might thus suppose that an extremely small withdrawal of heat from the sun might cause a copious condensation, and this change of molecular state would, through the alteration of reflection, &c., alter, to a great extent, the distribution over the various particles of the sun's surface of an enormous quantity of heat.

Again, convection is very strong in the sun, since the force of gravity is very strong, so that great mechanical changes might very easily result.

[B. S.]

Friday, May 24, 1867.

JOHN PETER GASSIOT, Esq. F.R.S. Vice-President, in the Chair.

A. S. HERSCHEL, Esq. B.A. F.R.A.S.

On the Shooting-stars of the Years 1866–67, *and on the probable Source of certain Luminous Meteors in the material Substance of the Zodiacal Light.*

REGARDED as an exhibition of a variable phenomenon, recurring at the end of every cycle of nearly thirty-three years, on a particular date of November, the meteoric shower witnessed on the morning of the 14th of November, 1866, appears to have been a fair example of the average scale of the November meteors at one of their principal returns. While it incomparably surpassed all the commoner displays of shooting-stars that are known to have occurred during the past period of more than thirty years, it nevertheless fell considerably short of the celebrated meteoric shower seen in America on the morning of the 13th of November, 1833. That shower, it will be remembered, took place quite unexpectedly, while one of the distinguishing features of the recent November star-shower was that it afforded a complete verification of the astronomical theory which made their return expected by Olbers, as about to take place in 1867; and more recently by Professor H. A. Newton, who anticipated the recurrence of the shower in 1866.

Although intended only to guide observation, due notice of the shower was timely given by Professor Newton,* to the effect that a considerable star-shower might be expected to take place on the morning of the 14th of November, 1866. On the eve of the occurrence the well-timed appeal was repeated in many places in the public papers, and a wide-spread and very intense popular interest in the phenomenon was excited. In England circular letters were addressed to the Members of the Royal Astronomical and Royal Meteorological Societies, by their respective Presidents, Mr. Pritchard and Mr. Glaisher, suggesting to observers the propriety of making concerted observations of the star-shower, during the second hour after midnight on the two mornings of the 13th and 14th of November, so that if

* 'American Journal of Science,' 2nd series, p. 60 *et seq.*

possible a number of the meteors might be simultaneously observed at distant places. On the first of those mornings the sky was completely overcast, and on the morning of the 14th of November the meteoric shower made its appearance with, certainly, great beauty, but so exactly during the hour named beforehand for simultaneous observations, that many, even of the most zealous of the confederated observers, as they expressed themselves, "gave up recording, and betook themselves to counting."

Accordant duplicate observations could hardly be expected to be obtained, when meteors were so very numerous, and were so strikingly uniform as to size, as seldom very greatly to exceed first magnitude stars in brightness. Of fireballs sufficiently brilliant to have attracted general attention (at least over Great Britain), there were very few. The shower, accordingly, met with less favourable observation, as regards ascertaining the absolute altitude of the meteors, than that which was observed more successfully for the same purpose in the previous year,* when the heights and velocities of several meteors were determined.

One observation of the kind, obtained in the recent meteoric shower, will, however, shortly be mentioned in detail. On the other hand, most important results were obtained by Mr. Glaisher at Greenwich,† Professor Adams and Dr. Challis at Cambridge,‡ and by other eminent observers and astronomers, both at home and abroad, who gave their best attention to the subject. The vexed question of where the meteors came from, was thus satisfactorily disposed of. The orbit of the meteoric group was finally determined; and, lastly, three recent comets, to which closed orbits have been assigned, now rank, almost certainly, as forming part of the material currents which give rise, respectively, to the meteors of the 10th of August, the 14th of November, and the 20th of April. The November star-shower of 1866, accordingly, both for the astronomical premonition which it fulfilled, and for the novel views which followed it, marks a new era in meteoric astronomy, not unlike that which dawned upon cometary astronomy when Clairault calculated the day of the return of Halley's comet, in the year 1759, and the comet appeared, almost punctually, at the appointed time.

In America the star-shower on the morning of the 14th of November last, was expected to be visible to the best advantage. Only 172 meteors, however, mostly of small size, were counted at the Washington National Observatory on that morning, during an interval of two hours and a half in which the sky was clear; indicating about the same rate of falling as on the preceding night. There was nothing peculiar, either in colour or in motion observed. At Newhaven,

---

* 'British Association Report,' 1866, p. 139; and these 'Notices,' vol. iv. p. 648.

† See Diagram showing the average number of meteors per minute at Greenwich. 'Monthly Notices, R.A.S.,' vol. xxvii. p. 54.

‡ Ibid. pp. 75 and 247.

U.S.A.,* the average number of meteors seen by one person of Professor Newton's staff of observers, on the same morning, with a beautifully clear sky, was about thirty-eight meteors in one hour, the average number on the preceding night having been from sixteen to twenty meteors in an hour. So great was the disappointment of astronomers in America, who confidently expected a successful view of the phenomenon, that a telegram, *viâ* the Atlantic cable, which appeared in the New York 'Herald,' announcing the appearance of the shower in England, could hardly be believed, until the arrival of the English newspapers at New York dispelled the doubts of its correctness. Owing to the seasonable notice of the shower finding its way by various channels, into every civilized quarter of the globe, observers in Europe, and elsewhere, were not a whit less expectant than in America; and here, at least, they were not destined to be disappointed, since the star-shower was everywhere conspicuously seen, and witnessed with admiration. The following extract from a letter from Syria, published in the New York 'Tribune' of the 29th of December, realizes the figurative language with which the early Arabian chroniclers sought to adorn their description of the great November star-shower of the year A.D. 1202:—

"I have just received the Arabic newspaper of this week's issue, and find in it the following news about the meteors, which, for your benefit, I will translate literally:—

"'*Beirut Domestic Intelligence.*—There has preceded this a notice, in No. 431 of our journal, of the falling meteors of the 12th and 13th of November, and there happened a marvellous thing of the kind on the night of the 13th. . . . People of Beirut saw thousands of these meteors, mixed in commotion and confusion, and they compared their extent in the heavens to the spreading out of locusts in the sky. And we have news from Damascus, that the same events were seen there, and they compared them to the mighty armies, joined in a fierce strife, from the four quarters of the sky. . . .'

"The Arabic journal then gives a very fanciful letter on the subject from one of the learned men of Damascus, the scholar Solyman Effendi Sooloh, who says:—

"'In this past night the stars began the war from the east to the west, and from the southern to the northern side. They dashed at the pace of fiery steeds and ghouls, so that you could not distinguish the Pleiades from the Hyades from the passing of the meteors across them, and the intensity of the brightness. But you now thought that the two stars in Leo's nose had been dispersed, and the two fishes were eclipsed and immersed, and the spearman of Arcturus had forgotten his spear, and was thinking only of his own safety, and the Adhal was complaining to the bright daughters of Ursa Major about the extent of his wound, and the lofty pole had fallen into the claws of the Eagle, and the Hedrah was prostrate, and the face of night like

* 'American Journal of Science,' vol. xliii. pp. 78 *et seq.*

a leopard's skin; and to sum up all the heavens were like a sphere of fire, or a gleaming of sparks, excepting that the fire and sparks were harmless, not touching the earth, or injuring our safety, as if night's daring horsemen, who continued till morning beating each other in single combat, gave us protection and peace. This I write for his Excellency our Prince, the Sultan Abdul Aziz Khan. May God perpetuate the seat of his government to the end of the world's revolution!'"

The same letter from Syria further mentions that, "on the morning of the 11th (Sunday), at a little after midnight, some young men in Beirut, who were out of doors, saw what they described as a rain of fire, the stars seeming to have got loose, and to be running about the sky in disorder." The occurrence of such a companion shower to the principal display, is not by any means improbable, and well deserves attention.* The meteors on the morning of the 14th of November were also seen in Persia, on the road to Ispahan.†

A letter from Mr. W. Masters,‡ Professor at the Kishnaghur College, about fifty-seven miles due north of Calcutta, in India, to Sir John Herschel, gives the following description of the shower:—

"I looked out about half-past four, or a quarter to five, and . . . after counting fifty in about five minutes, I woke up four other persons to witness the phenomenon, and to give aid in watching and counting. We arranged ourselves looking in different directions, and as each saw a meteor there was a distinct call of the next number, 51, 52, 53, &c.; the stars shooting out sometimes faster than they could be counted. Some were lost on this account, . . . yet in less than half-an-hour we counted 420; had we been all together during the half-hour we would certainly have counted more than 500."

The meteors were visible also at Sealkote, and at Lahore. Dawn appearing, however, put an end to the display, but a bright meteor was still seen at Kishnaghur after daylight had appeared.

A short note of the phenomenon at Yokohama, in Japan, and a private memorandum from Mr. B. V. Marsh, of Philadelphia, U.S.A., received by the speaker, are as follows, and concur in showing that the star-shower was not visible in the extreme east of Asia:—

---

* "It appears from observations of Captain Meiraldi, that another meteoric swarm passed over Italy during the night of the 12th–13th, and its characters did not differ from those of the display on the 14th at Urbino: that is to say, a great number in a short space of time, beginning unexpectedly, and ceasing very suddenly."—'Les Mondes,' 2nd ser. vol. xii. p. 644. Private letters inform the writer that a great and sudden outburst of "hundreds of shooting-stars," lasting only a short time, but so sudden and bright as to be terrifying, was seen at Norwich shortly before four o'clock A.M., on the 14th. The same was seen near Staplehurst, in Kent, "on the stroke of four o'clock," on the morning of the 14th, when the regular shower of shooting-stars had almost ceased to be watched. The notes, perhaps, refer to one and the same phenomenon, which must have taken place about four o'clock on the morning of the 14th.—(A. S. H.)

† 'Les Mondes,' 2nd ser. vol. xiii. p. 451.

‡ Royal Astronomical Society's 'Monthly Notices,' vol. xxvii. p. 202.

"Yokohama, 17th November, 1866.—I looked out for shooting stars, but it rained and blew so, that no stars could be seen till the morning of the 14th, and then nothing was to be seen but one or two odd ones. – (J. P. L. Maclear)."

"Philadelphia, 22nd March, 1867.—B. R. Lewis, Dep. U. S. Consul General at Shanghai, writes to me, under date of the 29th of November, that he had not heard that any unusual display had been observed there.—(B. V. Marsh)."

The easternmost limit of visibility of the shower is, indeed, very clearly defined, by the above description of the phenomenon at Kishnaghur, about the Bengal Presidency of India.

The star-shower was well observed at the Cape of Good Hope Observatory, in South Africa, by Mr. G. W. H. Maclear.* The shower was at its height at twelve minutes past two o'clock (Cape time), corresponding to two minutes before one o'clock Greenwich time; and a brief letter from Sir Thomas Maclear ('Edinburgh Quarterly Review,' Jan. 1867, p. 258) gives the following graphic account of its appearance:—

"In the early part of the night of the 13th, few meteors or shooting stars appeared. At 1h. 3m. A.M., on the 14th, the volcano burst forth, with awful grandeur, from the neighbourhood of Regulus; orange-coloured meteors, leaving streaks of green, mingled with ordinary looking shooting stars dashing along in a south-westerly direction. The scene was beyond description, and thus, with little variation, the projectiles continued till daylight. The total number counted amounts to two thousand seven hundred and forty-two."

At Athens,† Rome,‡ Turin,§ Paris,‖ Brussels,¶ and throughout the continent, good observations of the star-shower were made. At Malta the meteors appeared falling like a shower of hail. At Urbino, in Italy, they were compared to a flight of handgrenades, and at Saragossa, in Spain, they reminded the inhabitants of the bombardment of the town. At Haddingham, in England, Mr. Dawes compared the shower with that which he witnessed at Ormskirk near Liverpool, on the 13th of November, 1832.** But the meteors, in 1866, were not so large as on that occasion. At the Royal Observatory, Greenwich, the shower first reached its maximum at seven minutes past one o'clock, and it was most intense for the space of an hour and a half, from half-an-hour after midnight until two o'clock; before and after which times the scale of the phenomenon hardly exceeded a con-

* Royal Astronomical Society's 'Monthly Notices,' vol. xxvii. p. 65.

† Vienna Academy, Sitzungsbericht, vol. liv. pt. ii. 6th Dec. 1866.

‡ Bullettino Meteorologico del Collegio Romano, vol. v. p. 121.

§ Stelle Cadenti osservate in Piemonte, nel 1866. By P. Barnabita. Turin, 1867, p. 25.

‖ 'Comptes Rendus,' vol. xliii. p. 906.

¶ Bulletins de l'Académie Royale de Belgique, 2nd ser. vol. xxii. No. 12, 1866.

** 'Monthly Notices,' R.A.S. vol. xxvii. p. 48.

siderable August shower. The altitude of the radiant point at the last-named hour, was thirty-three degrees above the east horizon; and as the shower, as was already mentioned, was not visible in America, some thirty degrees of longitude west of Greenwich, in the Atlantic Ocean, probably terminated the visibility of the shower towards the west.

The geographical limits of visibility of the star-shower of 1866, it will at once be seen, coincide with the area over which the November meteors appeared in 1832. The latter shower was seen as far south as the Mauritius, as far east as Arabia and the Persian Gulf, and over the whole continent of Europe, with the British Isles, but it was not visible in America. It was, moreover, a moderate display, but it was followed, twelve months later in America, by the great storm of meteors which suddenly appeared on the morning of the 13th of November, 1833. The recent exhibition may therefore be regarded as the prelude of a similar meteor-rain in America, perhaps partially visible in Europe, as great and bright as the two star-storms seen in America, and partially visible in Europe, in the years 1799 and 1833.

Unless unforeseen curvatures of the meteoric current disturb the geographical boundaries of the display, the first symptoms of the approaching star-shower will be perceived at day-break in England, on the morning of the 14th of November, 1867, when the light of the moon, then three days past the full, and of dawn appearing, will detract something from the numbers and brightness of the meteors. But the same oscillation of the curves in an opposite direction, it should be borne in mind, will bring Great Britain into full view of the centre of the shower, and make the principal spectacle of the meteors visible in Europe *before* day-break, as well as in America.

A leading feature of the great display in 1866, was the surprisingly brief duration of the shower, and the almost sudden rapidity of its appearance and disappearance. Speaking of the remarkable scarcity of meteors on the morning of the 13th, Mr. Serpieri at Urbino writes to Mr. Secchi at Rome:—"I began to doubt whether even the ordinary meteors of the November epoch would not this year be altogether wanting. But in the end, it seemed as if all those ordinary meteors had gathered themselves together into one dense array, to make their transit in the shortest possible time!"* The definable character of the shower led to a large number of accurate observations being made, on the moment of its maximum abundance. This was generally observed in England to have taken place first, about ten minutes past one o'clock; while a second maximum, hardly less marked, was observed at twenty minutes past one. At the Cape of Good Hope Observatory the shower reached its maximum at two minutes before one o'clock (Greenwich time), and afterwards pretty steadily and very rapidly declined. The difference, which amounts to about a quarter

* 'Les Mondes,' 2nd ser. vol. xii. p. 641.

of an hour, is easily explained, if the oblique direction is considered, in which the earth at this juncture traversed the meteoric current. The radiant point of the shower was determined on this occasion with more than ordinary exactness, and few observers differ far from fixing it near the small star, x (Bode) at the centre of Leo's sickle, being the very position assigned to it by Professor Twining, at the last great appearance of the shower on the 13th of November, 1833.

It was early pointed out by Encke that this position of the radiant point is almost vertically over the point of the ecliptic, towards which the earth is moving at the moment. Supposing a pointer laid against a flat horizontal ring, to indicate the direction in which the earth is moving in its orbit on the 14th of November. If the pointer is then inclined a little upwards (about $10\frac{1}{2}$ degrees), it indicates the position of the radiant point, or the direction from which the meteors appear to come. But as the earth itself is advancing to meet the shower, the real slope of the meteoric current is less oblique (about 17°) than it appears to be, at the point of the ecliptic where the earth encounters it. The *southern side* of the earth must evidently meet the sloping current first, after which the equatorial parts, and lastly the northern side of the earth, will be plunged into the stream. Twenty-four minutes would be required for the whole earth to become immersed, and thirteen minutes should elapse (which was very nearly the interval observed) from the time when the maximum display was experienced at the Cape of Good Hope, until the like should be perceived in England. A more favourable opportunity could hardly have been expected for turning such observations to account; and the general mode of apprehending the phenomenon is shown to be substantially correct, by the satisfactory manner in which they answer to the test.

In twenty-four minutes the whole width of the earth's diameter would enter the stream. As the denser part of the shower lasted an hour and a half, its thickness was nearly equal to four diameters of the earth, or about 30,000 miles.

Only two or three of the brightest meteors observed at any one station, were brighter than the planets. One such was seen after sunrise at the Observatory at Athens. An equally bright meteor was seen by Mr. Crumplen as late as nine o'clock in the morning, at Primrose Hill, in London. The scale of magnitudes of eighty-one meteors, whose paths were recorded by the speaker, with the assistance of Mr. A. Macgregor, at the Glasgow Observatory, were as follows :—

As bright as Jupiter, or brighter	2	meteors	= 3	per cent.
As bright as Sirius	14	,,	= 17	,,
1st mag. stars	39	,,	= 48	,,
2nd mag. stars	26	,,	= 32	,,
	81		100	

The meteors were in general white, occasionally inclining to blue, and frequently to tints of orange-red. Every meteor left a peculiarly hard and solid-looking straight streak of light upon its whole course, *ahead* of which the head was occasionally observed to shoot in its flight, very seldom leaving any large sparks upon its track, and rarely, if ever, terminating with an explosion, but disappearing by degrees, as if the material substance of the meteor was expended. The streaks were lance-like, tapering to the extremities; and commonly faded in a few seconds, from the ends towards the centre, without losing their lance-like form. But in the case of a good many streaks which remained visible for some minutes, the most persistent part of the train diffused itself into spiral, snake-like, scymitar-shaped, and every possible variety of nebulous-looking clouds, of silvery white, or in some very long-enduring cases, of flame-coloured light. Of the latter kind was the stationary streak over Dundee, in Scotland, whose real height and position are described, approximately, on the next page. Dr. Schmidt describes a streak at Athens, which remained visible in the sky fifty-one minutes, like a red cumulus cloud, only effaced by the approach of daylight. The evident tendency of the most persistent streaks seen in England, was to drift with a pretty rapid motion towards the south, or to a few points west of south, until they disappeared. A very persistent light-cloud left by a large meteor of the shower *in America* on the 13th of November, 1833, was calculated by Professor Twining to have drifted from its place *eastwards* with a speed of three or four miles in a minute.

An inquiry projected last year,* with reference to analyzing the light of the meteors by their spectrum, was put in practice; and meteoric spectra were observed both on the nights of the 9th and 10th of August, and on the morning of the 14th of November, 1866. Meteor spectroscopes were constructed by Mr. Browning in good time for the first of those occasions, and seventeen views of the spectra of the meteor streaks and nuclei were obtained. The prevailing character, in the freshly deposited streaks, was a continuous spectrum of considerable width, but destitute of colour. But as soon as some of the streaks began to fade, in eight cases there remained nothing but an extremely slender and bright yellow line of light, manifestly the light of some self-luminous gas, of which the nearest analogue in terrestrial flames is the light of incandescent sodium vapour. The spectra of the nuclei generally presented all the brighter colours of the prismatic spectrum. When feeble, the spectrum was still continuous, although destitute of colour, but three examples of nuclei were observed, which presented nearly homogeneous yellow light; one entirely destitute, and the other two accompanied only by a very faint continuous spectrum.

The nuclei of the November meteors presented to Mr. Browning the same peculiar features in the spectroscope as those which the speaker had already noted in the spectra of the nuclei of the August meteors; namely, a marked preponderance of a line, or broad band of yellow light, which sometimes appeared to form the totality of their

* Proceedings, R. I. vol. iv. p. 650.

spectrum; with the addition that two nuclei were observed by Mr. Browning, among the November meteors, whose light was equally homogeneous, but green. The light of the streaks, which was mostly blue, green, or steel-grey, generally appeared homogeneous,* and this observation the speaker was able to confirm. Although the experiments presented peculiar difficulties from the rapidly-fading character of the light-streaks, yet the homogeneous appearance of the light of the November meteor-streaks in the spectroscope was, in certain cases, so unequivocal, that their bluish-green colour immediately suggested to him the suspicion, which was, at the time, nothing more, but which now appears highly probable, that an analogy exists between their light and that of the gaseous nebulæ, and, particularly, of the nucleus of Tempel's comet.

The principal result of the meteor-spectroscopic observations was, that the total absence of the presumed bright line of sodium in the spectra of the streaks of the November meteors establishes a specific difference between the meteoric substances of the two currents, to which the August and November shooting-stars belong.†

Observations of one of the largest meteors of the shower were obtained at Sunderland, in England, and at Edinburgh, Aberdeen, and Glasgow, in Scotland.‡ The meteor appeared at twenty minutes before three o'clock, and the streak, to which the observations principally refer, remained visible a quarter of an hour. The nebulous cloud of light very nearly marked the point of disappearance of the meteor, and its altitude was ascertained to be between fifty-one and fifty-seven miles above the earth's surface, over a spot a few miles distant from Dundee.

In a series of letters addressed to Padre Secchi, the celebrated astronomer of Rome,§ shortly before the recent appearance of the November shower, a fortunate theory of periodical meteors, announced by Sig. G. V. Schiaparelli, the astronomer of the Brera College at Milan, completely removes the obscurity of their origin, and claims for its distinguished author the praise of having, for the first time since the cosmical hypothesis of Chladni, laid a broad and sure foundation for a new science of meteoric astronomy.

Supposing that a cosmical cloud of particles should be drawn from stellar space by the sun's attraction, it is shown that it could not cross the earth's orbit in any other form than as a parabolic current.

---

* 'Monthly Notices,' R.A.S. vol. xxvii. p. 78.

† The orange tint of the nucleus of Comet II., 1862 (now supposed to be a member of the material current which furnishes the August meteors) occurred to the writer as being very remarkable to the naked eye, at the time of its appearance. It was noted with the telescope by Mr. Knott, in his 'Observations of the Comet at the Woodcroft Observatory, Cuckfield, Sussex.' See 'Monthly Notices,' R.A.S. vol. xxiii. p. 31.

‡ 'Proceedings of the Glasgow Philosophical Society,' vol. vi. p. 207.

§ 'Bullettino Meteorologico del Collegio Romano,' vol. v. Nos. 8, 10, 11, 12, and vol. vi. No. 2.

Assuming, for example, that such a cloud, of the sun's size, was originally moving at an aphelion distance of 20,000 times the earth's distance from the sun, with a velocity of only one hundred yards in a minute, and with no relative motions, or mutual attractions, existing among its particles, the cloud would represent at that distance a nebula of only one-tenth of a second of arc in angular width. In about one-and-a-half million years the cloud would arrive at less than the earth's distance from the sun, being then in the perihelion point of its orbit, which would there be undistinguishable from a parabola. By the inevitable laws of the sun's attraction, the cloud will be gradually deformed in its progress, until on the point of its perihelion passage it will become a current twenty-three miles broad, one hundred yards deep, in a direction measured from the sun, and extending upwards of six hundred millions of miles along the arc of the parabola. The perihelion passage of the current will occupy more than a year (387 days) and its particles will be four hundred million times more closely packed together than they were in the cosmical cloud before its deformation by the sun's attraction. As there are nebulæ in the sky larger than the sun's apparent disc, if that width only were adopted for the apparent magnitude of the cosmical cloud at its aphelion, it would be transformed into a parabolic current, which would occupy twenty thousand years in its perihelion passage. The transverse width of the current would at the same time be proportionately greater than in the previous case, yet not so large but that the earth will pass through it in a few hours, or at most, in one or two days. In this manner, avoiding all impossible assumptions, meteoric currents may clearly be accounted for, which, like that of August, have been visible for hundreds or thousands of years.

On the other hand a meteoric stream like that of November, which is visible for two or three years in succession at the end of every cycle of about thirty-three years, must be moving in a very much shorter ellipse, and must occupy a certain arc of that ellipse with the requisite materials for a meteoric shower. Mr. Le Verrier supposed * that a cosmical cloud, like those assumed to exist by Mr. Schiaparelli, was thrown into such an elliptic orbit by the action of the planet Uranus, and that the cloud must actually have passed close to Uranus in the year A.D. 126. The action of the planet having caused some of its parts to move faster than the rest, the cloud is gradually becoming transformed into a ring, but it is not yet such an ancient member of the solar system as, if the same kind of hypothesis could in that case be entertained, Mr. Le Verrier supposes the ring of the August meteors to represent.

It may be remarked that the earth passes periodically through the current of the November meteors, without sufficient attractive efficacy, while scattering some of the meteors, to deflect the main body of the current from its course. That the earth, however, in common with

---

* 'Comptes Rendus,' vol. xliv. p. 94.

the larger planets, must produce a sensible effect upon the direction of the stream, is certain; and Mr. Adams has shown,* as the result of very elaborate calculation, that assuming the orbit of the November meteoric current to be a long ellipse extending to near the orbit of Uranus, with a periodic time of 33·25 years, the joint effect of all the more important planets upon its course, must be to retard the date of the appearance of the shower one day at the end of every such cycle. The modern November star showers, of the 12th of November, 1799, the 13th of November, 1832, and 1833, and the 14th of November, 1866, as well as the much more decisive evidence derived from earlier displays of the same shower completely establish the correctness of this hypothetical form of the orbit; while other calculations show that no other possible orbit which the meteors might pursue consistently with the existence of a thirty-three-year period in the date of their returns, would agree, as the former orbit does, in a precise and accurate manner, with the very prominent variation of the date.

Periodical meteors are accordingly found to move in very eccentric orbits, inclined like the orbits of the comets at all possible obliquities to the ecliptic. Nor is this similarity of their orbits merely a conjecture. In the fourth of his series of letters to Padre Secchi,† Mr. Schiaparelli announces the important discovery, that a conspicuous and well-remembered comet (Comet III., 1862), which was visible to the naked eye for several weeks in August and September, 1862, and which Stämpfer calculated to have a periodic time of 113 years, coincides almost exactly in its path with the stream, supposed nearly parabolic, of the Perseids, as the 10th of August meteors are styled by Schiaparelli.

A period of revolution of 108 years is assigned to the August meteors by Schiaparelli, and one of 103 years was proposed by the speaker‡ last year, as bringing the August star-showers of A.D. 830, 933, 1243, and 1451, into conjunction with the remarkable star-shower of the 10th of August, 1863. The near coincidence of the time of the comet's return with the last-named great display of the "Perseids," affords a strong presumption that a connection of an intimate kind exists between the two important classes of bodies. Not less striking is the coincidence of the return of another periodical comet (Comet I., 1866), known as Tempel's comet, visible with the telescope, which passed its perihelion in January, 1866, with the recent reappearance of the November meteors. The November meteors and Tempel's comet, as calculated by Oppolzer, pass at the same time close to the earth's orbit, at the same ecliptic longitude, at the same obliquity to the ecliptic, and in the same direction round the sun. The periodic time of their return is also the same, so that their orbits must, evidently, coincide throughout their whole extent. Dr. Galle, the director of the Obser-

---

* 'Monthly Notices, R.A.S.' vol. xxvii. p. 247.

† 'Bullettino Meteorologico del Collegio Romano,' vol. v. No. 12.

‡ Proceedings R.I. vol. iv. p. 646.

vatory at Breslau, has recently pointed out a coincidence which exists between the elements of the orbit of a comet (Comet I., 1861) visible to the naked eye, and the orbit, supposed nearly parabolic, of the meteors of the 20th of April. The closeness of the coincidence makes it probable that the periodic time of 415 years, assigned to the comet, is also the periodic time of revolution of the April ring of meteors.

Of remarkable meteors visible in the past year, although only twice as bright as Sirius, one, having an extraordinary length of path, appeared at ten o'clock on the evening of the 11th of January, 1866. It shot almost horizontally across the English Channel from eighty-five miles over Paris to ninety-five miles over Cork, performing a path of not less than four hundred and fifty miles in about six or eight seconds.* The most experienced observers were astonished at the length of its luminous track, and compared it to "a bombshell fired from London into Ireland."

At midnight on the 10th of March, 1866,† a detonating meteor was seen in Hanoverian Prussia, which, the investigation of Dr. Heis informs us, shot from thirteen miles over Miete to three miles and a half over Lübbecke, a distance of thirty miles, in four or five seconds, with a speed of seven miles in a second. The detonation was so loud at Lübbecke as to awaken many from their sleep, with the noise, which was like a clap of thunder.

A similar meteor to the last appeared at eleven o'clock in the forenoon (Paris time) on the 20th of June, 1866, over Calais and Boulogne. It left a smoke-like train, that remained visible in broad daylight for several minutes.‡

The path of this fireball was from fifteen miles over Calais to four and a half miles over the neighbourhood of Boulogne and Montreuil-in-Somme. The meteor pursued a path of thirty miles in three and a half seconds, with a speed of eight or nine miles in a second, disappearing with a detonation which was heard, startlingly loud, as far as Maidstone in Kent, and Hastings in Sussex. At St. Omer it was at first believed that the neighbouring powder-mill of Esquerbes had exploded.

Three detonating meteors, to which allusion was made last year,§ made their appearance in England between the 19th and the 21st of November, during the five years 1861–65. The altitudes and other particulars of these meteors were defined. A fourth detonating fireball may now be added to the same list, if the following announcement in

---

* 'British Association Report,' 1866, pp. 84 and 126. As seen at Ticehurst, in Sussex, the meteor passed over 160° of the sky, appearing very near the horizon in the east, shooting overhead, and disappearing close to the west horizon.

† Twenty minutes after midnight, on the morning of the 11th of March, Münster time. A pamphlet by Prof. Heis, with map of meteor's course (8vo, Halle, 1866, H. W. Schmidt).

‡ British Association Report, 1866, pp. 126 and 128.

§ Proceedings R.I. vol. iv. p. 649.

the New York 'World' of the 24th of November, 1866, relates to a "meteorological phenomenon" which actually took place as there described:—

"At Nashville, about four o'clock last Tuesday morning (the 20th of November, 1866), a meteor, lighting the whole heavens, was seen in the direction of Rome, Ga., moving rapidly south-west. It appeared like a ball of fire as large as the sun. It exploded apparently ten miles off, with a tremendous report, like a 40lb. cannon, that shook the earth, and made the windows rattle."

In addition to two stonefalls, alluded to last year,* which happened, one at Shergotty, in India, and the other at Aumale, in Algeria, on the 25th of August, 1865, three other stonefalls took place, one an aërolite weighing about 5lbs., which fell at Dundrum,† in Ireland, with the usual accompaniment of a detonation, followed by a humming sound, on the 12th day of the same month and year. The second was an aërolite weighing about 3lbs., which fell at Bheenwall,‡ in Bengal, on the 20th of August; and the third, a fall of two stones at Muddoor,§ in India, on the 7th of September in the same year, making, with the two stonefalls already mentioned, a total of five aërolites in four weeks.

The next accounts of the same kind which have been received relate to a stonefall at St. Mesmin, in France, on the 30th of May, 1866, followed in ten days by the shower of stones at Knyahinya, in Hungary, on the 9th of June, 1866. At St. Mesmin, in the valley of the Seine, there fell, about four o'clock in the morning, three aërolites, one of them with such a clattering, and buzzing, and shrieking sound, that a signal man in the railway cutting, where it fell, remained for some time mute with affright.

The stone when picked up weighed 1lb., and a small indentation, about half-an-inch in width, with a fresh surface, covered with thin thread-like lines of the fused crust, marks a spot where an angle must have been broken off from the meteorite during its flight through the air.‖

The second and last stonefall recorded in the year is the shower of stones which happened between four and five o'clock in the afternoon at Knyahinya, near the northern frontier of Hungary.¶ It ranks as one of the greatest events of the kind, since the fall of about two hundred stones at Stannern, in Moravia, which strewed them-

* Proceedings R.I. vol. iv. p. 649.

† Scientific Papers from the Royal Irish Academy's Proceedings, vol. i. p. 230.

‡ British Association Report, 1866, p. 133.

§ 'Meteors and Aërolites.' By Dr. Phipson, p. 227.

‖ 'Comptes Rendus,' vol. xlii. 1866, June 18th.—A small lath of heavy wood astened to the end of a piece of string, and whirled round with the hand so as to describe a pretty wide circle, gives out a loud humming noise. The sounds which may be produced in this manner are evidently of the same kind as those which aërolites give out in falling through the air. Angular pieces of iron projected from a common sling produce in the same way a strange variety of humming, buzzing, and "shrieking" noises.—(A. S. H.)

¶ Vienna Academy, 'Sitzungsbericht,' vol. liv. 12th July, 1866.

selves over an area nine miles long by six miles wide. The largest stone at Knyahinya weighed six hundred-weights, and excavated a hole four feet deep and five feet wide, where it penetrated the ground to an extent of twelve feet in a slightly slanting direction from seventy-six degrees east of north to seventy-six degrees west of south. Other fragments, weighing from seven to thirty-seven, and one of them ninety pounds, were picked up near the larger mass, and about a thousand fragments, weighing together near upon a thousand pounds, were scattered over an area nine miles long, from north-east to south-west, and four miles wide. All the fragments are completely crusted over, and the meteor was seen, and the detonation was heard over a distance of sixty or eighty miles in a south-west direction.*

As fragments weighing a few pounds, or a few hundred-weights only, could not exist as a group with intervals of more than a hundred yards between them, at the earth's distance from the sun, without being disjoined by the sun's attractive force; and as it is difficult to imagine how such a compact group could be formed, even beyond the sphere of the sun's attraction, capable of resisting the sun's disturbing action at the earth, it is more probable that the shower of aërolites entered the atmosphere as a single stone. This was certainly the case with fragments of the meteorite of Butsura, two of which, picked up at a distance of more than two miles apart, were completely crusted over; and yet, when placed side by side, they have a nearly close-fitting junction, and are obviously seen to belong to one piece, by a vein of iron, which runs directly across the junction through the mass.

The most probable theory of the zodiacal light makes it to consist of numberless small bodies, revolving like planets round the sun, whose joint reflection of the sun's rays gives the appearance of a luminous haze stretching along the zodiac, and brightest in the neighbourhood of the sun. On particularly clear nights in the tropics, it sometimes appears to complete the circle of the midnight sky, from east to west, having been described, and frequently seen thus by Mr. G. Jones, on the long voyage of the Japan expedition from New York.†

The earth is, therefore, situated within its outer border. But if the bodies were spherical, and uniformly scattered between the earth and the sun, the theory of their phases, and maximum brightness, as seen from the earth, would lead us to expect a species of mock-suns, about fifty-five degrees distant from the true sun, one at each extremity of the light. As this is not observed, and as, on the contrary, both terminations of the light are diffuse, the bodies which compose it must, certainly, be more thinly scattered towards its edge, and it is even probable that they must consist of shapeless fragments, which is exactly what the circumstances of the falls of aërolites would lead us to expect.

* Vienna Academy, 'Sitzungsbericht,' vol. liv. 11th October, 1866.

† Observations on the Zodiacal Light, from April 2, 1853, to April 22, 1855. By Rev. George Jones, A.M. 'United States Japan Expedition,' vol. iii. p. 84.

The general presence of nickel, and the remarkable deficiency of oxygen in meteorites, cannot be urged against the probability of their originating in a boundless field of bodies, like the zodiacal light, because the sun itself, which is the largest body of the system, is shown by twenty or more lines of the solar spectrum to contain nickel; and the abundance of oxygen on the earth's surface, far from being a general proof of its redundancy in planetary and other stars, may be a peculiar circumstance of this planet, and perhaps of few other members of the material universe. The particular exhibition of an excess of free oxygen upon the surface of the earth will appear a much more exceptional condition, if we assign a wider field to the origin of aërolites, by accepting Mr. Schiaparelli's ninth postulate, in which he thus sums up his views regarding their extraction:—

"Since it may be regarded as certain that falling-stars, bolides and aërolites, differ from each other only in their magnitudes, we must conclude that the substance fallen from the sky *is a sample of that of which the stellar universe is composed.* And since in such substance there is no chemical element unknown upon the earth, the similarity of the composition of all the visible bodies of the universe, already rendered probable by researches with the spectroscope, receives a new argument of credibility." *

[A. S. H.]

* 'Bullettino Meteorologico del Collegio Romano,' vol. v. No. 11.

Friday, March 5, 1869.

SIR HENRY HOLLAND, Bart. M.D. D.C.L. F.R.S. President, in the Chair.

WILLIAM HUGGINS, Esq. F.R.S.

On some further Results of Spectrum Analysis as applied to the Heavenly Bodies.

THE speaker commenced by saying that four years ago he had the honour to give in the theatre of the Royal Institution an account of the results of an attempt to apply the method of analysis by the prism, for which science is indebted to Kirchhoff, to the light of the heavenly bodies. It was the speaker's purpose to describe, on the present occasion, some of the results which had been obtained in his observatory since the spring of 1865. The peculiar suitability of spectrum analysis as a mode of investigation of the bright objects in the heavens had been confirmed, not only by the gain of further information of the chemical and physical constitution of some of these immensely distant bodies, but also by knowledge of another kind which this elegant and searching method of analysis had revealed to us.

The speaker then described the three typical forms under which all spectra may be classed, and the interpretation which our present knowledge enables us to give of these different spectra when the light is emitted by bodies rendered luminous by heat. The spectra of fluorescent and phosphorescent bodies were not to be described.

1. *A continuous spectrum without dark or bright lines* shows, as a general rule, that the luminous source is in the solid or liquid state. In certain exceptional cases, however, a gas may give a spectrum which is apparently continuous. Dr. Balfour Stewart pointed out that as gases and vapours possess a power of general absorption, in addition to the selective absorption peculiar to each gas, a gas when luminous would emit light of all refrangibilities, producing a continuous spectrum, in addition to its spectrum of bright lines, and further that the intensity of this continuous spectrum would be in proportion to the opacity of the gas. The researches of Plücker and Frankland have shown that under certain conditions of density and temperature, the bright lines of hydrogen expand so as to produce a spectrum which is apparently continuous.

2. *A spectrum of bright lines* indicates that the luminous body is in the state of gas. Each gas and vapour has its own set of lines. The

lines may be greatly modified, or even altogether changed, under different conditions of temperature and density, as is well known in the case of nitrogen, the vapour of sulphur, and some other substances; but throughout all these changes each gas behaves in a way peculiar to itself. There appears to be one exception to the statement that a spectrum of bright lines is peculiar to luminous gas. Bunsen found that when solid erbia is heated to incandescence, the continuous spectrum contains bright bands.

3. *A continuous spectrum interrupted by dark lines* informs us that the light has passed through vapours at a lower temperature than the source of light. As the kinds of light absorbed by each vapour correspond precisely with the set of bright lines which that vapour emits when in the luminous state, it is possible to learn if the vapours are those of any of the substances with which we are acquainted.

The speaker said that following the arrangement adopted in the former discourse, the most important recent information obtained of the *fixed stars* results from the application of prismatic analysis in a new direction. Under certain conditions the spectrum of a luminous body is adapted to tell us whether that body is moving towards or from the earth. The importance of information on this point will be seen from the consideration that the proper motions of the stars represent that part only of their whole motion which is transverse to the line of sight; for any motion they might have in the visual direction, towards or from the earth, would not cause any visible displacement of the star, and could not therefore be ascertained by the ordinary methods of observation.

As it is upon the length of the waves, or upon the number contained in the series that enters the eye, or falls upon the prism, in a second that a judgment is formed of the colour of the light, or its place in the spectrum is determined, it follows that any circumstance which would alter the length of the waves *relatively to the observer*, or, in other words, cause a larger number of waves to enter the eye in a second of time, would cause a change in the colour or refrangibility of the light so far as the observer is concerned. It is obvious that if the observer advances to meet the light, a longer series of waves falls upon the retina in a second of time, each wave appears shorter, and he ascribes to the light a higher refrangibility than he would do if he were not advancing to meet the light. If he were receding from the star an alteration of refrangibility in the opposite direction would take place. The same effect would ensue, if the luminous source were in motion. Thus to a swimmer striking out from the shore, each wave appears shorter, and he passes a greater number of them in a given interval in proportion to his speed through the water.

Illustrations were given of this principle, which was first suggested in 1841 by Doppler, by means of an analogous change of pitch in sound. Two tuning-forks sounding in unison were moved rapidly towards and from the audience, when beats were heard, which told of a difference of pitch produced by the opposite motions of the forks.

As there exists beyond the visible spectrum, at both ends, a store of invisible waves, these would be advanced or degraded into visibility, in proportion as the colours of the spectrum were altered, and no change of colour would be perceived. It is therefore essential before we can apply this method to detect the radial motion of the stars, that we know the original refrangibility of some part of the light at the moment it left the star, and also that we are able to recognize this particular part of the light again in the spectrum of the star's light. When by means of a group of dark or bright lines we learn the presence of a terrestrial substance in the star, both these conditions are fulfilled.

Of all the stars which the speaker had compared with terrestrial elements, when working with his distinguished friend Dr. W. A. Miller, Treas. R.S., Sirius, which contains four very strong lines which are due to hydrogen, appeared the most suitable for this investigation. The apparatus employed, and the special precautions which were taken to ensure the perfect coincidence in his instrument of the stellar lines with those of the substance compared with it, were described by the speaker, who stated that after a prolonged comparison, extending over many weeks, of the line of hydrogen in Sirius in the green, at the place of F in the solar spectrum, with the line of terrestrial hydrogen, he found that the line in the star had undergone a shift in the spectrum equal to a difference of wave length, which would correspond to a motion of recession between the star and the earth of 41 miles per second. The speaker had obtained evidence from experiment that this shift was not due to unsymmetrical expansion of the line in hydrogen as the density is increased. The greater width of this line in Sirius than in the solar spectrum would show that the hydrogen in Sirius, though at a pressure considerably less than that of our atmosphere at the surface of the earth, is more dense than the hydrogen in the solar atmosphere by which the dark line F is produced. This conclusion is in accordance with the presumably enormous mass of Sirius, as suggested by its great intrinsic splendour.

The earth at the time of observation was moving from Sirius at about 11 miles per second, which would leave 30 miles as due to the star. A further connection is required for the solar motion in space, which is believed to be towards Hercules, with a velocity of 4 or 5 miles per second. The whole of this must therefore be deducted, leaving about 26 miles as the motion of Sirius from the earth in the line of sight. The true motion of the star would consist of this radial motion compounded with the transverse motion of from 24 to 40 miles per second, which is shown by its proper motion.

The speaker then described a further examination of the nebulæ (about fifty have been successfully observed) with a more powerful spectroscope, which confirms his previous conclusion that these bodies consist mainly of the gases nitrogen and hydrogen. He also found that when the spectra of these gases are made faint by the removal of the spark to a distance, all the lines are extinguished with the exception of the one line in each spectrum which is found in the nebulæ. If such an extinction takes place in the case of the nebulæ, since they are objects of sensible size, it must be attributed to a power of extinction of light existing in cosmical space.

Observations of four comets have been made. A large part of the light of these strange objects was found to be peculiar, and therefore emitted by the cometary matter. Brorsen's comet at its return in 1868, and a comet discovered by Winnecke, gave a spectrum of three bright bands. The spectrum of Winnecke's comet (comet II., 1868) was found to be identical with the spectrum of carbon as it appears when the induction spark is taken in olefiant gas, and in some other compounds of carbon. The spectrum of the comet was compared directly in the instrument with the spectrum of olefiant gas.

The speaker then described some observations of the sun. He found that while the solar lines are for the most part thickened when viewed in the light from the umbra of a spot, the lines C and F, due to hydrogen did not appear to be altered. This observation is of interest in connection with the constitution of the solar prominences as shown by the observations of the great eclipse of last August. The speaker nearly three years ago, at the same time that he had independently made attempts to see the prominences by means of the spectroscope, also tried the method of using absorbing media, by which the parts of the spectrum where the bright lines occur might remain, while all the rest of the spectrum was extinguished. In this way the faint prominences would be rendered visible, in consequence of the much greater relative diminution of the intensity of the illuminated screen of air, which on ordinary occasions conceals them from view. Recently he had succeeded in viewing the outline of these objects by means of a coloured glass combined with a spectroscope with a wide slit. He expected to be able to view these objects by means of coloured media alone.

[W. H.]

Friday, May 28, 1869.

SIR HENRY HOLLAND, Bart. M.D. D.C.L. F.R.S. President, in the Chair.

J. NORMAN LOCKYER, F.R.S.

On Recent Discoveries in Solar Physics made by means of the Spectroscope.

IN the year 1865 two very important memoirs dealing with all the telescopic and photographic observations accumulated up to that time on the subject of solar physics were given to the world. One of them was privately printed in this country, the other appeared in the 'Compte Rendu' of the Paris Academy of Sciences.

I shall not detain you with a lengthened notice of these remarkable papers. I shall merely refer to the explanation given in both of them of the reason that a sun-spot appears dark—the very keystone of any hypothesis dealing with the physical constitution of the sun.

English science, represented by Messrs. De la Rue, Stewart, and Loewy, said that a spot is dark because the solar light is absorbed by a cool, non-luminous, absorbing atmosphere, pouring down there on to the visible surface of the sun, in other words, on to the photosphere.

French science, represented by M. Faye, said that a spot is dark because it is a hole in the photosphere, and the feebly luminous and therefore radiating interior gases of the sun are there alone visible.

Now most of you will see in a moment that here was a clear issue, which probably the spectroscope, and possibly nothing else, could solve; for the spectroscope is an instrument whose special *métier* it is to deal with radiation and absorption. It tells us that the light radiated from different bodies gives us spectra of different kinds, according to the nature of the radiating body,—continuous spectra without bright lines in the case of solids and liquids, and bright lines, with or without continuous spectra, in the case of gases and vapours. It tells us also that absorption dims the spectrum throughout its length when the absorption is *general*, and dims it here and there only when the absorption is *selective*, the well-known Fraunhofer lines being, as you will readily see, an instance of the latter kind. So that we have general and selective radiation, and general and selective absorption.

Now then, with regard to the English theory, if there were more absorption in a spot than elsewhere, we might expect evidences of absorption; that is, the whole solar spectrum would be visible in the spectrum of a spot, but it would be dimmed, either throughout the length of the spectrum or in places only.

With regard to the French theory, radiating only gaseous matter to deal with, we should, according to the then generally received idea, get bright lines only in the spot spectrum.

Here then was a tempting opportunity, and one which I considered myself free to use; for, although the spectroscope had then been employed—and you all know how nobly employed—for four years in culling secrets from stars and nebulæ, there was not, so far as I know, either published or unpublished observation on the sun, the nearest star to us. The field was therefore open for me, and I was not entering into another man's labour, when, on the 4th of March, 1866, I attached a small spectroscope to my telescope, in order to put the rival theories to a test, and thus bring another power to bear on a question which had remained a puzzle since it was first started by Galileo some two-and-half centuries ago.

What I saw I will describe more fully by-and-by. It is sufficient here to mention that it was in favour of the English theory. There *was* abundant evidence of absorption in the spots, and there *was not* any indication of gaseous radiation.

Having then thus spectroscopically broken ground on the sun, a very natural inquiry was how next to employ this extension of a method of research, the discovery of which Newton had called, nearly two hundred years before, "the oddest, if not the most considerable, detection which hath hitherto been made in the operations of nature."

There seemed one question which the spectroscope should now put to the sun above all others, and it was this:—

"Assuming this absorbing atmosphere to encircle the sun, in accordance with the general idea and Kirchhoff's hypothesis, what are those strange red flames seen apparently in it at total eclipses, jutting here and there from beyond the sun's hidden periphery, and here again hanging cloudlike?"

The tremendous atmosphere, which apparently the spectroscope had now proved to be a cool absorbing one, was supposed to be indicated during eclipses by a halo of light called the "Corona," in which corona the red flames are visible. Now, as the red flames are always observed to give out more light than the corona, they were probably hotter than it; and reasoning thus on the matter with my friend Dr. Balfour Stewart one day, we came to the conclusion that they were most probably masses of glowing gas.

Now, this being so, the spectroscope *could* help us, and in this way.

The light from solid or liquid bodies, as you all I am sure know, is scattered broadcast, so to speak, by the prism into a long band of light, called a continuous spectrum, because from one end of it to the other the light is persistent.

The light from gaseous and vaporous bodies, on the contrary, is most brilliant in a few channels; it is *husbanded*, and, instead of being scattered broadcast over a long band, is limited to a few lines in the band—in some cases to a very few lines.

Hence, if we have two bodies, one solid or liquid and the other gaseous or vaporous, which give out exactly equal amounts of light, then the bright lines of the latter will be brighter than those parts of the spectrum of the other to which they correspond in colour or refrangibility.

Again, if the gaseous or vaporous substance gives out but few lines, then, although the light which emanates from it may be much less brilliant than that radiated by a solid or liquid, the light may be so localized, and therefore intensified, in one case, and so spread out, and therefore diluted, in the other, that the bright lines from the feeble light source may in the spectroscope appear much brighter than the corresponding parts of the spectrum of the more lustrous solid body. Now here comes a very important point: supposing the continuous spectrum of a solid or liquid to be mixed with the discontinuous spectrum of a gas, we can, by increasing the number of prisms in a spectroscope, dilute the continuous spectrum of the solid or liquid body very much indeed, and the dispersion will not seemingly reduce the brilliancy of the lines given out by the gas; as a consequence, the more dispersion we employ the brighter relatively will the lines of the gaseous spectrum appear.

The reason why we do not see the prominences every day in our telescopes is that they are put out by the tremendous brightness of our atmosphere near the sun, a brightness due to the fact that the particles in the atmosphere reflect to us the continuous solar spectrum. There is, as it were, a battle between the light proceeding from the prominences and the light reflected by the atmosphere, and, except in eclipses, the victory always remains with the atmosphere.

You will see, however, in a moment, after what I have said, that there was a possibility that if we could bring a spectroscope on the field we might turn the tide of battle altogether, assuming the prominences to be gaseous, as the reflected continuous spectrum might be dispersed almost into invisibility, the brilliancy of the prominence lines scarcely suffering any diminution by the process.

The first attempt was made in 1866, a Herschel-Browning spectroscope being attached to my telescope, and the first and many succeeding attempts failed; there was not dispersion enough to dilute the spectrum of the regions near the sun sufficiently, and as a consequence the tell-tale lines still remained veiled and invisible. Nature's secrets were not to be wrested from her by a *coup de main.*

The year 1868 brought us to the now famous eclipse, to see which scientific men hastened from all civilized Europe to India. To this eclipse and its results I need only refer, as they have already been dwelt on at some length in this theatre; suffice it to say that in the eclipse the spectroscope did its duty, and that the gaseous nature of the prominences was put beyond all question.

But there was a magnificent pendant to the eclipse, to which I must request your special attention. One of the observers, M.

Janssen—a spectroscopist second to none—the representative, in that peaceful contest, of the *Académie des Sciences* and of the *Bureau des Longitudes*, was so struck with the brightness of the prominences rendered visible by the eclipse that, as the sun again lit up the scene, and the prominences disappeared, he exclaimed, "*Je reverrai ces lignes là!*"; and, being prevented by clouds from putting his design into execution that same day, he rose next morning long before the sun, and as soon as our great luminary had risen from a bank of vapours, he succeeded in obtaining spectroscopic evidence of the protuberances he had seen surrounding the eclipsed sun the day before. During the eclipse M. Janssen had been uncertain even as to the number of lines he had observed, but he now by this new method at his leisure determined that the prominences were built up of hydrogen, this fact being indicated by the presence of two bright lines corresponding to the dark lines C and F in the ordinary solar spectrum.

Let me show you how this result was accomplished, by throwing an enlarged photograph of my telescope and spectroscope on the screen. We have first the object-glass of the telescope to collect the sun's rays and to form an image of the sun itself on a screen. In this screen is an excessively narrow slit, through which alone light can reach the spectroscope. This entering beam is grasped by another little object-glass and transformed into a cylinder* of light containing rays of all colours, which is now ready for its journey through the prisms. In its passage through them it is torn by each succeeding prism more out of its path, till at last, on emerging, it crosses the path it took on entering, and enters the little telescope you see, thoroughly dismembered but not disorganized.

Instead now of a cylinder of light containing rays of all colours, we have a cylinder of each ray which the little telescope compels to paint an image of the slit. Where rays are wanting, the image of the slit remains unpainted—we get a black line; and when the telescope is directed to the sun, so that the narrow slit is entirely within the image of the sun, we get in the field of view of the little telescope a glorious coloured band with these dark lines crossing it.

Of course it is necessary for our purpose to allow only the edge of the sun to fall on the slit, leaving apparently a large portion of the latter unoccupied. What is seen, therefore, is a very narrow band in the field of view of the little telescope, and a large space nearly dark, as the dispersion of the instrument is so great that the atmospheric light is almost entirely got rid of, for a reason you are already acquainted with.

Mr. Ladd will now show you on the screen what is seen when the slit reaches a prominence. First a line in the red, very obvious and brilliant, next a more delicate line in the yellow, then another in the green, and two others in the violet; all these lines, with the exception

* Cylindrical, that is, in the case of each pencil.

of the yellow line, are in the positions occupied by known lines of hydrogen.

As the height of these bright lines must vary with the height of the prominences, and as the lines will only be visible where there is any hydrogen to depict, it is obvious that the form of the prominences may be determined by confining the attention to one line, and slowly sweeping the slit over it.

The first fruits then of this new method of working with an uneclipsed sun was to tell us the actual composition of the prominences, and to enable us to determine their shapes and dimensions.

For the next steps you must permit me to refer more particularly to my own observations.

When I was first able to obtain results in this country similar to those previously obtained by M. Janssen, though unknown to us, my instrument was incomplete; when other adjustments had been added by Mr. Browning, I found that at whatever part of the sun's edge I looked I could not get rid of the newly discovered lines. They were not so long as I had seen them previously, but there they were, not to be extinguished, showing that for some 5000 miles in height all round the sun there was an envelope of which the prominences were but the higher waves. This envelope I named the "Chromosphere," as it is the region in which all the variously coloured effects are seen in total eclipses, and because I considered it of importance to distinguish between its discontinuous spectrum and the continuous one of the photosphere. And now another fact came out. The bright line F took the form of an arrow-head, the dark Fraunhofer line in the ordinary spectrum forming the shaft, the corresponding chromospheric line forming the head; it was broad close to the sun's edge, and tapered off to a fine point, an appearance not observed in the other lines.

Nature is always full of surprises, and here was a surprise and a magnificent help to further inquiry lurking in this line of hydrogen! MM. Plücker and Hittorf had already recorded that, under certain conditions, the green line of hydrogen widened out; and it at once struck me that the "arrow-head" was nothing but an indication of this widening out as the sun was approached,

I will now, then, for one moment leave the observatory work to say a word on some results recently obtained by Dr. Frankland and myself, in the researches on the radiation and absorption of hydrogen and other gases and vapours, upon which we have for some time been engaged.

First, as to hydrogen, what could laboratory work tell us about the chromosphere and the prominences?

It was obviously of primary importance—

1. To determine the cause to which the widening of the F line was due.

2. To study the hydrogen spectrum very carefully under varying conditions, with a view of detecting whether or not there existed a line in the orange.

We soon came to the conclusion that the principal, if not the only cause of the widening of the F line was *pressure.*

Having determined, then, that the phenomena presented by the F line were phenomena depending upon and indicating varying pressures, we were in a position to determine the atmospheric pressure operating in a prominence, in which the red and green lines are nearly of equal width, and in the chromosphere, through which the green line gradually expands as the sun is approached.

With regard to the higher prominences, we have obtained evidence that the gaseous medium of which they are composed exists in a condition of *excessive tenuity;* and that even at the lower surface of the chromosphere, that is, on the sun itself, in common parlance, the pressure is very far below the pressure of the earth's atmosphere.

Now I need hardly point out to you that the determination of the above-mentioned facts leads us necessarily to several important modifications of the received theory of the physical constitution of our central luminary—the theory which we owe to Kirchhoff, who based it upon his examination of the solar spectrum. According to his hypothesis, the photosphere itself is either solid or liquid, and it is surrounded by an extensive cool and non-luminous atmosphere composed of gases and the vapours of the substances incandescent in the photosphere.

We find, however, instead of this compound cool and non-luminous atmosphere outside the photosphere, one which is in a state of incandescence, is therefore luminous, and which gives us merely, or at all events mainly, the spectrum of hydrogen; and the tenuity of this incandescent atmosphere is such that it is extremely improbable that any considerable atmosphere, such as the corona has been imagined to indicate, exists outside it.

Here already, then, we find the "cool absorbing atmosphere" of the theorists terribly reduced in height, and apparently much more simple in its composition than had been imagined by Kirchhoff and others. Dr. Frankland and myself have shown separately—

1. That a gaseous condition of the photosphere is quite consistent with its continuous spectrum, whether we regard the spectrum of the general surface or of spots. The possibility of this condition has also been suggested by Messrs. De la Rue, Stewart, and Loewy.

2. That a sun-spot is a region of greater absorption.

3. That when photospheric matter is injected into the chromosphere, we see bright lines.

4. That there are bright lines in the solar spectrum itself.

All these are facts which indicate that the absorption to which the reversal of the spectrum and the Fraunhofer lines are due takes place in the photosphere itself or extremely near to it, instead of in an extensive outer absorbing atmosphere. And this conclusion is strengthened by the consideration that otherwise the newly discovered bright lines of hydrogen should themselves show traces of absorption on Kirchhoff's theory; but I shall show you presently that, so far from this being the case, they *appear bright actually in the very centre of the disc*, and, moreover, the vapours of sodium, iron, magnesium, and barium are often bright in the chromosphere, showing that they would always be bright there *if the vapours were always present*, as they should be on Kirchhoff's hypothesis; so that we may say that the photosphere *plus* the chromosphere is the real atmosphere of the sun, and that the sun itself is in such a state of fervid heat that the actual outer boundary of its atmosphere, *i.e.* the chromosphere, is in a state of incandescence.

With regard to the line in the orange I have nothing yet to tell. Dr. Frankland and myself are at the present moment working upon it.

I have next to take you a stage lower into the bowels, not of the earth, but of the sun.

As a rule, the chromosphere rests conformably, as geologists would say, on the photosphere, but the atmosphere (as I have just defined it) is tremendously riddled by convection currents; and where these are most powerfully at work the upper layers of the photosphere are injected into the chromosphere. Thus I have observed the lines due to the vapour of sodium, magnesium, barium, and iron in the spectrum of the chromosphere, appearing there as very short and very *thin lines*, generally much thinner than the black lines due to their absorption in the solar spectrum.

These injections are nearly always accompanied by the strangest contortions of the hydrogen lines, of which more presently. Sometimes during their occurrence the chromosphere seems full of lines, those due to the hydrogen towering above the rest.

At the same time we have tremendous changes in the prominences themselves, which I have recently been able to see in all their beauty. I attempted to accomplish this in the first instance by means of an oscillating slit, but hearing that Mr. Huggins had succeeded in doing the same thing by means of absorptive media, using an open slit, it struck me at once that an open slit was quite sufficient, and this I find to be the case. By this method the smallest details of the prominences and of the chromosphere itself are rendered perfectly visible and easy of observation, and for the following reason. As you already know, the hydrogen Fraunhofer lines (like all the others) appear dark because the light which would otherwise paint an image of the slit in the place they occupy is absorbed, but when we have a prominence on the slit, there is light to paint the slit, and as in the case of any one of the hydrogen lines we are working with light of one refrangibility only, on which the prisms have no dispersive power, we may consider the prisms abolished. Further, as we have the prominence image coincident with the slit, we shall see it as we see the slit, and the wider we open the slit the more of the prominence shall we see. We may use either the red, or yellow, or green light of hydrogen for the purpose of thus seeing the shape and details of the prominences; how far the slit may be opened depends upon the purity of the sky at the time. I have been perfectly enchanted with the sight which my spectroscope has revealed to me. The solar and atmospheric spectra being hidden, and the image of the wide slit and the part of the prominence under observation alone being visible, the telescope or slit is moved slowly, and the strange shadow-forms flit past, and are seen as they are seen in eclipses. Here one is reminded, by the fleecy, infinitely-delicate cloud-films, of an English hedge-row with luxuriant elms; here of a densely intertwined tropical forest, the intimately interwoven branches threading in all directions, the prominences generally expanding as they mount upwards, and changing slowly, indeed almost imperceptibly.

It does not at all follow that the largest prominences are those in which the intensest action, or the most rapid change is going on—the action as visible to us being generally confined to the regions just in, or above, the chromosphere; the changes arising from violent uprush or rapid dissipation—the uprush and dissipation representing the birth and death of a prominence. As a rule, the attachment to a chromosphere is narrow and is not often single; higher up, the stems, so to speak, intertwine, and the prominence expands and soars upward until it is lost in delicate filaments, which are carried away in floating masses.

Since last October, up to the time of trying the method of using the open slit, I had obtained evidence of considerable changes in the prominences from day to day. With the open slit it is at once evident that changes on the small scale are continually going on; but it was only on the 14th of March that I observed any change at all comparable in magnitude and rapidity to those already recorded by M. Janssen.

About 9h. 45m. on that day, with the slit lying nearly along the sun's edge instead of across it as usual, I observed a fine dense prominence near the sun's equator, on the eastern limb, with signs of intense action going on. At 10h. 50m., when the action was slackening, I opened the slit and saw at once that the dense appearance had all disappeared, and cloud-like filaments had taken its place. The first sketch, now exhibited, embracing an irregular prominence with a long perfectly straight one, was finished at 11h. 5m., the height of the prominence being 1′ 5″, or about 27,000 miles. I left the Observatory for a few minutes, and on returning, at 11h. 15m. I was astonished to find that the straight part of the prominence had entirely disappeared; not even the slightest rack appeared in its place. Whether it was entirely dissipated, or whether parts of it had been wafted towards the other part, I do not know, although I think the latter explanation the more probable one, as the other part had increased.

So much then for the chromosphere and the prominences, which I think the recent work has shown to be the last layer of the true atmosphere of the sun. I shall now invite your attention to spots.

Now, as a rule, precisely those lines which are injected into the photosphere by convection currents are most thickened in the spectrum of a spot, and the thickening increases with the depth of the spot, so that I no longer regard a spot simply as a cavity, but as a place in which principally the vapours of sodium, barium, iron, and magnesium occupy a lower level than they do ordinarily in the atmosphere.

I have told you before, that when these lines are observed in the chromosphere, they usually are thinner than their usual Fraunhofer lines.

I will now show a photograph of a spot spectrum on the screen. You will see a black band running across the ordinary spectrum; that black band indicates the general absorption which takes place in a sun spot. Now mark the behaviour of the Fraunhofer lines; see how they widen as they cross the spot, putting on a sudden blackness and width in the case of a spot with steep sides, expanding gradually in a shelving one. The behaviour of these lines is due to selective absorption.

We have, then, the following facts: mark them well—

1. The lines of sodium, magnesium, and barium, when observed in the chromosphere, are among those which are thinner than their usual Fraunhofer lines.

2. The lines of sodium, magnesium, and barium, when observed in a spot, are among those which are thicker than their usual Fraunhofer lines.

They show, I think, that a spot is the seat of a downrush or downsinking.

Messrs. De la Rue, Stewart, and Loewy, who brought forward the theory of a downrush before my observations of an actual downrush were made in 1865, at once suggested as one advantage of this explanation that all the gradations of darkness, from the faculæ to the central umbra, may be supposed to be due to the same cause, namely, the presence to a greater or less extent of a relatively cooler absorbing atmosphere; thus suggesting as one cause of the darkening of a spot—

1. The general absorption of the atmosphere, thicker here than elsewhere, as the spot is a cavity.

To which the spectroscope added in 1866, as you know—

2. Greater selective absorption.

I have Dr. Frankland's permission to exhibit an experiment connected with our researches on absorption which will show you that this increased selective absorption can be fairly grappled with in our laboratories. I will show you on the screen the absorption line due to sodium vapour, in one part as thin as it is in the ordinary solar spectrum; in another, almost if not quite as thick as it appears in a spot; and I accomplish this result in the following way:—Here I have an electric lamp, and by means of this slit I only permit a fine line of light to emerge from it; here the beam passes through a bisulphide of carbon prism, and there you see on the screen the glorious spectrum, due to the dismemberment of the fine line of polychromatic light. Mr. Pedler will now place a glass tube containing metallic sodium, sealed up with hydrogen, in front of the slit, and will heat it with a spirit lamp.

As the sodium vapour rises you see the dark line of absorption make its appearance as an extremely fine line, and finally you see that the light which traverses the upper layer of the sodium scarcely suffers any absorption—the line is thin; while, on the contrary, the light which has traversed the lower, denser layers has suffered tremendous absorption: the line is inordinately thick, such as we see it in the spectrum of a spot.

So much then for the selective absorption. My recent observations, to which I will shortly draw attention, show, I think, that it is of great importance, especially in connection with the fact that the passage from the penumbra to the umbra is generally less gradual than that from the photosphere to the penumbra. You see now how much is included in the assertion that the photosphere is gaseous.

You are all, I know, familiar with that grand generalization of Kirchhoff's, by which he accounted for the Fraunhhofer lines.

If we have a gas or a vapour less luminous than another light-source, and view that light-source through the gas or vapour, then we shall observe absorption of those particular rays which the gaseous vapour would emit if incandescent.

Let us confine our attention to the hydrogen Fraunhofer lines.

When I observe the chromosphere on the sun's limb, with no brighter light-source behind it, I observe its characteristic lines *bright*. But when I observe them on the sun itself—that is, when the brighter sun is on the other side of the hydrogen envelope, then, as a rule, its function is reduced—is toned down—the envelope acts as an absorber—the lines are observed black.

Now what must we conclude when I tell you that, at the present time, it is almost impossible to observe the sun for an hour without observing the hydrogen lines, every now and then, *bright upon the sun itself!*

Not only are the lines observed bright, but it would appear that the strongly luminous hydrogen is carried up by the tremendous convection currents at different pressures; and under these circumstances the bright line is seen to be expanded on both sides of its normal position. Moreover, at times there is a dim light on both sides the black line, and the line itself is thinned out, showing that, although there is an uprush of strongly luminous material, the column is still surmounted by some less luminous hydrogen, possibly separated from the other portion, which still performs the functions of an absorber.

This seems established by another fact, namely that at times the lines, still black, expand on both sides, as if, in fact, in these regions there were a depression in the chromosphere; you already know that the pressure is greater at the base of the chromosphere than at the summit.

For this reason it is best to observe these phenomena by means of the green line, which expands in a more decided manner by pressure than does the red.

I now come to a new field of discovery opened out by these investigations, a branch of the inquiry which I fear you will consider more startling than all the rest—a branch, however, which I have had many opportunities of studying, and which has required me to move with the utmost caution. I allude to the movements of the hydrogen envelope and prominences at which I have before hinted.

Any one who has observed the sun with a powerful telescope, especially in a London fog—all too great a rarity unfortunately for such work—will have been struck with the tremendous changes observed in spots. Now, change means movement, and as spot phenomena occur immediately below the level of the chromosphere we may easily imagine that the chromosphere and its higher waves, the prominences, will also partake of the movements, be they up- or down- rushes, cyclones, or merely lateral motions. I have thrown on the screen a photograph of a drawing of a sun-spot observed under the clear sky of Rome by Father Secchi—a drawing I regard as a most faithful counterpart of nature.

You see how the photosphere is being driven about and contorted; how here it seems to be torn to ribbons by the action of some tremendous force, how here it is dragged down and shivered to atoms.

The spectroscope enables us to determine the velocities of these movements with a considerable approach to accuracy; and at times they are so great that I am almost afraid to mention them to you.

Let me first endeavour to give you an idea how this result is arrived at, and I must here beg your indulgence for a gross illustration of one of the most supremely delicate of nature's operations.

Imagine a barrack out of which is constantly issuing with measured tread and military precision an infinite number of soldiers in single or Indian file; and suppose yourself in a street seeing these soldiers pass. You stand still, and take out your watch, and find that so many pass you in a second or minute, and that the number of soldiers, as well as the interval between them, is always the same.

You now move slowly towards the barrack, still noting what happens. You find that more soldiers pass you than before in the same time, and, reckoned in time, the interval between each soldier is less.

You now move still slowly from the barrack, *i.e.* with the soldiers.

You find that fewer soldiers now pass you, and that the interval between each is longer.

Now suppose yourself at rest, and suppose the barrack to have a motion now towards you, now from you.

In the first case the men will be paid out, so to speak, more rapidly. The motion of the barrack-gate towards you will plant each soldier nearer the preceding one than he would have been if the barrack had remained at rest. The soldiers will really be nearer together.

In the second case it is obvious that the interval will be greater, and the soldiers will really be further apart.

So that, generally, representing the interval between each soldier by an elastic cord, if the barrack and the eye approach each other by the motion of either, the cord will contract; in the case of recession, the cord will stretch.

Now let the barrack represent the hydrogen on the sun, perpetually paying out waves of light, and let the elastic cord represent one of these waves; its length will be changed if the hydrogen and the eye approach each other by the motion of either.

Particular wave lengths with the normal velocity of light are represented to us by different colours.

The long waves are red.

The short waves are violet.

Now let us fix our attention on the green wave, the refrangibility of which is indicated by the F line of hydrogen. If any change of wave length is observed in this line, *and not in the adjacent ones*, it is clear that it is not to the motion of the earth or sun, but to that of the hydrogen itself and alone that the change must be ascribed.

If the hydrogen on the sun is approaching us *the waves will be crushed together;* they will therefore be shortened, and the light will incline towards the violet, that is, towards the light with the shortest waves; and if the waves are shortened only by the $\frac{1}{10000000}$th of a millemeter we can detect the motion.

If the hydrogen on the sun is receding from us the waves will be drawn out, they will therefore be longer, and the green ray will incline towards the red.

I must next point out, that there are two different circumstances under which the hydrogen may approach or recede from the eye.

I have here a globe, which we will take as representing the sun. Fix your attention on the centre of this globe: it is evident that an uprush or a downrush is necessary to cause any alteration of wave length. A cyclone or lateral movement of any kind is powerless; there will be no motion to or from the eye, but only at right angles to the line of sight.

Next fix your attention to the edge of the globe—the limb, in astronomical language; here it is evident that an upward or downward movement is as powerless to alter the wave length as a lateral movement was in the other case, but that, should any lateral or cyclonic movement occur here of sufficient velocity, it might be detected.

So that we have the centre of the disc for studying upward and downward movements, and the limb for studying lateral or cyclonic movements, if they exist.

If the hydrogen-lines were invariably observed to broaden out on both sides, the idea of movement would require to be received with great caution; we might be in presence of phenomena due to greater pressure, both when the lines observed are bright or black upon the sun; but when they widen out, sometimes on one side, sometimes on the other, and sometimes on both, this explanation appears to be untenable, as Dr. Frankland and myself in our researches at the College of Chemistry have never failed to observe a widening out, equally or nearly so, on both sides the F line when the pressure of the gas has been increased.

You see now on the screen a diagram showing the strange contortions which the F hydrogen line undergoes at the centre of the sun's disc. Not only have we the line bright, as I have before told you, but the dark one is twisted in places, generally inclining towards the red; and often when this happens we have a bright line on the violet side. You see it, sometimes, stopping short of one of the small sun-spots; swelling out prior to disappearance; invisible in a facula between two small spots; changed into a bright line, and widened out on both sides two or three times in the very small spots; becoming bright near a spot, and expanding over it on both sides; very many times widened out near a spot, sometimes considerably, on the less refrangible side; and, finally, extended as a bright line without any thickening over a small spot.

Now the other Fraunhofer lines on the diagram may be looked upon as so many milestones telling us with what rapidity the uprush and downrush take place; for these twistings are nothing more or less than alterations of wave length, and thanks to Ångström's map we can map out distances along the spectrum from F in $\frac{1}{100000000}$ths of a millimeter from the centre of that line; and we know that an alteration of that line $\frac{1}{100000000}$th mm, towards the violet means a velocity of 38 miles a second towards the eye, *i.e.* an uprush; and that a similar alteration towards the red means a similar velocity from the eye, *i.e.* a downrush. The fact that the black line inclines to the red shows that the less bright hydrogen descends; the fact that the bright line—where both are visible side by side—inclines to the violet, shows that the more vivid hydrogen ascends; and the alteration of wave length is such that 20 miles a second is very common.

Now, observations of the lateral motions at the limb are of course made by the chromospheric bright lines seen beyond the limb. Here the velocities are very much more startling; not velocities of uprush and downrush, as you now know, but swinging and cyclonic motions of the hydrogen.

I will first show you a cyclone observed on the 14th of March, but before I do so let me make one remark. Although the slit used is as narrow as I can make it, let us say $\frac{1}{500}$th—I have not measured it—of an inch, a strip of this breadth, of the sun's image, is something considerable, as the glorious sun himself is painted by my object-glass only about ·94 inch in diameter, so that after all the slit lets in to be analyzed a strip some 1800 miles wide.

Now, suppose we have a cyclone of incandescent hydrogen some 1500 miles wide tearing along with a very rapid rotatory motion, it is clear that all this cyclone could fall within the slit; and that if the rotatory motion were sufficiently rapid the spectroscope should separate the waves which are carried towards us from those which are receding. It does this: as you see, we have an alteration of wave-length both towards the red and violet, amounting to something like 40 miles a second. Now it should be clear to you that, by moving the slit first one way and then the other, we may be able to bring it in turn to such positions that only the light proceeding from either side of the cyclone can enter it. Then we shall have changes of wave-length in one direction only, in each case precisely as you see was observed.

Now, let us suppose that instead of a cyclone, we have a motion of some portions of the prominence towards the eye; and that, moreover, the rate of motion varies excessively in some portions. What we shall see will be this. The portion of the prominence at rest will give us no alteration of wave-length; its bright line will be in a line with the corresponding black one in the spectrum. The portion moving towards the eye, however, will give us an alteration of wave-length towards the violet. You are now in a position to grasp the phenomena revealed to me by my spectroscope on the 12th instant, when at times the F line was triple! the extreme alteration of wave-length being such that the motion of that part of the prominence giving the most extreme alteration of wave-length must have exceeded 120 miles per second, if we are to explain these phenomena by the only known possible cause which is open to us.

By moving the slit it was possible to see in which part of the prominence these great motions arose, and to follow the change of wave-length to its extremest limit.

By the kindness of Dr. Balfour Stewart I am able to exhibit to you some of the Kew sun-pictures which show you how these spectroscopic changes are sometimes connected with telescopic ones.

On the 21st April there was a spot very near the limb which I was enabled to observe continuously for some time. At 7.30 A.M. there was a prominence visible in the field of view, in which tremendous action was evidently going on, for the C, D, and F lines were magnificently bright in the ordinary spectrum itself, and as the spot-spectrum was also visible it was seen that the prominence was in advance of the spot. The injection into the chromosphere surpassed anything I had seen before, for there was a magnesium cloud quite separated from the limb, and high up in the prominence itself.

By 8.30 the action had quieted down, but at 9.30 another throb was observed, and the new prominence was moving away with tremendous velocity. While this was going on, the hydrogen lines suddenly became bright on the other side (the earth's side) of the spot, and widened out considerably—indeed to such an extent that I attributed their action to a cyclone, although, as you know, this was a doubtful case.

Now, what said the photographic record? The sun was photographed at 10h. 55m. A.M., and I hope you will be able to see on the screen how the sun's surface was disturbed near the spot. A subsequent photograph at 4h. 1m. P.M. on the same day shows the limb to be actually broken in that particular place: the photosphere seems to have been absolutely torn away behind the spot, exactly when the spectroscope had afforded me possible evidence of a cyclone!

In connection with the last branches of the research I have brought to your notice, I may remark that we have two very carefully prepared recent maps of the solar spectrum, one by Kirchhoff, the other by Angström, made a few years apart and at different epochs with regard to the sun-spot period. If you look at these maps you will see a vast difference in the relative thicknesses of the C and F lines, and great differences in the relative darkness and position of the lines; and if I had time I could show you that we now may be supplied with a barometer, so to speak, to measure the varying pressures in the solar and stellar chromospheres; for, depend upon it, every star has, has had, or will have, a chromosphere, and there are no such things as "worlds without hydrogen," any more than there are stars without photospheres. I suggested in 1866 that possibly a spectroscopic examination of the sun's limb might teach us somewhat of the outburst of the star in Corona, and already we see that all that is necessary to get just such an outburst in our own sun is to increase the power of his convection currents, which we know to be ever at work. Here, then, is one cataclysm the less in astronomy—one less "World on Fire," and possibly also a bright light thrown on the past history of our own planet.

I might show you further that we now are beginning to have a better hold on the strange phenomena presented by variable stars, and that an application of the facts I have brought to your notice this evening, taken in connection with the various types of stars which have been indicated by Father Secchi with admirable philosophy, opens out generalizations of the highest interest and importance; and that having at length fairly grappled with some of the phenomena of the nearest star, we may soon hope for more certain knowledge of the distant ones.

At present, however, we may well leave speculation for those who prefer it to acquiring facts; let us rather, emboldened by the work which this new method of research has enabled us to accomplish in this country, under the worst atmospheric conditions, in seven short months, go on quietly deciphering one by one the letters of this strange hieroglyphic language which the spectroscope has revealed to us—a language written in fire on that grand orb which to us earth-dwellers is the fountain of light and heat, and even of life itself.

[J. N. L.]

Friday, May 6, 1870.

SIR HENRY HOLLAND, Bart. M.D. D.C.L. F.R.S. President, in the Chair.

RICHARD A. PROCTOR, B.A. F.R.A.S.

On Star-grouping, Star-drift, and Star-mist.

NEARLY a century has passed since the greatest astronomer the world has ever known,—the Newton of observational astronomy, as he has justly been called by Arago,—conceived the daring thought that he would gauge the celestial depths. And because in his day, as indeed in our own, very little was certainly known respecting the distribution of the stars, he was forced to found his researches upon a guess. He supposed that the stars, not only those visible to the naked eye, but all that are seen in the most powerful telescopes, are suns, distributed with a certain general uniformity throughout space. It is my purpose to attempt to prove that—as Sir Wm. Herschel was himself led to suspect during the progress of his researches—this guess was a mistaken one; that but a small proportion of the stars can be regarded as real suns; and that in place of the uniformity of distribution conceived by Sir Wm. Herschel, the chief characteristic of the sidereal system is *infinite variety.*

In order that the arguments on which these views are based may be clearly apprehended, it will be necessary to recall the main results of Sir Wm. Herschel's system of star-grouping.

Directing one of his 20-feet reflectors to different parts of the heavens, he counted the stars seen in the field of view. Assuming that the telescope really reached the limits of the sidereal system, it is clear that the number of stars seen in any direction affords a means of estimating the relative extension of the system in that direction, provided always that the stars are really distributed throughout the system with a certain approach to uniformity. Where many stars are seen, there the system has its greatest extension; where few, there the limits of the system must be nearest to us.

Sir Wm. Herschel was led by this process of star-grouping to the conclusion that the sidereal system has the figure of a cloven disc. The stars visible to the naked eye lie far within the limits of this disc. Stars outside the relatively narrow limits of the sphere including all the visible stars, are separately invisible; but where the system has its greatest extension these orbs produce collectively the diffused light which forms the Milky Way.

Sir John Herschel, applying a similar series of researches to the southern heavens, was led to a very similar conclusion. His view of the sidereal system differs chiefly in this respect from his father's, that he considered the stars within certain limits of distance from the sun to be spread less richly through space than those whose united lustre produces the milky light of the galaxy.

Now it is clear that if the supposition on which these views are based is just, the three following results are to be looked for.

In the first place the stars visible to the naked eye would be distributed with a certain general uniformity over the celestial sphere; so that if on the contrary we find certain extensive regions over which such stars are strewn much more richly than over the rest of the heavens, we must abandon Sir Wm. Herschel's fundamental hypothesis and all the conclusions which have been based upon it.

In the second place, we ought to find no signs of the aggregation of lucid stars into streams or clustering groups. If we should find such associated groups we must abandon the hypothesis of uniform distribution and all the conclusions founded on it.

Thirdly, and most obviously of all, the lucid stars ought not to be associated in a marked manner with the figure of the Milky Way. To take an illustrative instance. When we look through a glass window at a distant landscape we do not find that the specks in the substance of the glass seem to follow the outline of valleys, hills, trees, or whatever features the landscape may present. In like manner, regarding the sphere of the lucid stars as in a sense the window through which we view the Milky Way, we ought not to find these stars, which are so near to us, associated with the figure of the Milky Way, whose light comes from distances so enormously exceeding those which separate us from the lucid stars. Here again, then, if there should appear signs of such association, we must abandon the theory that the sidereal system is constituted as Sir Wm. Herschel supposed.

It should further be remarked that the three arguments derived from these relations are independent of each other. They are not as three links of a chain, any one of which being broken the chain is broken. They are as three strands of a triple cord. If one strand holds, the cord holds. It may be shown that all three are to be trusted.

It is not to be expected, however, that the stars as actually seen should exhibit these relations, since far the larger number are but faintly visible; so that the eye would look in vain for the signs of law among them, even though law may be there. What is necessary is that maps should be constructed on a uniform and intelligible plan, and that in these maps the faint stars should be made bright, and the bright stars brighter.

The maps exhibited during this discourse have been devised for this purpose amongst others. There are twelve of them, but they overlap, so that in effect each covers a tenth part of the heavens. There is first a north-polar map, then five maps symmetrically placed around it; again, there is a south-polar map, and five maps symmetrically placed round that map; and these five so fit in with the first five as to complete the enclosure of the whole sphere. In effect every map of the twelve has five maps symmetrically placed around it and overlapping it.*

Since the whole heavens contain but 5932 stars visible to the naked eye, each of the maps should contain on the average about 593 stars. But instead of this being the case, some of the maps contain many more than their just proportion of stars, while in others the number as greatly falls short of the average. One recognizes, by combining these indications, the existence of a roughly circular region, rich in stars, in the northern heavens, and of another, larger and richer, in the southern hemisphere.

To show the influence of these rich regions, it is only necessary to exhibit the numerical relations presented by the maps.

The north-polar map, in which the largest part of the northern rich region falls, contains no less than 693 lucid stars, of which upwards of 400 fall within the half corresponding to the rich region. Of the adjacent maps, two contain upwards of 500 stars, while the remaining three contain about 400 each. Passing to the southern hemisphere, we find that the south-polar map, which falls wholly within a rich region, contains no less than 1132 stars! One of the adjacent maps contains 834 stars, and the four others exhibit numbers ranging from 527 to 595.

It is wholly impossible not to recognize so unequal a distribution as exhibiting the existence of special laws of stellar aggregation.

It is noteworthy, too, that the greater Magellanic cloud falls in the heart of the southern rich region. Were there not other signs that this wonderful object is really associated with the sidereal system it might be rash to recognize this relation as indicating the existence of a physical connection between the Nubecula Major and the southern region rich in stars. Astronomers have indeed so long regarded the Nubeculæ as belonging neither to the sidereal nor to the nebular systems, that they are not likely to recognize very readily the existence of any such connection. Yet how strangely perverse is the reasoning which has led astronomers so to regard these amazing objects. Presented fairly, that evidence amounts simply to this: The Magellanic clouds contain stars and they contain nebulæ; therefore they are neither nebular nor stellar. Can perversity of reasoning be pushed farther? Is not the obvious conclusion this, that since nebulæ and stars are *seen* to be intermixed in the Nubeculæ, the nebular and stellar systems form in reality but one complex system.

As to the existence of star-streams and clustering aggregations, we have also evidence of a decisive character.† There is a well-

* It will be understood that the description here, and all which immediately follows, replaces portions of the discourse which would only be intelligible when illustrated by means of the diagrams and illuminated maps actually employed.

† Here, again, without the illustrative maps the argument is necessarily rendered imperfect.

marked stream of stars running from near Capella towards Monoceros. Beyond this lies a long dark rift altogether bare of lucid orbs, beyond which again lies an extensive range of stars, covering Gemini, Cancer, and the southern parts of Leo. This vast system of stars resembles a gigantic sidereal billow flowing towards the Milky Way as towards some mighty shore-line. Nor is this description altogether fanciful; since one of the most marked instances of star-drift presently to be adduced refers to this very region. These associated stars *are* urging their way towards the galaxy, and that at a rate which, though seemingly slow when viewed from beyond so enormous a gap as separates us from this system, must in reality be estimated by millions of miles in every year.

Other streams and clustering aggregations there are which need not here be specially described. But it is worth noticing that all the well-marked streams recognized by the ancients seem closely associated with the southern rich region already referred to. This is true of the stars forming the River Eridanus, the serpent Hydra, and the streams from the water-can of Aquarius. It is also noteworthy that in each instance a portion of the stream lies outside the rich region, the rest within it; while all the streams which lie on the same side of the galaxy tend towards the two Magellanic clouds.

Most intimate signs of association between lucid stars and the galaxy can be recognized,—(i.) in the part extending from Cygnus to Aquila; (ii.) in the part from Perseus to Monoceros; (iii.) over the ship Argo; and (iv.) near Crux and the feet of Centaurus.

Before proceeding to the subject of Star-drift, three broad facts may be stated. They are, I believe, now recognized for the first time, and seem decisive of the existence of special laws of distribution among the stars:—

First, the rich southern region, though covering but a sixth part of the heavens, contains one-third of all the lucid stars, leaving only two-thirds for the remaining five-sixths of the heavens.

Secondly, if the two rich regions and the Milky Way be considered as one part of the heavens, the rest as another, then the former part is three times as richly strewn with lucid stars as the second.

Thirdly, the southern hemisphere contains one thousand more lucid stars than the northern, a fact which cannot but be regarded as most striking when it is remembered that the total number of lucid stars in both hemispheres falls short of 6000.

Two or three years ago, the idea suggested itself to me that if the proper motions of the stars were examined, they would be found to convey clear information respecting the existence of variety of structure, and special laws of distribution within the sidereal system.

In the first place the mere amount of a star's apparent motion must be regarded as affording a means of estimating the star's distance. The nearer a moving object is, the faster it will seem to move, and *vice versâ*. Of course in individual instances little reliance can be placed on this indication; but by taking the average proper motions of a set of stars, no untrustworthy measure may be obtained of their average distance, as compared with the average distance of another set.

For example, we have in this process the means of settling the question whether the apparent brightness of a star is indeed a test of relative nearness. According to accepted theories the sixth-magnitude stars are ten or twelve times as far off as those of the first magnitude. Hence their motions should, on the average, be correspondingly small. Now, to make assurance doubly sure, I divided the stars into two sets, the first including the stars of the 1st, 2nd, and 3rd, the second including those of the 4th, 5th, and 6th magnitude. According to accepted views, the average proper motion for the first set should be about five times as great as that for the second. I was prepared to find it about three times as great; that is, not so much greater as the accepted theories require, but still considerably greater. To my surprise I found that the average proper motion of the brighter orders of stars is barely equal to that of the three lower orders.

This proves beyond all possibility of question that by far the greater number of the fainter orders of stars (I refer here throughout to lucid stars) owe their faintness not to vastness of distance, but to real relative minuteness.

To pass over a number of other modes of research, the actual mapping of the stellar motions, and the discovery of the peculiarity to which I have given the name of star-drift, remains to be considered.

In catalogues it is not easy to recognize any instances of community of motion which may exist among the stars, owing to the method in which the stars are arranged. What is wanted in this case (as in many others which yet remain to be dealt with) is the adoption of a plan by which such relations may be rendered obvious to the eye. The plan I adopted was to attach to each star in my maps a small arrow, indicating the amount and direction of that star's apparent motion in 36,000 years (the time-interval being purposely lengthened, as otherwise most of the arrows would have been too small to be recognized). When this was done, several well-marked instances of community of motion could immediately be recognized.

It is necessary to premise, however, that before the experiment was tried, there were reasons for feeling very doubtful whether it would succeed. A system of stars might really be drifting athwart the heavens, and yet the drift might be rendered unrecognizable through the intermixture of more distant or nearer systems having motions of another sort, and seen accidentally in the same general direction.

This was found to be the case, indeed, in several instances. Thus the stars in the constellation Ursa Major, and neighbouring stars in Draco, exhibit two well-marked directions of drift. The stars β, γ, δ, ε, and ζ of the Great Bear, besides two companions of the last-named star, are travelling in one direction, with equal velocity, and clearly form one system. The remaining stars in the neighbourhood are travelling in a direction almost exactly the reverse. But even this relation, thus recognized in a region of diverse motions, is full of interest. Baron Mädler, the celebrated German astronomer, recognizing the community of motion between ζ Ursæ and its companions, calculated the cyclic revolution of the system to be certainly not less than 7000 years. But when the complete system of stars showing this motion is considered, we get a cyclic period so enormous, that not only the life of man, but the life of the human race, the existence of our earth, nay, even the existence of the solar system, must be regarded as a mere day in comparison with that amazing cycle.

Then there are other instances of star-drift where, though two directions of motion are not intermixed, the drift character of the motion is not at once recognized, because of the various distances at which the associated stars lie from the eye.

A case of this kind is to be met with in the stars forming the constellation Taurus. It was here that Mädler recognized a community of motion among the stars, but he did not interpret this as I do. He had formed the idea that the whole of the sidereal system must be in motion around some central point; and for reasons which need not here be touched on, he was led to believe that in whatever direction the centre of motion may lie, the stars seen in that general direction would exhibit a community of motion. Then, that he might not have to examine the proper motions all over the heavens, he inquired in what direction (in all probability) the centre of motion may be supposed to lie. Coming to the conclusion that it must be towards Taurus, he examined the proper motions in that constellation, and found a community of motion which led him to regard Alcyone, the chief star of the Pleiades, as the centre around which the sidereal system is moving. Had he examined farther he would have found more marked instances of community of motion in other parts of the heavens, a circumstance which would have at once compelled him to abandon his hypothesis of a central sun in the Pleiades, or at least to lay no stress on the evidence derivable from the community of motion in Taurus.

Perhaps the most remarkable instance of star-drift is that observed in the constellations Gemini and Cancer. Here the stars seem to set bodily towards the neighbouring part of the Milky Way. The general drift in that direction is too marked, and affects too many stars, to be regarded as by any possibility referable to accidental coincidence.

It is worthy of note that if the community of star-drift should be recognized (or I prefer to say, *when* it is recognized), astronomers will have the means of determining the relative distances of the stars of a drifting system. For differences in the apparent direction and amount of motion can be due but to differences of distance and position, and the determination of these differences becomes merely a question of perspective.*

* Here no account is taken of the motions of the stars within the system; such motions must be minute compared with the common motion of the system.

Before long it is likely that the theory of star-drift will be subjected to a crucial test, since spectroscopic analysis affords the means of determining the stellar motions of recess or approach. The task is a very difficult one, but astronomers have full confidence that in the able hands of Mr. Huggins it will be successfully accomplished. I await the result with full confidence that it will confirm my views.

Turning to the subject of Star-mist, under which head I include all orders of nebulæ, I propose to deal but with a small proportion of the evidence I have collected to prove that none of the nebulæ are external galaxies. That evidence has indeed become exceedingly voluminous.

I shall dwell, therefore, on three points only.

First, as to the distribution of the nebulæ :—They are not spread with any approach to uniformity over the heavens, but are gathered into streams and clusters. The one great law which characterizes their distribution is an avoidance of the Milky Way and its neighbourhood. This peculiarity has strangely enough been regarded by astronomers as showing that there is no association between the nebulæ and the sidereal system. They have forgotten that marked contrast is as clear a sign of association as marked resemblance, and has always been so regarded by logicians.

Secondly, there are in the southern heavens two well-marked streams of nebulæ. Each of these streams is associated with an equally well-marked stream of stars. Each intermixed stream directs its course towards a Magellanic Cloud, one towards the Nubecula Minor, the other towards the Nubecula Major. To these great clusters they flow, like rivers towards some mighty lake. And within these clusters, which are doubtless roughly spherical in form, there are found intermixed in wonderful profusion, stars, star clusters, and all the orders of nebulæ. Can these coincidences be regarded as accidental? And if not accidental, is not the lesson they clearly teach us this, that nebulæ form but portions of the sidereal system, associating themselves with stars on terms of equality (if one may so speak), even if single stars be not more important objects in the scale of creation, than these nebulous masses, which have been so long regarded as equalling, if not outvying, the sidereal system itself in extent?

The third point to which I wish to invite attention is the way in which in many nebulæ stars of considerable relative brightness, and belonging obviously to the sidereal system, are so associated with nebulous masses as to leave no doubt whatever that these masses really cling around them. The association is in many instances far too marked to be regarded as the effect of accident.

Amongst other instances* may be cited the nebula round the stars c^1 and c^2 in Orion. In this object two remarkable nebulous nodules centrally surround two double stars. Admitting the association here to be real (and no other explanation can reasonably be admitted), we are led to

* Eight pictures of nebulæ were exhibited in illustration of this peculiarity.

interesting conclusions respecting the whole of that wonderful nebulous region which surrounds the sword of Orion. We are led to believe that the other nebulæ in that region are really associated with the fixed stars there; that it is not a mere coincidence, for instance, that the middle star in the belt of Orion is involved in nebula, or that the lowest star of the sword is similarly circumstanced. It is a legitimate inference from the evidence that all the nebulæ in this region belong to one great nebulous group, which extends its branches to these stars. As a mighty hand this nebulous region seems to gather the stars here into close association, showing us in a way there is no misinterpreting, that these stars form one system.

The nebula around the strange variable star, Eta Argûs, is another remarkable instance of this sort. More than two years ago I ventured to make two predictions about this object. The first was a tolerably safe one. I expressed my belief that the nebula would be found to be gaseous. After Mr. Huggins' discovery that the great Orion nebula is gaseous, it was not difficult to see that the Argo nebula must also be so. At any rate, this has been established by Captain Herschel's spectroscopic researches. The other prediction was more venturesome. Sir John Herschel, whose opinions on such points one would always prefer to share, had expressed his belief that the nebula lies far out in space beyond the stars seen in the same field of view. I ventured to express the opinion that those stars are involved in the nebula. Lately there came news from Australia that Mr. Le Sueur, with the great reflector erected at Melbourne, has found that the nebula has changed largely in shape since Sir John Herschel observed it. Mr. Le Sueur accordingly expressed his belief that the nebula lies *nearer* to us than the fixed stars seen in the same field of view. More lately, however, he has found that the star Eta Argûs is shining with the light of burning hydrogen, and he expresses his belief that the star has consumed the nebulous matter near it. Without agreeing with this view, I recognize in it a proof that Mr. Le Sueur now considers the nebula to be really associated with the stars around it. My belief is that as the star recovers its brilliancy observation will show that the nebula in its immediate neighbourhood becomes brighter (*not* fainter through being consumed as fuel). In fact, I am disposed to regard the variations of the nebula as systematic, and due to orbital motions among its various portions around neighbouring stars.

As indicative of other laws of association bearing on the relations I have been dealing with, I may mention the circumstance that red stars and variable stars affect the neighbourhood of the Milky Way or of well-marked star-streams. The constellation Orion is singularly rich in objects of this class. It is here that the strange 'variable' Betelgeux lies. At present this star shows no sign of variation, but a few years ago it exhibited remarkable changes. One is invited to believe that the star may have been carried by its proper motion into regions where there is a more uniform distribution of the material whence this orb recruits its fires. It may be that in the consideration of such causes of variation affecting our sun in long past ages a more satisfactory explanation than any yet obtained may be found of the problem geologists find so perplexing—the former existence of a tropical climate in places within the temperate zone, or even near the Arctic regions.*

It remains that I should exhibit the general results to which I have been led. It has seemed to many that my views tend largely to diminish our estimate of the extent of the sidereal system. The exact reverse is the case. According to accepted views there lie within the range of our most powerful telescopes millions of millions of suns. According to mine the primary suns within the range of our telescopes must be counted by tens of thousands, or by hundreds of thousands at the outside. What does this diminution of numbers imply but that the space separating sun from sun is enormously greater than accepted theories would permit? And this increase implies an enormous increase in the estimate we are to form of the vital energies of individual suns. For the vitality of a sun, if one may be permitted the expression, is measured not merely by the amount of matter over which it exercises control, but by the extent of space within which that matter is distributed. Take an orb a thousand times vaster than our sun, and spread over its surface an amount of matter exceeding a thousandfold the combined mass of all the planets of the solar system :—So far as living force is concerned, the result is —*nil*. But distribute that matter throughout a vast space all round the orb :—That orb becomes at once fit to be the centre of a host of dependent worlds. Again, according to accepted theories, when the astronomer has succeeded in resolving the milky light of a portion of the galaxy into stars, he has in that direction, at any rate, reached the limits of the sidereal system. According to my views, what he has really done has been but to analyze a definite aggregation of stars, a mere corner of that great system. Yet once more, according to accepted views, thousands and thousands of galaxies, external to the sidereal system, can be seen with powerful telescopes. If I am right, the external star-systems lie far beyond the reach of the most powerful telescope man has yet been able to construct, insomuch that perchance the nearest of the outlying galaxies may lie a million times beyond the range even of the mighty mirror of the great Rosse telescope.

But this is little. Wonderful as is the extent of the sidereal system as thus viewed, even more wonderful is its infinite variety. We know how largely modern discoveries have increased our estimate of the complexity of the planetary system. Where the ancients recognized but a few planets, we now see, besides the planets, the families of satellites; we see the rings of Saturn, in which minute satellites must

---

* Sir John Herschel long since pointed to the variation of our sun as a possible cause of such changes of terrestrial climate.

be as the sands on the sea-shore for multitude; the wonderful zone of asteroids; myriads on myriads of comets; millions on millions of meteor-systems, gathering more and more richly around the sun, until in his neighbourhood they form the crown of glory which bursts into view when he is totally eclipsed. But wonderful as is the variety seen within the planetary system, the variety within the sidereal system is infinitely more amazing. Besides the single suns, there are groups and systems and streams of primary suns; there are whole galaxies of minor orbs; there are clustering stellar aggregations, showing every variety of richness, of figure, and of distribution; there are all the various forms of nebulæ, resolvable and irresolvable, circular, elliptical, and spiral; and lastly, there are irregular masses of luminous gas, clinging in fantastic convolutions around stars and star-systems. Nor is it unsafe to assert that other forms and varieties of structure will yet be discovered, or that hundreds more exist which we may never hope to recognize.

But lastly, even more wonderful than the infinite variety of the sidereal system, is its amazing vitality. Instead of millions of inert masses, we see the whole heavens instinct with energy,—astir with busy life. The great masses of luminous vapour, though occupying countless millions of cubic miles of space, are moved by unknown forces like clouds before the summer breeze; star-mist is condensing into clusters; star-clusters are forming into suns; streams and clusters of minor orbs are swayed by unknown attractive energies; and primary suns singly or in systems are pursuing their stately path through space, rejoicing as giants to run their course, extending on all sides the mighty arm of their attraction, gathering from ever new regions of space supplies of motive energy, to be transformed into the various forms of force,—light, and heat, and electricity,—and distributed in lavish abundance to the worlds which circle round them.

Truly may I say, in conclusion, that whether we regard its vast extent, or its infinite variety, or the amazing vitality which pervades its every portion, the sidereal system is of all the subjects man can study, the most imposing and the most stupendous. It is as a book full of mighty problems,—of problems which are as yet almost untouched by man, of problems which it might seem hopeless for him to attempt to solve. But those problems are given to him for solution, and he *will* solve them, whenever he dares attempt to decipher aright the records of that wondrous volume.

[R. A. P.]

Friday, March 17, 1871.

Sir Henry Holland, Bart. M.D. D.C.L. F.R.S. President, in the Chair.

J. Norman Lockyer, Esq. F.R.S.

On the Recent Solar Eclipse.

My duty to-night, a pleasant one, although it is tinged with a certain sense of disappointment, is to bring before you the observations which were made of the recent eclipse in Spain and Sicily, to connect them with our former knowledge, and to show in what points our knowledge has been extended. In these observations, as you know, we had nothing to do with the sun as ordinarily visible, but with the most delicate phenomenon which becomes visible to us during eclipses. I refer to the *Corona.*

General Notions of the Corona.

Let me, in the first place, show you what is meant by this term, and state the nature of the problems we had before us. I have here some admirable drawings, which I will show by means of the lamp, of the eclipse that was observed in 1851 by several astronomers who left England in that year to make observations in Sweden, where the eclipse was visible. You must bear in mind that the drawings I shall bring to your notice were made in the same region, at places not more than a few miles apart.* The first drawing was made by an observer whose name is a sufficient guarantee for its accuracy—I refer to Mr. Carrington—*and when the sky was absolutely free from clouds.* In the next diagram you will see the corona is changed. The bright region round the sun is no longer limited to the narrow border of light round the dark moon, as seen by Mr. Carrington, but it is considerably expanded. The third gives still a greater extension, although that picture was drawn within a quarter of a mile of the place where Mr. Carrington's was taken. And lastly, we have a drawing made by the present Astronomer Royal, of that same eclipse, *through a cirro-stratus cloud,* as unlike Mr. Carrington's as anything can possibly be. So that you see we began with a thin band of light about the moon,

---

* Mr. Carrington observed at Lilla Edet, on the Gota River. The Astronomer Royal observed at Göttenburg. The second drawing referred to was made by Pettersen, at Göttenburg; the third by a friend of the Rev. T. Chevallier, at the same place; and I might have added another by Fearnlay, taken at Rixhöft, in which the corona is larger than in any of the others. The series is most instructive. See Mem. R. A. S., vol. xxi.

which would make the corona a few thousand miles high, and we end with a figure which Mr. Airy graphically likens to the ornament round a compass-card, and which gives the corona a height equal to about once and a half the sun's diameter.

I will next bring before you some drawings made during the eclipse of 1858, which was not observed in European regions, but in South America by two first-rate observers—one, M. Liais, a French astronomer, who was stationed at Olmos, in Brazil; the other, Lieutenant Gilliss, who was also there as a representative of the American Government, and observed some thousand miles away in Peru.

I will throw on the screen the appearances observed by these gentlemen, and I think you will acknowledge the same variations between their results, as to degree, while in one case we get a perfectly new idea of the phenomena—a difference in kind. I would especially call attention in the Olmos drawing to those extraordinary bundles of rays of wonderful shapes, which you see are so much brighter than the other portions of the corona. Such forms have been seen in other eclipses, but they are somewhat rare. The drawing made by Lieutenant Gilliss bears the same relation to that made by M. Liais as Mr. Carrington's did to the Astronomer Royal's; so that we may say that we not only get variations in the dimensions of the corona as seen at different stations, but that we furthermore get a strange structure introduced now and then in our drawing in regions where absolutely no corona at all exists in the other.

So much by way of defining the phenomena and giving an idea of the eye observations generally.

Let me now attempt to show you how the phenomena observed in the last eclipse bear upon the results which had been previously accumulated by means of telescopic and naked-eye observations, and by means of the polariscope and spectroscope.

I.—Telescopic and Naked-Eye Observations.

a.—A Part of the Corona is undoubtedly Solar.

The first use I propose to make of the telescopic and naked-eye observations of last year, is to show you a photographic copy of an admirable drawing made by Mr. Brett, who, though unfortunate enough to see the sun only for a very short time, was yet sufficiently skilled to make good use of that brief period. This drawing will bring before you the fact, that even when a large portion of the sun remained unobscured by the moon, Mr. Brett was enabled to see a dim ring of light round the unobscured portion, which since the year 1722 has been acknowledged, beyond all question, I think I may say, to represent something *at the sun.* It was observed in 1722 round the uneclipsed sun, and in more recent times by Mrs. Airy in 1842, and by Rumker $1\frac{1}{2}$ minute before totality in 1860, not to mention other instances. Therefore, we have one observation made during this eclipse, confirming the old one, that in the corona there is a region of some small breadth at all events which is absolutely solar, and which it only requires a diminution of the solar light to enable us to see. This, then, we may look upon as the known; now let us feel our way gradually outwards.

b.—Rays, or Streamers, are added at Totality.

The drawings made in all the eclipses which have been carefully recorded bring before us quite outside this narrow, undoubtedly *solar region*, observed before totality, as I have shown, and also by Mr. Carrington and by Lieutenant Gilliss during totality in 1851 and 1858, extraordinary appearances of a different order. While in fact we have a solar ring from 2′ to 6′ high, we have rays of all shapes and sizes visible outside, in some cases extending as far as 4°, and in all cases brighter than the outer corona on which they are seen, the rays being different in different eclipses, and appearing differently to different observers of the same eclipse, and even at the same station. Here is a copy of a drawing made by M. Rumker of the eclipse of 1860, and I show it for the purpose of calling your attention to the fact that the two curious rays represented in it belong to a different order of things from those which we see in the rest of the corona. From the beginning to the middle of the eclipse the east rays were the most intense. In the next drawing, which was made by the same observer, you see something absolutely new: and now the western side of the corona is the most developed; we have a new series of bright rays, and altogether it is difficult to believe that it is a drawing made by the same observer of the same eclipse.

The third drawing is a representation of the same eclipse by M. Marquez, who observed with a perfection of minute care which has scarcely ever been equalled: I bring it before you to show that the rays he saw were altogether differently situated. We may conclude then that the rays, although extremely definite and bright—as bright or brighter than the other portions of the corona which are visible before totality, they being *invisible* before totality—appear different to different observers of the same eclipse, and to the same observer during different *phases.*

c.—They change from Side to Side.

I have already said that M. Rumker observed that from the beginning to the middle of totality the rays on the *east* side of the sun were longest and brightest, and that from the middle to the end of totality the rays on that side of the sun where the totality ended were longest and brightest.

We will now carry this observation a step further, by referring to three drawings made by M. Plantamour in the same eclipse, that of 1860. In the first drawing we have the beginning of the total eclipse as seen in the telescope: with the naked eye naturally we should get the sun disappearing at the east or left-hand side, the moon moving from west to east; in the telescope things are reversed, and we have it right instead of left: and here we have the same thing that M. Rumker observed, namely, that when the eastern limbs were in contact bright rays (M. Plantamour saw three) were visible on the side at which the contact took place. When the moon was half way over the sun, *two* rays of reduced brilliancy were observed on that side, not necessarily in the same position as *those* first observed, but one of *these* has been abolished altogether; and on the other side of the sun, where totality was about to end, we have three rays gradually suggesting themselves: at the end of totality the rays visible at the commencement are abolished, and now instead of them and of those seen at the middle of the eclipse, we have a bran new set of rays on the side of the moon from whence the sun is about to emerge.

This observation I need hardly say is of considerable importance in connection with the fact that from the year 1722 almost every observer of a total eclipse has stated that there is a large increase of brilliancy, and an increase of the size of the corona on the side where the sun has just been covered, or is just about to emerge.

Now, what was there bearing on this point in the recent observations? I have here three drawings, which, though roughly done, you will see are of great importance side by side with those of M. Plantamour. These are drawings which have been sent in to the Organizing Committee by Mr. Gilman, who lives in Spain, and who took considerable interest in the eclipse, and sent the results of his observations to England with the eclipse party when they came home; and it is of importance that you should see everything that Mr. Gilman has done. If you agree with this explanation of the square form of the corona, which was observed in Spain this year, it will explain the quadrangular form observed in the corona in a good many previous eclipses. Mr. Gilman says that at the commencement of totality—let me remind you, the commencement was determined by the disappearance of the sun at the east limb of the moon, which is east in Mr. Gilman's drawing, as he was observing with the naked eye — the commencement, he says, was determined by the corona flashing out very much like a capital D. You see on the black board exactly the outline, and you will at once mentally associate one half of the diagram with the rays observed by M. Plantamour, and the other half in which there is a nearly perfect ring of light round the moon, with the corona observed by Mr. Carrington all round it in a cloudless sky. At mid-eclipse Mr. Gilman also observed the corona, sketched out its outline carefully, and found rays coming out on the opposite side, adding themselves on to the perfect ring first seen there. Opposite the two salient angles he observed at the commencement of totality—represented by the top and bottom of the upright stroke of the capital D—there were two others; *the corona now appeared square*, and then, just before the end of totality came on, the two corners first seen were observed to disappear altogether, leaving nothing but a perfect ring, and where, at the beginning of the eclipse, nothing was seen but a perfectly round ring, the two exactly similar forms on the opposite side shot forth, and you got a D reversed (ᗡ). Mr. Warrington Smyth, who drew a square corona, saw the light flash out into the corona before the end of totality, and believes that all the angles of the square were not visible at one and the same time.

Here, then, you have observations of exactly the same character as those of M. Plantamour, to which I have referred. In the drawings of both are shown the inner part of the corona, which you saw growing in the observations of 1851, to which were added the strange forms observed in 1858. You have these strange variations positively growing at the same place and the same time, in the same and in different eyes. Obviously there must be very much that is non-solar, call it personality, atmospheric effect, or what you will, connected with it. We have added to the stable the unstable. The question is, to what is this unstable portion due?

d.—They are very variously represented.

I will now refer to other drawings of the late eclipse, which were made in Sicily. For some reason or other, which I do not profess to understand, the corona, which appeared in Spain to be square, and to Mr. Gilman like a D at the beginning, and like a D reversed (ᗡ) at the end,—to all those with whom I have conversed who saw it in Sicily, it appeared as round as you see it here, in this drawing made by Mr. Griffiths; and, instead of being square, we had sent to us all sorts of pictures, a large number of them representing a stellate figure. Here is a drawing made by a Fellow of the Royal Society, on board one of Her Majesty's ships (the 'Lord Warden') which were trying to save the poor 'Psyche' at Catania. In this we have perfectly regular rays drawn from every region of the sun, some long, some short, but similar rays are almost invariably opposite each other; but in the interior, inside these rays, the corona is just as it was observed by Mr. Griffiths at Syracuse. I now show you a drawing made by an American gentleman at sea, between Catania and Syracuse, with one ridiculously long ray, a ray as long as was seen by Otto Struve in 1860. Other drawings were made, even on board the same ship, so unlike each other, and so bizarre, that I need only refer to them as showing that there at all events must be some personality. We have then to account for the variations between the observations made in Spain and those made in Sicily. I regret that we have not a third order of difficulties to contend with, as doubtless we should have had if observations had been made by Mr. Huggins' party in North Africa.

e.—The Rays are accompanied by a Mass of Light.

These changes of the rays from side to side are accompanied by, and are perhaps to a certain extent due to, the bursting forth of brilliant light in their neighbourhood, where the limbs are nearest in contact. This was first observed by Miraldi in the eclipse of 1724, and has frequently been recorded since. Mr. Warrington Smyth, to whom I have before alluded, states that he noticed this in the last eclipse, and the photographs, I think, have recorded it; but as there is some uncertainty on this point, I need only suggest it.

f.—Long Rays are seen extending from the Cusps before and after Totality.

So far I have referred only to the rays visible during totality, but long rays were seen when a crescent of the sun was visible in 1860 and 1868 by Mr. Galton and Mr. Hennessy. Mr. Brett caught the same phenomenon last year; but as the sky was cloudy the commencements of the rays only were seen, appearing like delicate brushes in prolongation of the cusps. These observations are of great value, *as no one for one moment imagines that these rays are solar*, and yet they are very like those seen during totality.

g.—Sometimes Dark Rays, called Rifts, are seen instead of Bright ones.

These rays to which I have referred are, however, not the only kind of rays that are observed. At times are seen, as it were, openings in the corona; the openings being of the same shape as the rays, that is, expanding as they leave the dark moon, and opening more or less exactly as the rays do. Like the rays also they are sometimes very numerous; in other eclipses they are few in number. Let us take the eclipse observed in India in 1868. Several drawings made there showed the corona as square as it was drawn in Spain last year; others as round as it was seen in Sicily; but the eclipse was not observed only in India, it was observed at Mantawalok-Kelee by Captain Bullock, and at Whae-Whan, on the east coast of the Malayan Peninsula, by Sir Harry St. George Ord, Governor of the Straits Settlements. In the former place we had rifts expanding rapidly as they left the sun—one forms an angle of 90°, the sides of another being *parallel*—separating patches of corona, which in some places extends 2½ diameters of the moon from the sun.

At Whae-Whan we are told that at one particular moment of the eclipse "it was noticed that from several points in the moon's circumference darker rays emanated, extending to a considerable distance into space, and appearing like shadows cast forth into space by something not very well defined;" these dark rays afterwards "diminishing."

Now let us pass on to the eclipse of 1869. In two drawings made by Dr. Gould, in which the changes in the bright bundles of rays come out in a most unmistakable way, we get similar rifts, which changed as violently as did the rays; while in another drawing made by Mr. Gilman the whole corona is furrowed by narrow rifts in all regions lying between violet, mauve-coloured, white, and yellowish white rays!

Now, what have we bearing on this point in the recent observations? No rift was *seen* in Sicily; one rift was recorded by the sketchers in Spain, but more than one rift was photographed in both places. We must remember, however, in thus bringing eye-sketches and photographs into comparison, first that the eye too often in such observations retains a general impression of the whole phenomenon, while the plate records the phenomenon as it existed at the time at which it was exposed; and secondly, that we know that the plates record chemically, while the eye records visually. We are dealing with two different kinds of light.

I will show you two photographs on the screen. Although the lucid intervals were very rare, we were fortunate enough to get one photograph of the coronal regions in Syracuse, and one in Spain. I now show you the photograph made by the American party in Spain. You see here that, probably owing to a cloud, we get a certain amount of light driven on to the dark moon, and you also see the indications of the rifts. This photograph was taken with an instrument with a small field of view, so that the most important parts of the corona were rendered invisible by the instrument itself.

Lord Lindsay, who also photographed in Spain, recorded no rifts.

In the other photograph, taken at Syracuse, the result is better. We have the equivalent of the rift in the photograph I showed you before. The instrument was extremely unsteady, and the definition not so good as it would have been if Mr. Brothers had had a good opportunity of displaying his skill. We get other fainter indications of other rifts here and there, and the question whether these rifts agree in the photograph taken in Spain with those in that taken in Syracuse is one of great importance; and it is to be hoped that before long it will be set at rest. Some observers think they agree; others think they do not.

But there is an important consideration based on that photograph, to which I must draw your particular attention. I have shown you the photograph as it may be thrown on the screen; but in the photograph itself there are delicate details which it is impossible to reproduce. The dark portions in the corona indicated in the copy I have shown you are merely the bases of so many dark wedges driving out into space, like their prototypes in the Indian eclipse. It is Mr. Brothers's opinion, I believe, that all you see on the screen round the dark moon, all that enormous mass of light, nearly uniform in texture, and these beautiful broad rays between the rifts are really and absolutely parts of the solar corona. I confess I do not wish to commit myself to such an opinion. We want more facts, and the *onus probandi* lies with those who insist upon that view, and I have yet to hear an explanation of them on that basis.

h.—The Corona sometimes seems to be Flickering or Rotating.

We now come to the next point. Time out of mind, that is, for the last two centuries, the corona has been observed to be flickering, wavering, or rotating, moving in every conceivable way and direction. In 1652 it was described as "a pleasant spectacle of rotatory motion."

Don Antonio Ulloa remarked of the corona observed in the eclipse of 1788, "It seemed to be endued with a rapid rotatory motion, which caused it to resemble a firework turning round its centre." The terms whirling and flickering were applied in the eclipse of 1860. This extraordinary condition of things was also thoroughly endorsed by the late observations. It certainly exists, and is among the observations we have to take into account. When I saw an officer of one of the ships at Catania, I asked him if he had taken a drawing of the corona. "No," he said. I asked him, "Did you see any rays?" "Yes." "Then why did you not make any drawing of them?" His answer was, "How on earth could you draw a thing that was going round and round like a firework?" This was not the only observation of the kind, and the tendency of such observations I need hardly say is to strengthen a belief in the unstable, and therefore uncosmical, nature of their rays.

Is this variation of light due to the brilliancy of the corona, and the rapid change of the rays, which is one of the results which comes out clearest? In 1842 the brilliancy of the corona was stated to be insupportable to the naked eye. A similar remark was made to me by several of those officers who saw the last eclipse in Sicily.

II.—Polariscopic Observations.

With regard to the polarization experiments, by the kindness of Mr. Spottiswoode I am enabled to show you, in a very clear way, the *raison d'être* of the polariscopic observations made during this and former eclipses; but the polariscopic ground is a wide one, and it is not my intention to cover it to-night.

I have had this arrangement of lamp, reflector and prisms made so that you may see how the polariscope can determine the percentage of reflected light at different angles, and the direction of reflexion. Assume this lamp to represent the sun, let this reflector close to the lamp represent a particle near the sun, reflecting light to us, we shall naturally have the light reflected at a much larger angle than if the reflector were close to the screen representing a particle in our own air. Having this idea of the angle of reflexion in your minds, and the fact that the larger the angle under these conditions the more the polarization, if you take this lamp, as I have said, to represent the sun, and this mirror to represent any particle, of whatever kind you choose to imagine, it is clear, that in order to get the maximum polariscopic effect from that particle, that you must have it so situated that it will reflect light at a considerable angle to the beam coming from this lamp.

Now it is clear, that in order to polarize the beam most strongly, I must place the reflector close to our imaginary sun. If I so place it as to represent a particle in our own atmosphere, the angle will be so small that the polarization of the light will hardly be perceptible.

Here is our sunlight, which we will polarize at as great an angle as we can, by placing the reflector close to the imaginary sun, and send it through this magnificent prism, which Mr. Spottiswoode has been good enough to place at our disposal; and in the path of the beam I will place an object so that you may determine whether there is polarized light. [Experiment.] You see there is considerable brilliancy in those colours; their brilliancy depending upon the amount of polarization.

Now if, instead of having our reflector close to our imaginary sun to represent a particle in the sun's atmosphere, we place it near the screen to represent a particle in our own, in which case the angle is extremely small, the brilliancy of the colours will entirely disappear. You see it has disappeared. The colours, as colours, are distinguishable, but their brilliancy has gone.

That is the rationale of the polariscopic observations, which have been made on the occasion of the last eclipse with more elaboration than they ever were before. If we found the corona to be strongly polarized, this was held to be a great argument in favour of the corona being a real solar appendage, an argument strengthened if the polarization was also found to be radial. At present, however, a great many of the observations that have been made have not been received, and those that have been received are as discordant as those obtained in former eclipses, and therefore my account is an imperfect one, because I have not had an opportunity of discussing all these observations. Indeed, if I had, I should hesitate to give an opinion on the subject. When Mr. Carrington saw that small corona in 1851, and Mr. Gilliss saw that small corona in 1858, neither of them traced any polarization whatever; but when M. Liais saw that large corona in 1868 which was invisible to Mr. Gilliss, he in his turn saw an immense amount of polarization, which led him to believe that the corona was solar, the whole of it, rays and everything included, and that we had an indication of a solar atmosphere two or three times higher than the diameter of the sun; that is, an atmosphere two or three millions of miles in height. This observation is not in accordance with the general conclusions from the drawings I have shown you; and let me add that the assumption of reflexion at the sun is not without its difficulties, and that we have not yet traced reflected sunlight, even when the strongest polariscopic effects have been observed.

III.—Airy's and Mädler's Conclusions as the Results of the Pre-spectroscopic Observations.

Before passing to the spectroscopic observations. I will state the conclusions at which the Astronomer Royal and M. Mädler arrived after the observations of 1860 had been gathered together.

The Astronomer Royal, in a lecture delivered before the British Association at Manchester in 1861, stated that the assumption of an atmosphere extending to the moon explained the observation of Plantamour, which could, he thought, be explained in no other way, and he held also that the polarization experiments seemed to show the same thing. The Astronomer Royal was content to find the reflexion, which so many now insist must be at the sun, taking place somewhere between the earth and moon.

M. Mädler's verdict is in the same direction, and though he does not perhaps express so decided an opinion, he maintains that the atmosphere plays a principal part in the phenomenon; and after detailing experiments to show this, he remarks of the solar and atmospheric portions, "Both cover each other and unite in one phenomenon, so that the corona is a mixed phenomenon."

I shall shortly show you that the spectroscope, leaving the telescope out of consideration, has taught us that this is true; though I shall not be able to show you that it is the whole truth; we are not yet in a position to do that. Mädler concludes his observations by remarking, "We cannot share the doubts of those who are afraid to surround the sun with too many envelopes; neither do we find anything unnatural in the statement that the sun has as many atmospheres as Saturn has rings; but we gladly admit that we cannot yet say anything positive. We have here a large field of probabilities, and the decision may yet be distant."

We can speak with more certainty now!

IV.—Spectroscopic Observations.

a.—Spectrum of the Corona first observed by Tennant, Pogson, and Rayet.

We now come to the consideration of those observations in which we are aided by a most powerful and our most recent ally, the spectroscope, first used in the eclipsed sun, as you know, in the eclipse of 1868. You all know that in that year the question of the nature of red flames was for ever settled by M. Janssen, Major Tennant, Capt. Herschel, and others, who observed that eclipse in the most admirable manner; but we have nothing to do with the red flames now, we have to do with something outside them.

Now, most of you are under the impression, and it was mine until the day before yesterday, that the only thing we learnt about the corona in the eclipse of 1868, was that its spectrum was a continuous one; and I need not tell anyone in this theatre that the assertion that it was continuous was one that was extremely embarrassing, and implied that we had something non-gaseous outside the red flames, which seemed very improbable to those who know anything about the subject. But some of you will no doubt remember that, besides Major Tennant, who made this observation, we had a French observer, M. Rayet, who gave us a diagram of the spectrum of one of the prominences, and Mr. Pogson, who has now been for some time in India, and is a well-known observer, who gave us, nominally as the spectrum of a prominence, a spectrum with some curious variations from M. Rayet's diagram.

I exhibit a copy of M. Rayet's diagram of the spectrum of a prominence, as he called it. At the bottom is what he considered as the spectrum of the lower portion of the prominence, while in the higher portion, where we get fewer lines, as he considered, is the spectrum of the higher portion of the prominence; the spectrum of the lower portion contains the lines B, D, E, and F, and some other lines, in all nine, while the spectrum of the upper part of the prominence, as he thought it, only contains three lines. It was at first difficult to account for these observations. In the first place, one could not understand the line B being given, because I soon found that the line B was not seen as a bright line in the chromosphere spectrum; it was clearly the line C that was intended. Hence doubt was thrown on the other lines; it seemed as if M. Rayet was wrong about his elongated lines D, E, and F, and probably meant C near D and F. And so it was explained---I am ashamed to say by myself—that there was no particular meaning in these elongated lines except that the spectrum of the prominence some distance away from the sun was simpler than it was nearer the sun, as happens in all prominences, as we may now determine any day we choose to look at the sun by means of the spectroscope.

Now let us hear Mr. Pogson. He gave a diagram showing five lines in the spectrum of what he thought a prominence, and he writes:—" A faint light was seen [in the spectroscope], scarcely coloured, and certainly free from either dark or bright lines. While wondering at the dreary blank before me, and feeling intensely disappointed, some bright lines came gradually into view, reached a pretty considerable maximum brilliancy, and again faded away. Five of these lines were visible, but two decidedly superior to the rest. . . . . . The readings of the two brightest were secured. It struck me as strange that these brightest lines should appear at a part of the spectrum not corresponding to any very conspicuous dark lines in the solar spectrum. . . . . [These lines are a little less refrangible than E.] The third line seen in order of brilliancy must have been either coincident with, or very near the place of the sodium line D, but it was much fainter than the two measured, while the fourth and fifth lines were extremely faint." [They were very faint and DOUBLED, and near F. I have seen F give way to a double line in our hydrogen experiments, though I am not prepared to say this is an explanation of Mr. Pogson's observations.]

The fact that we have here the first observations of the spectrum of the sun's corona is one beyond all doubt; and why M. Rayet and Mr. Pogson thought they were observing prominences when they were observing above them, is explained by a remark made by Captain Tupman, of the Royal Marine Artillery, who acted as jackal to Professor Harkness, and picked out the brighter spots of the corona for his observation. Professor Harkness observing the prominence bright lines, said to Captain Tupman, " You have turned the telescope on to a prominence; I want the corona." " No," said Captain Tupman, " I am giving you the corona as well as I can." It was certainly the corona in both cases. Here you see, dimly and darkly, the first outcome of the spectroscope on the nature of the corona; a record as fairly written as anything at the sun can write it; and I am more anxious to lay stress on these observations, since they have lain fallow for two years, and show the importance of observations, not only in extending our knowledge, but in explaining prior observations; and it is an additional reason for never rejecting an observation. What was, however, dim and dark in 1868 shone out brightly in 1869, thanks to the skill of the American observers of the eclipse of that year.

b.—Laboratory Experiments bearing on these Observations.

But before I proceed to refer to the admirable observations made in America during this eclipse, I wish to introduce you to some work which was commenced in 1868, and has been done quite independently of eclipses. In a lecture which I delivered here about two years ago, I described to you some of the facts observed by the spectroscope in the bright-line region which had been spectroscopically determined to exist all round the sun, and which, as in it all the various coloured effects are seen in total eclipses, I had named the Chromosphere. It was clear that by the new method of observing this without any eclipse, by partially killing, so to speak, the atmospheric light, we got a percentage only of the phenomenon, as the atmospheric light could only be killed by an amount of dispersion which enfeebled and shortened the chromospheric lines; so that although we could say that an envelope of some 5000 or 6000 miles in height existed round the sun, we could not fix this as a maximum limit. Further, when we examined the spectrum of this envelope we got long lines and short lines; and I told how the short lines indicated a low stratum, and how a long line indicated a higher one. To explain this, I will show you an observation made long before the new method was thought of. Even before that time we had abundant evidence of such strata, if we could not determine their nature: we had distinct evidence either of one thing *thinning* out, and then another, or that various substances were situated at different levels, under different conditions; on the first hypothesis, at the extreme outside of the chromosphere the last thing would thin out, and then there would be an end of all things as respects the sun.

I will show you a drawing made by Professor Schmidt of the eclipse of 1851. I do not wish to call your attention to the strange shape of the large prominence, but to the fact, that as the moon passed over this region we get a thin red band, first along the edge of the dark moon, and after the moon had passed over still further we see this red layer, *suspended as it were in the chromosphere*, with a white layer below it. This is the explanation of the long and short lines visible in the spectrum of the chromosphere; in the red layer we have hydrogen almost alone; below, its red light was conquered by other light with bright lines in all parts of the spectrum, and we get white light.

Lord Lindsay tells me he has a distinct indication, written by the sun himself, that in one particular part of the chromosphere, as recorded photographically in Spain, there were three such layers. And over and over again we find recorded white light close to the sun, then red alone, or red mixed with yellow, then violet, and lastly green. And M. Mädler remarks on this very admirably, "The violet band is the link between the prominences and the corona."

Before going further, I will show you the difference in the appearance of what we may term hot hydrogen and cold hydrogen, that is, hydrogen which we drive into different degrees of incandescence by means of the spark. After Dr. Frankland and myself were able to determine that the pressure in these solar regions was small, we came to the conclusion, that outside the hot hydrogen there must be some cooler hydrogen, in order that the phenomena we observed, both in the laboratory and in the observatory, should agree.

I have in this tube hydrogen at a certain pressure, and here we have a coil which will enable us to send a spark through it; you see we get a certain amount of redness in that tube, and if you look on one side or above you will see a sort of bluish-greenish light. Now that redness represents the condition of the hydrogen in the region of the sun where Dr. Schmidt gave us that extremely thin red ring, and the combination of the blue and red would give you something very like violet.

But here I have hydrogen under a different condition. In the tube its rareness is not excessive; but in this globe, of which I am about to speak, you have the nearest approach to a vacuum ever obtained through which a spark will pass; and I beg to call your attention to what will now happen. This globe contains the same chemical element, prepared at the same time as the chemical element you have in the tube, but you see that, so far as colour goes, we have something perfectly different in this case. Now we send the spark through it. I would beg Professor Tyndall, if he will be good enough, to observe the spectrum of this hydrogen in this globe. [Professor Tyndall did so.] You will see that there is one line? [*Professor Tyndall.* Yes.] And a continuous spectrum? [*Professor Tyndall.* And a continuous spectrum.] Cool hydrogen gives us only the bright line F, plus a continuous spectrum, and many of you will know the extreme importance of that observation. It accounts for the F line being observed without the C line in 1868 and last year, and also for the continuous spectrum observed in the Indian eclipse.

c.—The American Eclipse.

When we come from the Indian to the American eclipse with the considerations to which I have drawn your attention, namely, the existence of these different layers due to the different elements and conditions of the same element thinning out, we shall see the extreme importance of the American observations, for they establish the fact that outside the hydrogen layer there was a layer giving only a line in the green, the line which Rayet and Pogson had observed associated with the hydrogen spectrum and the spectrum of the yellow substance. Here obviously we have, I think, merely an indication of another substance thinning out, in spite of the extraordinary suggestion which was put forward that the corona was nothing but a *permanent solar aurora.*

I need hardly tell you that the idea of a permanent aurora anywhere was startling, and that of a permanent solar aurora more startling still; but what I claim is that during last year's observations we made this very startling idea into a most beautiful fact, namely, that this outer layer of the chromosphere is in all probability nothing more nor less than an indication of an element lighter than hydrogen, although this is not yet absolutely established, for the line is coincident with one of the lines in the spectrum of iron.

d.—The Layers increase very rapidly in Density. Reproduction of the Coloured Phenomena.

Dr. Frankland and myself were early drawn to consider the solar nature of the large coronas, to which I have called your attention, as extremely questionable, even on the supposition of cool hydrogen, because we did not see how, with its temperature and pressure, it could extend very far: and an experiment which I have to make here will probably make that clearer.

We have in these glass vessels hydrogen a little more brilliant now the spark passes through it than that you saw in the globe, because I have been compelled to mix with it a certain amount of mercury vapour. Below, we have at the present moment sodium vapour being generated from metallic sodium in one tube, and mercury vapour in the other. I hope, if the experiment succeeds, you will see that a good many of the coloured phenomena seen in the chromosphere during eclipses may be easily reproduced by such experiments as this; and not only the coloured phenomena but *the increase of brilliancy* accompanied by changes of colour recorded. You can now all see the yellow tinge at the bottom of one tube, and the green tinge at the bottom of the other; and if there were time to continue this experiment by increasing the density of the vapours now associated with the hydrogen, I could make the bottom portion of each tube where the vapours are densest shine out almost like the sun, while the cool hydrogen at the top would remain not more brilliant than it is at present. We should have as it were a section of the chromosphere.

V.—Conclusion.

I will proceed now, if you will allow me, to some of the general results obtained during the last eclipse.

I think that, although the work has been very unfortunately interrupted, still the result has been most satisfactory. By putting together observations here and observations there, I consider our knowledge of the sun is enormously greater than it was a few months ago. For instance, we are enabled to understand the long-neglected observation of Rayet, and the equally long-neglected observation of Pogson ; and we know that outside the hydrogen there is, in all probability, a new element existing in a state of almost infinite tenuity. And we are sure of the existence of cool hydrogen above the hot hydrogen, a fact which seemed to be negatived by the eclipse of 1869.

I think if we had merely determined that there was this cool hydrogen, all our labour would not have been in vain, as it shows the rapid reduction of temperature ; but there is more behind. I told you that M. Mädler, in summing up the observations made up to 1860, came to the conclusion that part of the corona was certainly solar, and that whether the outer portions were or were not solar, was a matter of doubt. I do not say that we have settled that absolutely, but we have firm evidence that *some* of the light of the corona is due to reflexion between the earth and the moon. The outer corona was observed to have a rosy tinge over the prominences, and the spectrum of the prominences was detected many minutes above them, as well as on the dark moon. It could not have got this colour *at the sun*, for its intrinsic colour is green, and the red light of the hydrogen supplied at the sun is abolished altogether, is absorbed, and can only reach the corona *at the sun*, so to speak, as dark light.

It is a great fact that we are sure, as far as observation can make us sure, that there is a glare round the hydrogen which gives us the spectrum of hot hydrogen on the corona, *where we know that hot hydrogen does not exist.* Assume the hot hydrogen which gives us the red light to be only two minutes high, the spectroscope has picked it up eight minutes from the sun! The region of cool hydrogen is exaggerated in the same way. We get it where there is no indication of the cool hydrogen existing. And then with regard to the element which gives us the line of the green, we get that twenty minutes or twenty-five minutes away from the sun. Well, no man who knows anything about the matter will affirm that it is certain that the element exists at that distance from the sun.

Therefore I think we have absolutely established the fact that as the sun—the uneclipsed sun—gives us a glare round it, so each layer of the chromosphere gives us a glare round it. That is exactly what was to be expected, and that it is true is proved by the observation —a most important observation made in Spain — that the air, the cloud, ever between us and the dark moon, gives us the same spectrum that we get from the prominences themselves.

Given, however, the layers and elements in the chromosphere extended as far as you will, and apparently increased or not by reflexion *not at the sun*, we have still to account for rays, rifts, and the like. If anyone will explain either Mr. Brothers's photograph or Mr. Gilman's picture of the eclipse of 1869, containing those dark bands starting from the moon and fading away into space, and the bright variously-coloured rays between them on any solar theory, he will render great service to science. But in the meantime I must fall back upon M. Mädler's opinion of 1860, with the addition to it that I have stated that we have found, at all events, *that some* of the doubtful light is now solar; we have turned the opinion into a fact.

Bear in mind that close to the sun you have a white layer composed of vapours of many substances, including all the outer ones; outside this is a yellow region; above that a region of hydrogen, incandescent and red at the base, cooler, and therefore blue, higher up, the red and blue commingling and giving us violet; and then another element thinning out and giving us green. Take these colours in connection with those which are thrown on our landscapes or on the sea during eclipses, each region being lit up in turns with varying, more or less monochromatic light, and that light of the very colour composing the various layers, each layer being, as I have shown, so much brighter than the outer ones that its light predominates over them. Is it too much to suggest to those who may be anxious to attempt to elucidate this subject, that probably if they would consider all the conditions of the problem presented by that great screen, the moon, allowing each of these layers by turn to throw its light earthwards, the inequalities of the edge of the *globular* moon allowing here light to pass from a richer region, here stopping light from even the dimmer ones, they would be able to explain the rays, their colours, variations, apparent twistings, and change of side? I do not hesitate to ask this question, because it is a difficult one to answer, since the whole question is one of enormous difficulty. But difficult though it be, I trust I have shown you that we are on the right track, and that in spite of our bad weather, the observations made by the English and American Government Eclipse Expedition of 1870 have largely increased our knowledge.

With increase of knowledge generally comes a necessity for changing the nomenclature belonging to a time when it was imperfect. The researches to which I have drawn your attention form no exception to this rule. A few years ago our science was satisfied with the terms prominences, sierra, and corona, to represent the phenomena I have brought before you, the nature of both being absolutely unknown, as is indicated by the fact that the term *sierra* was employed, and aptly so, when it was imagined the prominences might be solar mountains! We now know many of the constituent materials of these strange things; we know that we are dealing with the exterior portion of the solar atmosphere, and a large knowledge of solar meteorology is already acquired, which shows us the whole mechanism of these prominences. But we also know that part of the corona is not at the sun at all. Hence the terms *leucosphere* and *halo* have been suggested to designate in the one case the regions where the general radiation, owing to a reduced pressure and temperature, is no longer subordinate to the selective radiation, and in the other, that part of the corona which is non-solar. Neither of these terms is apt, nor is either necessary. All purposes will be served if the term corona be retained as a name for the exterior region, including the rays, rifts, and the like, about which doubt still exists, though it is now *proved* that some part is *non-solar*, while for the undoubted solar portion the term Chromosphere—the bright-line region—as it was defined in this theatre now two years ago, exactly expresses its characteristic features, and differentiates it from the photosphere and the associated portion of the solar atmosphere.

Here my discourse would end, if it were not incumbent on me to state how grateful I feel to Her Majesty's Government for giving us the opportunity of going to the eclipse; to place on record the pleasure we all felt in being so closely associated in our work with the distinguished American astronomers who from first to last aided us greatly; and to express our great gratitude to all sorts of new friends whom we found wherever we went, and who welcomed us as if they had known us from our childhood.

[J. N. L.]

Friday, March 22, 1872.

SIR HENRY HOLLAND, Bart. M.D. D.C.L. F.R.S. President, in the Chair.

J. NORMAN LOCKYER, Esq. F.R.S. M.R.I.

On the Eclipse Expedition, 1871.*

I UNDERSTAND my duty to-night to be to give an account of the observations made, not by all who observed the eclipse of last December, but by the members of the party which went out under the auspices of the British Association, and it is extremely fortunate that nothing more is required of me: first, because most valuable work was done by the other parties, which of itself would require more time to state than I have at my disposal; and secondly, because the amount of material obtained by the members who were sent out from England, and by the friends who met them at every point, is so great, that it would be impossible in one discourse to give anything like an exhaustive account of it. Here are some of the records in this portfolio. You will see at once that even for one party I can only make a selection, and I am perfectly aware of the extreme responsibility which attaches to anyone who may venture to make a selection out of such an enormous mass of material as we have collected.

Before I proceed to discuss the work done by the different parties, it will be desirable to give an idea of the arrangements, and for this purpose I have prepared several maps, which will enable you to see what the British Association parties did.

In the first instance I may remark that the weather conditions were somewhat problematical. Another point of great importance was that much of the ground was fortunately occupied, and it was essential when placing the parties to bear these two considerations in mind—the possibility of bad weather, and then the importance of so arranging matters that if some of the observers were clouded out, belonging to our parties, then the story might be continued by other observers.

Here we have a map of India, which gives you a general idea of

* The chief results obtained by the expedition have been taken from the *ad interim* Report presented to the British Association Meeting at Brighton. The lecture itself dealt mainly with the methods and instruments employed.—J. N. L.

the path of the shadow during the eclipse. The shadow, you see, strikes India on the western coast, and it runs down in a south-westerly direction, and cuts the northern portion of Ceylon.

When we arrived in India we found that the Indian observers, consisting of those well-known men Tennant, Herschel, Hennessy, Pogson, and others, had determined, from their knowledge of the climatic conditions of India at that time of the year, to occupy the central part of the line, and also a station at a low level; the eminent French physicist M. Janssen taking up his position at the top of the Nielgherries. We were to station ourselves either east or west, or both, of these parties. Whether east or west would depend upon the monsoon, and the great question that was being discussed on our arrival was, Was the monsoon favourable?

I have not time to go into the many interesting points touching the answer to this question; but I may say shortly that what we heard was, that if the weather was likely to be bad on the east side of the hill range, generically called the Ghauts, there was a good chance for anyone occupying a position west of those hills. What happened was that we did occupy the positions marked by blue wafers on the map, namely, Bekul on the west coast, Manantoddy on the western slope of the Ghauts, Poodocottah in the eastern plain, and in the island of Ceylon, first, Jaffna; and secondly, Trincomalee.

Such were our arrangements. The parties were stationed along the line of totality. Very different were the arrangements of the Sicilian party of the former year. In Sicily we were compelled to throw ourselves *across* the line of totality in the direction which I have indicated on this map of Sicily.

Now what was the work we had to do? If you will allow me to refer to two or three results of the former Eclipse Expedition, I will endeavour to put them before you without taking up too much of your time.

One of the most important among the results obtained in the eclipse of 1870 was this: far above the hydrogen which we can see every day without an eclipse—far above the prominences, the spectrum of hydrogen had without doubt been observed by two or three of the American observers, who were more fortunate than we were. Among them Professor Young stated, that the spectrum of hydrogen was observed to a distance of 8 minutes from the sun; he then adds, "far above any possible hydrogen atmosphere." This is point number one.

Another of the points was this: the unknown substance which gives us a line coincident, according to Young, with a line numbered 1474 by Kirchhoff, had been observed by the American observers to a height of 20 minutes above the limb of the dark moon.

Now, it was a very obvious consideration that if we got a spectrum of hydrogen 8 minutes from the dark moon, when we thought we knew that the hydrogen at the sun did not really extend more than 10 seconds beyond the dark moon, there was something at work which had the effect of making it appear very much more extensive than it really was; and it was fair to assume that if this happened in the case of the hydrogen, it might also happen in the case of the unknown stuff which gives us the line 1474.

In support of this view we had one of the few observations which were made in Sicily, in the shape of a drawing of the corona, as seen by Professor Watson, who observed at Carlentini. He saw the corona magnificently; and being furnished with a powerful telescope, he made a most elaborate drawing of it, a rough copy of which I will throw on the screen. You will see at once that we had in this drawing something which seemed to militate against the idea that the 1474 stuff at the sun did exist to a height of 20 minutes. According to Professor Watson the boundary of the real corona was clearly defined, its height being far under that stated.

Next, we had another observation of most important bearing on our knowledge of the base of the corona. I refer to the announcement of the observation by Professor Young of a stratum in which all the Fraunhofer lines were reversed. It was asserted that there was undoubtedly a region some 2 seconds high all round the sun, which reversed for us all the lines which are visible in the solar spectrum. We had in fact in a region close to the photosphere the atmosphere of the sun demanded by Kirchhoff at some distance above the photosphere.

Last, not least, we had the photographic evidence. There was in Sicily a photographic station at Syracuse, and the Americans had another in Spain. I now show on the screen a drawing—it is not the photograph itself—but a drawing of a photograph made by the party in Sicily; what we have on this photograph, is a bright region round the dark moon, which is, undoubtedly, solar, but stretching out right away from this, here and there are large masses of faint light, with dark spaces between them, which have been called rifts. Now the question is, Is this outer portion solar?

Having thus brought rapidly before you some of the questions which we had principally to bear in mind, and, if possible, settle (though that is too much to hope for in any one Eclipse Expedition) in the work we had to do in India, I will next bring to your notice some new methods of inquiry which had been proposed, with the object of extending former observations.

I may here remark that the Royal Astronomical Society, in the first instance, invited me to take charge of an expedition to India merely to conduct spectroscopic observations; but although this request did me infinite honour, I declined it, because the spectroscope alone, as it had been used before, was, in my opinion, not competent to deal with all the questions then under discussion. I have told you that some of the most eminent American observers had come to the conclusion that the spectrum of hydrogen observed in the last eclipse round the sun, to a height of 8 minutes, was a spectrum of hydrogen "far above any possible hydrogen" at the sun. Hence it was in some way reflected. Now with our ordinary spectroscopic methods it was extremely difficult, and one might say impossible, to determine whether the light which the spectroscope analyzed was really reflected or not; and that was the whole question.

It became necessary, therefore, in order to give any approach to hopefulness, to proceed in a somewhat different way in the 1871 expedition; and in order to guard against failure, to supplement such new observations by photographs; and fortunately we were not long in coming to a conclusion that this might be done with some considerable chance of success.

I have here a train of prisms. I will for one moment take one prism out of the train, and we will consider what will happen if we illuminate the slit of the lantern with a monochromatic light and observe it through the prism. If we render sodium vapour incandescent, we know we get a bright yellow image of the slit, due to the vapour of the metallic sodium only giving us yellow light. But why is it that we get a line? Because we always employ a line for the slit. But suppose we vary the inquiry. If, instead of a straight line we have a crooked line for the slit, then we ought to see a crooked line through the prism. Now, allow me to go one step further: suppose that instead of a line, whether straight or crooked, we have a slit in the shape of a ring, shall we see a ring through the prism? You will see that we shall. And then comes this question: If, when we work in the laboratory we examine these various slits, illuminated by these various vapours, why should it not happen that if we observe the corona in the same way, we shall also get a ring built up by each ray of light which the corona gives to us; since we know, from the American observations, that there were bright lines in the spectrum of the corona, as observed by a line slit? In other words, the corona examined by means of a long train of prisms, should give us an image of itself painted by each ray which the corona is competent to radiate towards us.

Now let us pass to the screen, the screen merely replacing the retina. We will first begin with the straight slit with which you are familiar—we now have our slit fairly focussed on the screen—we then in the path of the beam interpose one of these prisms, and there we get on the screen a bright line.

Now, to continue the argument, we replace the straight slit by a crooked one, and you see we get a crooked image on the screen. We now replace this crooked slit by a ring. We have now a ring-formed image on the screen. So that you see we can use any kind of narrow aperture we choose, and as long as we are dealing with light which is monochromatic, or nearly so, we get an image of the aperture on the screen.

If we consider the matter further it will be evident that we may employ a mixture of vapours, and extend this result.

We will now, for instance, instead of employing sodium vapour, employ a mixture of various vapours. You see now that each ray given by these substances, instead of building up a line image, is building up for us a ring image—that we have now red, green, yellow, blue, and violet rings.

Now that was the consideration which led to the adoption of one of the new attempts to investigate the nature of the corona used this time. It was, to use a train of prisms, pure and simple, using the corona as the slit, a large number of prisms being necessary to separate the various rings we hoped to see, by reason of their strong dispersion. On the screen the rings to a certain extent intersect each other; and it would have been easier to show you the ring-form of the images if we could have used more prisms than one.

If this is good for a train of prisms such as I have referred to, it is good for a single prism in front of the object-glass of a telescope. Such was the method adopted by Professor Respighi, the distinguished Director of the Observatory of the Capitol of Rome, who accompanied the expedition.

Now you may ask how would this method, if it succeeded, be superior to the ordinary one? In this way. If we were dealing merely with reflexion, then all the rings formed by vapours of equal brilliancy at the base of the chromosphere would be of the same height, while if reflexion were not at work the rings would vary according to the actual height of the vapours in the sun's atmosphere, and the question would be still further advanced if the spectrum did not contain a ring representing the substance which underlies the hydrogen.

Our *new* spectroscopic equipment then was as follows :—

1. A train of five prisms.
2. A large prism of small angle placed before the object-glass of a telescope.
3. Integrating spectroscopes *driven by clockwork.*
4. A self-registering integrating spectroscope, furnished with telescopes and collimators of large aperture, and large prisms. (This instrument was lent by Lord Lindsay.)

Now a word about the polariscopic instruments, referring you to my lecture given last year for a general notion of the basis of this class of observation.

A new idea was that observations to determine the polarization of the corona might be made with the same telescope and eye, both with the Biquartz and the Savart.

By the kindness of Mr. Spottiswoode, who has placed his magnificent polarizing apparatus at our service, I hope to be able to show you on the screen the mode of examining the corona by means of those two instruments, so as to enable you pretty well to follow what was actually done.

Let me begin with the Biquartz polariscope. In the first instance I will throw on the screen a representation of the corona itself, and we will then insert a Biquartz, and see its effect when I flood the screen with polarized light. You now see an indication of what would be observed supposing the polarization was due to polarized light diffused in the region between us and the dark moon and eclipsed sun, in which case the polariscopic effect would be observed generally over the dark moon, the corona and the region of the sky outside the corona. But this is not all; not only does this arrangement enable us to determine the existence of such a general polarization, but the vertical line in the Biquartz called the line of junction indicates the plane of polarization, when the colours on both sides of it are the same; so that we have two colours strongly contrasted in either half of the field when we are away from the plane of polarization, and a uniform colouring of the whole field when in or at right angles to that plane. By turning this prism through 90 degrees, you see I entirely change the colours.

But we are not limited to the Biquartz in this inquiry. We can apply the Savart polariscope. Having still our image of the corona on the screen, I now replace the Biquartz by a Savart.

We now no longer see a line of junction with the similar or different colours on either side of it, but lines of colour running across the image. I turn the prism. We first see the lines with a white centre, then with a dark one; while at times they are altogether absent. And as a departure from the plane, when we use the Biquartz, gives us the strongest contrasts of colour, so you observe that with the Savart under these circumstances all indications of polarization vanish.

Now, if we assume polarization to be general, and the plane of polarization vertical, we should get those coloured bands, as you see them there, crossing the corona and dark moon, the lines being vertical and dark-centred. If the plane of polarization were horizontal, we should find the lines horizontal and the central one white.

But so far as we have gone, we have been dealing with polarization which is general, and we have not attempted to localize polarization at the corona itself. But I have here an apparatus, by means of which, quietly, in this theatre one can see as admirable an example as we should desire, of polarization, assumed to be particular to the sun and not general—I mean radial polarization. We have simply a circular piece of mahogany, or something else which polarizes light equally well, with a hole in the middle with sloping sides, cut as you see this cut, and then we place behind it a candle, so that the light of this candle, after falling on oiled tissue paper stretched across the aperture, can be reflected to the eye by the sides, the direct light of the candle being stopped by a central metallic diaphragm. We have now a source of polarized light of a different kind from the last. The next thing we have to do is to introduce into a small telescope exactly the same kind of apparatus we have there, though of course on a much smaller scale, and examine the ring of light seen when we put the candle behind the aperture. On examining the ring of light which is now visible by means of this telescope, which contains a Biquartz and analyzer, I see the most exquisite gradations of colour on either side the line of junction which cuts the field of view and the bright ring in the centre into two.

Now, instead of the candle, we will employ the electric lamp; and instead of the eye, the screen; but I must inform you that the great heat of the electric lamp prevents the appearance being perfectly successful on the screen, as the reflecting varnish is melted.

In this experiment we cannot work with an image of the corona. We must make our corona out of the image of the ring we hope to get on the screen; and then, by employing the Biquartz in the same way as before, instead of getting similar colours on either side of the line of junction, as we did when we were working in the plane of polarization, and getting the greatest contrasts, as we did when we worked 45 degrees away, you observe we get different colours in each part of the ring.

On the screen we now have a highly-magnified image of the hollow cone of iron which I am compelling to reflect the light from the lamp; and by inserting this Biquartz I throw various colours over different portions of that ring, which I beg you to consider for one moment as the solar corona, and the colours change as I rotate this prism. You will at once be able to explain the different actions of this Biquartz in this instance. The reflexion and therefore the plane of polarization is no longer general, but varies from point to point of the reflecting surface. It is in fact radial, and hence the delicate radiate arrangement of colour.

Such then were some of the new methods and new instruments we used for the first time in our researches. And I hope you will allow me to use this term, although our work was conducted a long way from the Royal Institution, the natural home of research in England.

I must now state very briefly some of the results of our work; and first, the certain results.

We were able to make out the structure of the corona. We know all about the corona so far as the structure of its lower brighter strata, that portion, *viz.* which I referred to in my lecture last year as being visible both before and after totality, is concerned. You may define it as consisting of cool prominences; that is to say, if you examine a prominence any day, without waiting for an eclipse, and then go to an eclipse and examine the lower portion of the corona you will find the same phenomena, minus the brightness. You find the delicate thread-like filaments which you are now all so familiar with in prominences—filaments which were first thrown on a screen in this theatre; the cloudy light masses, the mottling, the nebulous structure, are all absolutely produced in the corona, as far as I could see it with a telescope with an aperture of $6\frac{1}{4}$ inches; and I may add that the portion some five minutes round the sun, reminded me forcibly in parts of the nebula of Orion, and of that surrounding η Argus, as depicted by Sir John Herschel, in his Cape observations.

We have shown that the idea that we did not get hydrogen above 10 seconds above the sun is erroneous; for we obtained evidence that hydrogen exists to a height of 8 or 10 minutes at least above the sun; and I need not tell you the extreme importance of this determination. One of the proofs we have of that lies in this diagram, showing the observations made by Professor Respighi, armed with an instrument, the principle of which I hope you are now familiar with.

Just after the sun disappeared Professor Respighi employed this prism to determine the materials of which the prominences which were then being eclipsed were composed; and he got the prominences shaped out in red, yellow, in blue, and in violet light; a background of impure spectrum filling the field, and then as the moon swept over the prominences these images became invisible; he saw the impure spectrum and the yellow and violet rings gradually die out, and then three bright and broad *rings* painted in red, green, and blue, gradually form in the field of view of his instrument; and as long as the more brilliant prominences were invisible on both sides of the sun, he saw these magnificent rings, which threw him in a state of ecstasy. And well they might.

These rings were formed by C and F, which shows us that hydrogen extends at least 7 minutes high, for *had we not been dealing with hydrogen we should have got a yellow ring as well*, because the substance which underlies the hydrogen is more brilliant than the hydrogen itself, and in addition to the red ring and the blue ring, which indicate the spectrum of hydrogen, he saw a bright green ring, much more brilliant than the others, built up by the unknown substance which gives us the Kirchhoff line, 1474.

Now at the time that Professor Respighi was observing these beautiful rings by means of a single prism and a telescope of some four inches aperture, some 300 miles away from him—he was at Poodocottah and I was at Bekul—I had arranged the train of prisms which you see here so that the light of the sun should enter the first prism, and after leaving the last one should enter my eye. And what I saw is shown, side by side with Respighi's observations, in this diagram, in which I have separated the rings somewhat, so that there should be less confusion than in the actual observation. Here is Professor Respighi's first observation. He gets indications of C, D^3, F, and the hydrogen line near G. He was observing the very lowest, brightest region of all, and therefore 1474 was obliterated by the brightness of the continuous spectrum; but as the eclipse went on D^3 was entirely obliterated, and afterwards he got C and F building up rings together with 1474, which was not represented in the lower regions of the prominence—not because it was not there, but because, as I have already insisted, of the extreme brilliancy of the background. Now my observation was made intermediately, as it were, between the two observations of Professor Respighi's. Let me show the observations together.

Respighi	..	C	D^3		F G	Prominences at beginning of eclipse.
Lockyer	..	C		1474	F G	Corona at 80 seconds from commencement.
Respighi	..	C		1474	F	Corona at mid eclipse.

Note that I had no object-glass to collect light, but that I had more prisms to disperse it; so that with me the rings were not so high as those observed by Respighi, because I had not so much light to work with: but such as they were I saw them better because the continuous spectrum was more dispersed, and because, with my dispersion, the rings—the images of the corona—therefore did not so much overlap. Hence doubtless Respighi missed the violet ring which I saw, so faint, however, that both that and 1474 were almost invisible, while C shot out with marvellous brilliancy, and D^3 was absent.

These observations thus tend to show, therefore, that instead of the element—the line of which corresponds with 1474—existing alone just above the prominences, the hydrogen accompanies it to what may be termed a great height above the more intensely heated lower levels of the chromosphere, including the prominences in which the lower vapours are thrown a greater height. With a spectroscope of small dispersion attached to the largest mirror of smallest focus which I could obtain in England, the gaseous nature of the spectrum, as indicated by its *structure*, that is bands of light and darker intervals as distinguished from a continuous spectrum properly so called, was also rendered evident.

These are results of the highest importance, which alone are worth all the anxiety and labour connected with the expedition.

But there is more behind.

The photographic operations (part of the expense of which was borne by Lord Lindsay) were most satisfactory, and the solar corona was photographed to a greater height than it was observed by the spectroscope, and with details which were not observed in the spectroscope.

Mr. Davis was fortunate enough to take an admirable series of five photographs at Bekul, and Captain Hogg also obtained some at Jaffna; but I am sorry to say the latter lack somewhat in detail.

I have prepared two lamps, because I am anxious to exhibit the photographs two at a time, that you may compare one with the other. [This was done.] You see that so far as the camera goes—and mark this well—the corona was almost changeless during the whole period of totality; this is true, not only for one place, but for all the places at which it was photographed.

I now exhibit two other photographs—one taken at Jaffna and the other at Ootacamund. Actinically the corona was the same and practically changeless at all the stations. You see that, though not so obvious as in the other case, there is the same similarity.

Before I leave the actinic corona, I am anxious to show you an image of it, taken during the American eclipse of 1869 in a camera exposed to the sun during the whole of the totality; to a certain extent in our recent photographs we have reproduced what was photographed in 1869.

The solar nature of most, if not all, of the corona recorded on the plates is established by the fact that the plates, taken in different places, and both at the beginning and end, of totality, closely resemble each other, and much of the exterior detailed structure is a continuation of that observed in the inner portion independently determined by the spectroscope to belong to the sun.

While both in the prism and the $6\frac{1}{4}$ inch equatorial the corona seemed to form pretty regular rings round the dark moon, of different heights according to the amount of light utilized by the instrument, on the photographic plates the corona, which, as I have before stated, exceeds the limits actually seen in the instrument I have named, has a very irregular, somewhat stellate outline, most marked breaks or rifts (*ignored by the spectroscope*), occurring near the sun's poles, a fact perhaps connected with the other fact that the most active and most brilliant prominences rarely occur there.

From the photographs in which the corona is depicted actinically we pass to the drawings in which it is depicted visually. I would first call attention to two drawings made by Mr. Holiday, who formed part of the expedition, and in whose eye every one who knows him will have every confidence.

First there is a drawing made at the commencement of the totality, and then a drawing made at the end. There is a wonderful difference between the drawings; the corona is in them very much more extensive than is represented actinically on our plates.

Here is another drawing, made by Captain Tupman, in which again we have something absolutely different from the photographs and from Mr. Holiday's sketches, inasmuch as we get an infinite number of dark lines extending down to the moon, and a greater extension than in the photographs, though in radial places the shape of the actinic corona and some of its details are shown.

Now the corona, as it appeared to me with the naked eye, was nothing but an assemblage of bright and dark lines, it lacked all the structure of the photographs and appeared larger; and I have asked myself whether these lines do not in some way depend on the size of the telescope, or the absence of a telescope. It seems as if observations of the corona with the naked eye, or with a telescope of small power, may give us such lines; but that when we use a telescope of large power, it will give, close to the moon, the structure to which I have referred, and abolish the exterior structure altogether, leaving a ring round the dark body of the moon such as Professor Respighi and myself saw in our prisms, and in the 6-inch telescope, in which the light was reduced by high magnification so as to bring the corona to a definite ring some 5 minutes high, while Professor Respighi, using a

4-inch telescope and less magnifying power, brought the corona down to a ring something like 7 minutes high.

And here we have an important connection between spectroscopic and telescopic work. If we employ a telescope in which the light is small or is reduced by high magnification, we bring the corona to a definite ring, and perhaps here we have the origin of the 'ring-formed' coronas.

Many instances of changing rays, like those seen by Plantamour in 1860, were recorded by observers in whom I have every confidence. One observer noted that the rays revolved and disappeared over the rifts.

We have next to deal with the polariscopic observations.

Mr. Lewis, in sweeping round the corona at a distance of 6′ or 7′ from the sun's limb, using a pair of compensating quartz wedges as an analyzer, which remained parallel to itself while the telescope swept round, observed the bands gradually change in intensity, then disappear, bands of a complementary character afterwards appearing, thereby indicating radial polarization.

Dr. Thomson at Bekul saw strong traces of atmospheric, but none of radial polarization, with a Savart. With the same class of instrument the result obtained by myself was precisely similar; while on turning in the Biquartz, at the top and bottom of the image of the corona, *i. e.* near the sun's equator, faint traces of radial polarization were perceptible for a short distance from the moon's limb. Captain Tupman, who observed with the polariscope after totality, announces strong radial polarization extending to a very considerable distance from the dark moon.

Leaving the extreme outside of the corona as a question to be determined at some future time—and it can well wait—let us come to the base of the corona, and deal with the region to which I have already referred, close to the sun.

What was the general conclusion at which we arrived on this important point? Before I state it, let me tell you the instrumental conditions of the inquiry. We can use such a spectroscope as the one with which you are all familiar, and so arrange matters that the slit shall be carried by a clock, so that it may follow accurately the edge of the moon; but if the least variation in the rate of motion takes place, the observation is rendered almost valueless. But if we employ a spectroscope, in which we sum up the light—do not localize the light, but throw it together—it does not matter whether your clock goes well or not, you are certain to have a result worthy of credit. But if you employ such an instrument as Professor Respighi employed, and abolish the slit altogether, the weight of any observations made with such conditions is very great.

Captain Maclear, who was observing with me at Bekul, has undoubtedly shown that when the light of our atmosphere is cut off by the interposition of the dark moon, we see very many more bright lines than we do when this is not the case, the lines being of unequal height.

Mr. Pringle, also at Bekul, showed that, at the end of totality, many lines flashed into one of these instruments, carried under these difficult conditions.

Captain Fyers, the Surveyor-General of Ceylon, observing with a spectroscope of the second kind, saw something like a reversal of all the lines at the beginning, but nothing of the kind at the end.

Mr. Fergusson, observing with a similar instrument, saw reversal neither at the beginning nor the end.

Mr. Moseley, whose observations are of great weight, says that at the beginning of the eclipse he did not see this reversal of lines. Whether it was visible at the end he could not tell, because at the close the slit had travelled off the edge of the moon.

Professor Respighi, using no slit whatever, and being under the best conditions for seeing the reversal of the lines, certainly did not see it at the beginning, but he considers he saw it at the end, though about this he is doubtful.

From the foregoing general statement of the observations made on the eclipse of last year, it will be seen that knowledge has been very greatly advanced, and that most important data have been obtained to aid in the discussion of former observations. Further, many of the questions raised by the recent observations make it imperatively necessary that future eclipses should be carefully observed, as periodic changes in the corona may then possibly be found to occur. In these observations the instruments above described should be considered normal, and they should be added to as much as possible.

I had intended, if time had permitted me, to point out how much better we are prepared for the observation of an eclipse now than we were when we went to India, and how a system of photograph record should be introduced into the spectroscopic and polariscopic work; but time will not allow me to do more than suggest this interesting topic. I am anxious, however, that you should allow me one minute more to say how very grateful we feel for the assistance rendered by all we met, to which assistance so much of our success must be ascribed. I wish thus publicly to express the extreme gratitude of everyone of our Expedition to the authorities in India and in Ceylon for the assistance we received from them; and our sorrow that Admiral Cockburn, a warm and well-known friend to Science, who placed his flagship at the disposal of the expedition, and the Viceroy, whose influence in our favour was felt in every region of India whither our parties went, and to whom we gave up our ship, are now, alas! beyond the expression of our thanks. We are also anxious to express our obligations to the directors and officers of the Peninsular and Oriental Company for the magnificent way in which they aided us. If they had not assisted us as they did, Science would have gained very much less than she has done from the observations of the last eclipse.

[J. N. L.]

Friday, May 10, 1872.

SIR HENRY HOLLAND, Bart. M.D. D.C.L. F.R.S. President, in the Chair.

NEVIL STORY-MASKELYNE, Esq. M.A. F.R.S.

On Meteoric Stones.

THE substantial unity of the celestial objects distinguished in common language by the names shooting or falling stars, fire-balls, and meteorites, and further the coincidence in many important respects of these with comets, and possibly with the zodiacal light, were suggestions made by Humboldt in the 'Cosmos,' which have received much confirmation from the subsequent advance of science.

The greater apparent velocity with which the ordinary meteors traverse the atmosphere as compared with that with which the less frequent larger bodies are seen to move, the marked periodicity that attends the recurrence of the former in several, and especially in two, notable cases of meteor-showers, offer an apparent contrast between these classes of meteors; it is not, however, in all probability, a real contrast, for the one class passes into the other by every gradation in the magnitude of the mass or masses of which the meteor consists, and consequently in the grandeur of the phenomena which accompany its advent. If of the material composing the ordinary falling star we have never yet been able to recognize any vestiges as reaching the earth, of the meteorite, on the other hand, the mineral collections of Europe contain numerous carefully collected specimens, which are the fragments that have escaped the fiery ordeal of the transit through our earth's atmosphere, and in these we recognize masses composed either of iron (siderites), or of stone (aerolites), or of a mixture of the two (siderolites). The phenomena associated with such falls of meteoric matter have been described in very similar language by those who have witnessed them in various parts of the world, and these accounts, whether coming from European observers or from Hindoo herdsmen (of which some were read by the lecturer), concur generally in the approach of the meteorite as a fiery mass, emanating from a cloud when seen by day and exploding often with successive detonations that are heard over a great extent of country, even in certain cases at points more than 60 miles distant, but finally reaching the earth with a velocity little higher than what might be due to the motion of a falling body. Externally these meteoric masses are generally hot when they fall; sometimes, however, they are not so: the discrepancies in the accounts being explained by one authenticated case in which the mass was internally intensely cold, though at first hot externally. The fallen meteorite is invariably coated with an incrustation, sometimes shining as an enamel, generally black, but occasionally colourless where the aerolite is free from ferrous silicates; and this incrustation is seen to have been formed in the atmosphere, since it is found coating surfaces of fragments that have been severed by the explosions in the air.

Aerolites frequently fall simultaneously in large numbers, many thousands of them being in such cases spread over a surface of the country some miles in extent; and such showers of stones seem to have entered the atmosphere as a group, though their numbers must subsequently have been greatly increased by the division accompanying their detonation.

The explanation of the incrustation and of the cloud left by the meteorite, or out of which it seems to emerge, is found in the transformation into heat of the energy actuating a body that enters our atmosphere with a motion of 12 to 40 miles in a second. The velocity of the body is almost instantaneously arrested by the atmospheric resistance, and in a very few seconds the mass becomes, comparatively speaking, stationary. Its surface must, as a consequence, be immediately fused, and the melted matter would be flung off from it into the surrounding air, fresh surfaces continually affording new fused material to form the cloud of, so to say, silicious spray that lingers along and around the path of the meteorite.

When the mass is small,—and in the case of meteoric showers and ordinary falling stars it cannot exceed a few ounces, and may often be but a few grains,—the whole material is thus consumed, and must ultimately fall as an unperceived, because widely-scattered, dust. The meteorite is the residue that survives this wasting action where the magnitude of the mass is more considerable. The cause of the violent and often successive explosions is probably to be sought in the expansion of the outer portions of the mass while the interior retains the contracted volume due to the intense cold of space with which the meteorite enters the atmosphere.

From time to time these contending conditions of volume may, as in a Prince Rupert's drop, produce explosion, the heated shell in the case of the meteorite flying off in fragments from the internally cold inner core, which if sufficient velocity remain to the mass will undergo a recurrence of the same conditions of surface fusion and explosion. The loudness of the detonation is also probably enhanced by the simultaneous collapse of the air on the vacuum that would follow the rapidly moving mass.

The pitted surface characteristic of meteorites probably bears witness to a similar effect of unequal dilatation operating more especially in the freshly-broken surfaces of the mass, small fragments splintering off in this way from the cold and brittle stone under the sudden influence of intense heat.

A remark made by Humboldt, that light and meteorites are the only sources of our knowledge regarding the universe external to our world, points to the true ground for our interest in the waifs and strays of extra-telluric matter that thus fall upon our globe.

In physical as well as in chemical characters aerolites resemble at the first aspect some terrestrial volcanic rocks.

The minerals of which they are composed are nearly entirely crystalline, as is evidenced by the colours in polarized light of such as are transparent. These minerals are usually aggregated with slight cohesion, and they present in by far the greatest number of cases a peculiar spherular or "*chondritic*" structure.

In these the spherules are composed of similar minerals to those which enclose them, and even contain metallic iron sometimes in microscopically fine grains disseminated through them.

A section of an aerolite was exhibited by the microscope in which some of the spherules had been broken before being cemented by the surrounding mass, and in another fissures were seen which had been filled with a fused material after one side of the fissure had slidden along the other: facts pointing to events in the history of the meteorite subsequent to its first formation.

The chemical composition and the mineral constitution of aerolites were illustrated by tables showing the elements met with in these bodies, and the minerals in which they were distributed. The former comprised about one-third of the known elements; among them magnesium, iron, silicon, oxygen, and sulphur were conspicuous; calcium, aluminium, nickel, carbon, and phosphorus coming next in importance, the basic elements of most importance by their amount being the same as those which are found by spectroscopic analysis to be present in the sun—and in those stars which have been the best examined.

The minerals most frequent in aerolites besides nickeliferous iron or troilite (iron monosulphide) and graphite, are bronzite (a ferriferous enstatite) and olivine, both the latter being essentially magnesium silicates. Augite and anorthite also occur (more particularly in the eukritic aerolites of Rose) and some minerals unknown in terrestrial mineralogy have also been met with; such are the different varieties of Schreibersite (phosphides of iron and nickel): calcium sulphide, asmanite (a form of silica crystallizing in the orthorhombic system and having the specific gravity of fused quartz), and a cubic mineral with the composition of labradorite. The crystalline form of bronzite was first determined from the crystals in a meteorite, and was found to confirm the conclusion Descloizeaux had arrived at as regards its system from observations on the distribution of the optic axes in the terrestrial bronzite and enstatite.

The question as to whence the meteorites come is one that we are not yet in a position to answer with certainty. The various hypotheses which suppose for them an origin in lunar volcanoes, or in our atmosphere, or again in a destroyed telluric satellite, or that would treat them as fragments of an original planet of which the asteroids are parts, or as masses ejected from the sun; all these hypotheses seem to be more or less precluded by the known velocities, the retrograde motion so frequently characterizing meteors and meteorites, or else by the chemical conditions that for instance are involved in the passage of the meteorite through the sun's chromosphere. Whether meteorites move or do not move in circumsolar orbits is at present impossible to say; because, while with our incomplete knowledge we cannot to-day attach the character of periodicity to any known class of meteorites, we are not justified in founding any conclusion on a negative result with so limited a foundation.

But even if all or some of them may have been, on their encountering the earth, members temporarily or permanently of the solar system, we may with considerable probability consider them as having originally entered our system from the interstellar spaces beyond it. Such at least must be our conclusion if we are to admit the unity of the whole class of phenomena of meteorites and falling stars. For, since the orbits of the two best known meteoric streams, those namely of August and November, have been identified with the orbits of two comets, and since in regard to one of these (that of November) Leverrier has shown, with great probability, that as a meteoric cloud it entered and became a member of our system only some 1700 years ago in consequence of the attraction of Uranus, while the August meteoric ring only differs in this respect from it, that it had at a much more remote period found an elliptic orbit round the sun; we are constrained on the assumption with which we started to recognize also in a meteorite a visitor from the regions of remote space. And so far as it goes, the observation by Secchi that the November falling stars exhibit the magnesium lines is in harmony with this view.

It may, however, further be said, that the tendency of scientific conviction is in the direction of recognizing the collection towards and concentration in definite centres, of the matter of the Universe, as a cosmical law, rather than the opposite supposition of such centres being the sources whence matter is dispersed into space. In the meteorites that fall on our earth (certainly in considerable numbers) we have to acknowledge the evidence of a vast and perpetual movement of matter in space, about which we can only reason as part of a great feature in the Universe which we have every ground for not supposing to be confined within the limits of the solar system.

That this matter, whether intercepted or not by the planets and the sun, should to an ever-increasing amount become entangled in the web of solar and planetary attraction, and that the same operation should be collecting round other stars and in distant systems such moving "clouds" of star-dust as have been treated by Schiaparelli, Leverrier, and other astronomers, or individual masses of wandering stone or iron is a necessary deduction from the view that we have assumed regarding the tendency of cosmical matter to collect towards centres. But in order to trace the previous stages of the history of any meteorite, and, in particular, to determine the conditions under which its present constitution as a rock took its origin, we have only for our guide the actual record written on the meteoric mass itself; and it is in this direction that the mineralogist is now working.

But the progress is necessarily a gradual one. We may indeed assert that the meteorites we know have, probably all of them, been originally formed under conditions from which the presence of water or of free oxygen to the amount requisite to oxidize entirely the elements present were excluded; for this is proved by the nature of the minerals constituting the meteorites and by the way in which the metallic iron is distributed through them.

And one suggestive and significant fact remains to be alluded to; the presence, namely, in some few meteorites of combinations of hydrogen and carbon, which if met with in a terrestrial mineral would with little hesitation be assigned to an organic origin. A few grains were exhibited to the audience of such a body, crystallized from ether, which solvent had extracted it to the amount of about 0·25 per cent. from six ounces of the Cold Bokkveldt meteorite.

Similar substances have been extracted by Wöhler, Roscoe, and other chemists from this and other meteorites. It was, however, observed, as pointing to the probability of the comparatively porous meteoric stone having in this case taken up the hydrocarbon as a substance extraneous to it (possibly when in the state of a vapour), that ether extracted it entirely from the solid lumps of the meteorite; pulverization not in any way adding to the amount obtained, or facilitating in any appreciable degree the separation of the substance.

[N. S. M.]

Friday, May 30, 1873.

Sir Henry Holland, Bart. M.D. D.C.L. F.R.S. President, in the Chair.

The Earl of Rosse, D.C.L. F.R.S. M.R.I.

On the Radiation of Heat from the Moon, the Law of its Absorption by our Atmosphere, and its Variation in Amount with her Phases.

In commencing his discourse, Lord Rosse remarked that of the three modes of action by light at a distance, which are believed always, to a greater or less extent, to exist together, one only—that of the direct action of the luminous rays on the eye—had been available to astronomers until very recent times; and he read a passage from Arago's works, in which he mentions that up to the time at which he wrote (1832)* no chemical action of the Moon's light on chloride of silver, nor heat effect on the most sensitive thermometer, had been detected in the concentrated rays of the Moon; also one from Sir J. Herschel's 'Outlines of Astronomy,'† to the same effect as regards absence of perceptible heat in the Moon's rays. He then described in order the successive attempts which have since been made to employ the thermopile where the thermometer had proved of inadequate sensitiveness. First, Melloni, having had placed at his disposal a segmental lens of 3 feet diameter and 3 feet focal length, after one or two unsuccessful attempts, was able to detect a heating effect due to the Moon's rays collected at its focus. Then about ten years later (Aug., 1856), Mr. Piazzi Smyth, while on Teneriffe, on directing his pile, armed with its polished conical reflector, alternately towards and away from the Moon, satisfied himself that the radiant heat was appreciable.

In 1861 Professor Tyndall experimented with a thermopile armed with a large tin conical reflector; but the smoke and heated air from the surrounding chimneys, which were rather less numerous in the direction of the Moon than in that to which he turned away his pile, completely masked, or rather overpowered, the heating effect from the Moon.

Professor Joule, in 1863, by means of a cylindrical glass vessel,

---

* 'Annuaire du Bureau des Longitudes' for 1833.

† Page 285 (5th edition).

divided in a vertical direction by a blackened pasteboard diaphragm, which extended to within one inch of the cover and of the bottom of the vessel, and in the upper of which spaces was delicately suspended a magnetized sewing needle furnished with a glass index, was able to detect the heat from a pint of water heated 30°, placed in a pan at 9 feet distance, also that of a moonbeam, admitted through an opening in a shutter, as it passed across the apparatus.

In the autumn of 1869 M. Marie-Davy, with a refractor of 9 inches aperture, detected the Moon's heat in the focus, and estimated the rise of temperature of the face of his thermopile at about twelve-millionths of a degree (Celsius); and a little earlier Mr. Huggins worked with the thermopile, but directed his attention more to the Fixed Stars than to the Moon.

During the course of the year 1868 * an arrangement was prepared for measuring the heat in the image of the Moon, formed by the mirror of the 3-foot reflector at Parsonstown. For the purpose of further concentrating the heat of this image of 2·9 inches mean diameter on to the face of a thermopile $\frac{1}{3}$rd of an inch diameter, a concave mirror of $3\frac{1}{2}$ inches diameter and 3 inches focal length was employed; a rock-salt lens being objectionable from its condensing moisture on its surface, and, moreover, being hardly procurable of sufficient size.

To secure greater steadiness of the needle than was otherwise obtainable, a second similar concave mirror and thermopile were placed by the side of the first, the similar poles of the thermopiles being connected with each other, and the others with the terminals of the galvanometer. Thus the deviations due to the Moon's heating effect were proportional to the *sum* of the effects due to each pile separately, and those arising from disturbing causes, acting on the two piles, to the *difference* of the effects due to each pile. To secure still greater steadiness of the needle, the two piles of four pairs each, which, having been made at *different* times by Messrs. Elliott, were not of equal power, were replaced by two more equal thermopairs constructed on the spot.† The apparatus was enclosed on all sides, except on that towards the mirror of the telescope, with a box of tin and glass, and the lattice-tube was covered with a cloth to keep draughts of air from the piles. Two covered wires led from the thermopiles to the galvanometer in the observatory, and the heating effect was determined by directing the telescope so that the Moon's image fell alternately, for the space of one minute, on each of the two small concave mirrors.

---

* See Proc. Royal Soc. 1869.

† In the construction of the thermopairs which were substituted for the thermopiles previously employed, the requisite sensitiveness was obtained by making the cross-section of the two elements small compared with the area of face, so that a smaller proportion of heat might be lost by conduction along the bars. On being compared with two of the ordinary construction they were found to be of about equal power, and therefore from the simplicity of construction they would appear to be deserving of the attention of physicists. (See Proc. Royal Soc. 1870.)

The observations made during the seasons 1868–9 and 1869–70, were found to follow pretty well Lambert's law for the variation of light with phase. It was found also that a piece of glass which transmitted 80 per cent. of the Sun's rays suffered only about 10 per cent. of the Moon's rays to pass through; thus a large amount of absorption before radiation from the Moon's surface was shown to take place.*

In the earlier experiments no attention had been paid to the correction to be applied for absorption of heat by the Earth's atmosphere; but, as the apparatus was gradually improved, it became indispensable to determine the amount of this correction before attempting to approach more nearly to the law of variation of the Moon's heat with her Phases than had been done in the earlier investigation.

By taking long series of readings for lunar heat through the greatest ranges of zenith distance available, a table expressing the law for decrease of heat with increase of zenith distance, closely following that deduced by Seidel for the corresponding decrease of the *light* of the stars, was obtained.† By the employment of this table, the determinations of the Moon's heat at various moments of the lunation were rendered comparable and available for laying down a more accurate "Phase-curve" than had been previously obtained.‡ This curve was found to agree more nearly with Professor Zöllner's law for the Moon's light, on the assumption that her surface acts as if it was *grooved* meridionally, the sides of the grooves being inclined at the uniform angle of 52° to the surface, than with Lambert's law for a perfectly *smooth* spherical surface.

To supply the want of the diagrams, which in the lecture exhi-

---

* Assuming Wollaston's estimate of 800,000 : 1 to be the correct value of the proportion between sunlight and that of Full Moon, and the percentage of heat transmitted by glass to be 10 per cent. for the Moon's rays, and 80 per cent. for the Sun's (the difference being due to absorption before radiation from the Moon's surface), the corresponding ratio for solar and lunar heat deduced was 80,000 : 1. The later observations on lunar heat would require some modifications of this value, but in the present uncertainty as to the value for the proportion between sunlight and moonlight for which

Wollaston gives the value	801,072
Bouguer ,, ...	300,000
Bond ,, ...	470,980
Zöllner, 1st method ,, ...	618,000
,, 2nd ,, ,, ...	619,000

further examination of the question would appear unprofitable.

† Following the reasoning from which Pouillet concluded that the Earth's atmosphere absorbs solar heat as if it were a medium of uniform density of height equal to $\frac{1}{80}$ of the Earth's radius, we deduced from the measurement of the Moon's heat, a height of $\frac{1}{60}$ of the earth's radius.

‡ In deducing the observations previous to employing them in the formation of a Phase-curve, corrections were applied for "Augmentation of the Moon's semidiameter," for change in the value of the Moon's distance, for reduction of the readings of the galvanometer from tangent to arc and variation of the Sun's distance. The last was not necessary in deducing the law of atmospheric absorption. Lambert's law for change with Phase was assumed in deducing the absorption law.

bited the laws of absorption in the atmosphere, and of variation of heat and light, with the phases, abbreviated tables are appended below.*

The distribution of light on two white globes, constructed in accordance with Lambert's and Zöllner's hypotheses, on which a beam from the electric light was thrown, was shown to be very different in the two cases; the brightest spot on the former being at the centre, and

---

*

Zenith Distance.	Light of Stars transmitted by Atmosphere.	Moon's Heat transmitted by Atmosphere.
0°	1·000	1·000
30	0·984	0·988
40	0·962	0·958
50	0·902	0·907
60	0·800	0·836
70	0·642	0·698
80	0·407	0·465
85	0·208	—

N.B.—Before entering the Atmosphere the Moon's heat = 1·262, so that at the Zenith fully $\frac{1}{8}$th is absorbed before it reaches the Earth's surface.

Distance from Full Moon.	Lambert's Formula.	Phase-curve for Heat (Observed).	Phase-curve for Heat transmitted by Glass.	Curve representing Zöllner's Photometric Observations.	Zöllner's Formula for Moon's Light.
100°	96	44	—	—	—
90	128	62	—	—	—
80	165	89	—	—	—
70	205	117	11·4	88	—
60	246	149	16·7	109	—
50	286	186	22·0	132	154
40	324	228	27·3	166	212
30	355	276	33·5	212	278
20	381	335	46·3	271	346
10	398	394	64·3	342	417
0	404	403	69·5	390	488
10	398	367	56·7	327	417
20	381	323	44·5	269	346
30	355	278	33·5	218	277
40	324	234	24·4	167	213
50	286	191	18·1	122	157
60	246	155	14·5	84	109
70	205	127	11·8	58	71
80	165	103	9·2	49	—
90	128	78	6·5	—	—
100	96	54	3·8	—	—
I.	II.	III.	IV.	V.	VI.

N.B.—To compare the heat transmitted by glass with Zöllner's photometric observations (Column V.) the quantities in Column IV. must be multiplied by 5·792.

on the latter at 52° on each side from the centre at the time of Full Moon, and at other times on the former at the bright limb, from which it gradually decreases towards the terminator; while on the latter there is a rapid decrease from the bright limb to a minimum about half-way to the terminator, after which it increases again, and then fades away on approaching the terminator.

On examining the Phase-curve which had been obtained, a certain want of symmetry on the two sides of Full Moon was perceived, which was ascribed to the unequal distribution of mountain and plain on the lunar surface, as was shown by a rough diagram of the lunar surface with its so-called "seas." It had also been found that the percentage of the Moon's heat transmitted by a sheet of glass diminished from 17·3 per cent. at Full Moon to about 13·3 per cent. at 22½°, 11 per cent. at 45°, and 10 per cent. at 67½° distance from Full Moon; a circumstance which might have been accounted for by supposing that there is a constant amount of radiant heat coming from the Moon in addition to that which, like the light, varies with the Phase, had it not been found that as the Moon approached tolerably near the Sun, as for instance, on March 27, 1871, when her distance from full was 138°, no perceptible amount of heat radiated from her surface.

The less rapid decrease of the Moon's heat than of her light on going farther from Full Moon, and the increase of percentage of heat transmitted by glass towards the time of Full Moon, may probably be explained on the assumption that when the Sun's heat and light strike the Moon's surface the whole of the former and only a certain proportion of the latter, depending on the intrinsic reflecting power or "Albedo" of the surface, leave it again, and consequently the shaded portions, which are inclined more towards the position of the Earth at Quadrature than at Full Moon, reflect a larger amount of heat as compared with the light at the former than at the latter time, and a greater flatness of the Heat- than of the Light- Phase-curve is the result.

With the view of obtaining a decisive result on the question, whether or not the Moon's surface requires an appreciable time to acquire the temperatures due to the various amounts of radiant heat falling on it at different moments, simultaneous determinations of the amount of the Moon's heat and of her light were made, whenever the state of the sky allowed of it, during the Eclipse of November 14, 1872. The Eclipse was a very partial one, only about $\frac{1}{40}$ of the Moon's diameter being in shadow; but although this circumstance, coupled with the uncertain state of the sky, rendered the observation far less satisfactory than it would otherwise have been, yet it was sufficient to show that the decline of light and heat as the penumbra came over the lunar surface and their increase after the middle of the Eclipse were sensibly proportional.*

Rather with the view of finding the value of the galvanometer

---

* Both were reduced to about one-half what they were before the eclipse.

readings in terms of the radiation from a surface of known temperature, and capable of being reproduced by anyone at any future time, than in the expectation of getting more than the roughest approximation to the temperature of the lunar surface, readings of the galvanometer were taken when the faces of the piles were alternately exposed for periods of one minute to the radiation from two blackened tin vessels filled with water of different temperatures. It was thus found that were the atmosphere removed, the surfaces of New and Full Moon might be respectively replaced by blackened tin vessels of equal apparent area to that of the Moon, and at temperatures of 50° and 247° Fahrenheit.*

In conclusion, Lord Rosse expressed a hope that among the many subjects which engage the attention of astronomers that of radiation of heat from the Moon might not be entirely lost sight of, more especially by those who live in a more favourable climate for such observations than that of the British Isles.

Friday, February 25, 1876.

WARREN DE LA RUE, Esq. D.C.L. F.R.S. Manager, in the Chair.

The REV. STEPHEN JOSEPH PERRY, F.R.S.

The Transit of Venus.

THE history of the transit of Venus leads us back only a few centuries. Whilst engaged in the study of planetary laws, Kepler discovered that both Mercury and Venus might be seen as black spots on the bright background of the sun in the year 1631, the former in November, and the latter in December. Gassendi witnessed this transit of Mercury, but that of Venus was not seen, as the sun rose in the west of Europe just after the egress of the planet.

For the transit of a planet it is only necessary that its latitude should be less than the sun's semidiameter when the planet is in inferior conjunction, or that the earth and planet should have the same longitude near a node of the planet's orbit. Now Venus and the earth perform their sidereal revolution in 224·701 and 365·256 days respectively; hence in 243 years Venus would go round the sun almost exactly 395 times, the two bodies returning to the same relative positions. A transit in 1631 must therefore be followed by another in 1631 + 243 = 1874, &c. But the time of thirteen revolutions of Venus amounts very nearly to eight years, and Jeremiah Horrocks found that Venus would only change her latitude by 22′ between the conjunctions eight years apart, and that, as the sun's diameter is more than 30′, Venus might be visible eight years after 1631. Horrocks and Crabtree were the only observers of this transit of 1639.

Both the above transits occurred near the ascending node, and 243 years later a similar pair must follow at the same node; but the orbits of the larger planets being almost circular, half that period should bring Venus and the earth into similar relative positions at the opposite, or descending node, and therefore $121\frac{1}{2}$ years after December, 1639, viz. in June, 1761, another transit might be expected.

In 1677 Halley observed a transit of Mercury at St. Helena, and this observation suggested to his mind the possibility of employing the transit of Venus as a most accurate means for determining the exact distance of the sun. A few years later, in a paper read before the Royal Society, he proposed a method of observation, which still bears his name. This consists in recording the exact local time of the ingress and the egress of the planet at two stations, whose difference of latitude, and position with regard to the earth's axis of rotation, render the total durations of the transit as unequal as possible.

The inapplicability of Halley's method to transits that are too centrical, caused Delisle to propose another method, which is equally suited to all transits. His two observers are placed rather east and west, than north and south, and, as far as rotation and obliquity of orbit will permit, almost opposite the extremities of that solar diameter, which passes through the point of contact either at ingress or egress. The observations required at each station are a single contact, and a very accurate longitude determination, so as to obtain the absolute difference of time of contact at the two stations. This difference should of course be made as great as possible, so that any small error in observed time of contact, or in longitude, may bear the smallest possible proportion to the whole.

In June, 1761, the conditions were not favourable for Halley's method, and Delisle's failed from want of accurate longitudes.

The transit which followed eight years later was well observed, and the solar parallax and distance deduced. The observers were numerous, and the computed values differed considerably. Encke, in 1837, combined the results of various nations, and obtained 8″·55 as the solar parallax, which makes the sun's distance more than 95,000,000 miles.

This value was accepted universally as a fair approximation, until other methods of determining the solar parallax raised a most serious doubt as to the accuracy of Encke's result. Foucault's determination of the velocity of light by a revolving mirror, and that of Fizeau by a rotating wheel, combined with the observed difference in time of the phenomena of Jupiter's satellites, or with Bradley's constant of aberration, gave severally 8″·86 and 8″·88 as the solar parallax. Hansen, from the lunar inequalities found 8″·92, and Stone 8″·85. Le Verrier, from 106 years of meridional transits of Venus, from her observed latitude in 1761 and 1769, and from an occultation of a star by Mars, deduced respectively 8″·86, 8″·85, and 8″·87. And finally, the discussion of the parallactic displacements of Mars led Sir Thomas Maclear, Mr. Stone, and Dr. Winnecke to the values 8″·90, 8″·94, and 8″·96. The close agreement of these numbers induced Mr. Stone to undertake the laborious task of a re-examination of Encke's value. The result of his labours, in which great attention was paid to the description given by each astronomer of what he observed as true contact, changed the solar parallax given by the transit of 1769 from 8″·55 to 8″·91, which is nearly the mean of the values found previously by methods differing so completely in their nature.

The sun's distance enters into the computation of the lunar tables, and this fact alone makes it a matter of the highest moment for a great naval power and a commercial people. Hence early attention was drawn to the preparations necessary for the accurate observation of the transit of 1874 by the astronomical representative of the Admiralty. There were two principal reasons that made it advisable to secure stations well suited for Delisle's method. 1. Only two observations of contact are required, and therefore, supposing the chances of good and of bad weather to be equal, the odds in favour of a complete Delislean observation would be as four to one, compared with Halley's method. 2. Our transit instruments and altazimuths are now so perfectly made that much greater reliance can be placed on the accuracy of our longitudes, even where the telegraph is wanting, than on our contact observations.

For ingress the British stations were the Sandwich Islands and Kerguelen, strengthened by Rodriguez; and for egress, New Zealand and Egypt. For Halley's method, Professor Struve combined with Sir G. Airy; the Russians occupying the north, and the British the south at the three Halleyan stations of Kerguelen, New Zealand, and Rodriguez.

Besides the indirect methods of observed contacts used in former transits, the solar parallax can now be found more directly by actual measures whilst Venus is on the sun, or photographs can be taken during the transit and measured afterwards. The heliometer, for actual measurement of greatest and least distance of limbs, seems to have been used only by the Germans and Russians, and by Lord Lindsay. The British expeditions were provided with double-image micrometers for measuring cusps, and diameters, and least distance of limbs near contact.

Our photographic preparations owe much to Dr. De la Rue and to Captain Abney. We adopted the equatorial mounting, with an image enlarged to four inches, and dry-plate photography, whilst the Americans preferred a fixed telescope and heliostat, without enlargement by eye-piece; and the French chose daguerreotype in preference to photography. An excellent arrangement was devised by Mr. Christie, for carrying out the idea of Dr. Janssen, of taking on the same sensitized plate a succession of photographs, with only second intervals, of the position of Venus near contact.

The only spectroscopes taken out for the observation of external contact at ingress were, as far as is known at present, those of Dr. Janssen, Tacchini, Lord Lindsay, and Captain Tupman, and an instrument constructed specially for our Kerguelen station by Mr. Browning.

The most advantageous moment for taking direct measures with the heliometer, or for obtaining photographs of Venus on the sun, is when the two stations to be compared together are situated on the same radius of the circular section, at the earth's distance, of the right cone enveloping the sun and Venus. Mr. Proctor has clearly pointed out the advantages of this in multiplying the good stations of observation, though many of the best mid-transit stations are equally good for Halley's method.

The active preparations for the transit of 1874 occupied several years. Large telescopes had to be collected, and new instruments devised, before the observatories were constructed, and the observers trained; but thanks to the untiring zeal and energy of Sir G. Airy, Captain Tupman, and the whole Greenwich staff, all was in readiness by the time appointed for departure. The observers for distant stations left England in the May and June that preceded the transit.

The Kerguelen expedition, to which I will now direct special attention for a short time, left in two detachments, which met at the Cape, where Commodore Hewett, V.C., and the Admiralty astronomer, Mr. Stone, did all in their power to assist us. H.M.S. 'Encounter' and 'Supply' had been appointed by the Admiralty to take the astronomers from the Cape to Kerguelen, but an accident had happened to the lifting piece of the screw of the 'Encounter' during the Ashantee war, and when this was known in England, H.M.S. 'Volage' was at once put into commission to replace the 'Encounter.' We left the Cape of Good Hope in the 'Volage' and 'Supply' on September 18th, passed the Crozets on October 2nd, and came almost within sight of Kerguelen on the 6th, when a fearful storm, that carried away a cutter, and drowned many of our live stock, made it impossible to advance for nine-and-forty hours. When the wind abated and the sky cleared, the maps of H.M.S. 'Challenger' enabled us to steer straight into Royal Sound without any difficulty. It was the end of spring, and yet the country was completely covered with snow. We found two sealing schooners in Royal Sound, waiting for a favourable wind to take them to Heard Island. Captain Bailey kindly offered to be our guide, and we soon found a site, which served excellently for our primary station. The anchorage for our vessels was perfect, the landing safe, the foundations for our instruments the solid rock, the supply of water good, all our huts could be placed on the same level, and the spot was protected from the N. and W. gales, without interfering with the horizon. When the heavy work of draining, landing stores, erecting huts, building piers, and fixing the instruments, was well advanced at our principal station, we selected a site for a second observatory, six or seven miles south of the first. Two observers were placed there with a transit and small altazimuth for finding local time and latitude, and with two four-inch telescopes for observing contacts.

The Germans and Americans, who at first intended to occupy Heard Island, were now stationed at Kerguelen; the latter in Royal Sound near the Prince of Wales' Foreland, some 8 miles N.E. of our first station, and the former at Betsy Cove, 10 miles N. of the Americans. The Crozets had been also abandoned, and thus a large cloud over the east end of Kerguelen might destroy at once all the best Delislean stations for retarded ingress, and the south stations for Halley's method. I determined therefore, if it was at all advisable, to attempt the occupation of Heard Island. Lieutenant Coke generously volunteered to take the observations, and Captain Fairfax prepared the

'Volage' for the trip, which was likely to be a stormy one. We had already made every possible inquiry from the sealers, and their reports were discouraging. Captain Fuller, who is best acquainted with those seas, we had not as yet met, but we expected to see him about two weeks before the transit. His account of the island left us finally no hope of a successful trip, as he told us we should probably have to wait many days before it would be possible to land at all, and even then our instruments, chronometers, &c., would have to be conveyed on shore in the sealing boats through a heavy surf. We were forced, therefore, however reluctant we might be, to give up Heard Island, and to occupy instead a third position on Kerguelen. This proved in the end to be particularly fortunate, as the whole of the transit was observed at the third station, and we learned afterwards from the sealers that the sun was not visible on December 8th at Whisky Bay, the only landing place on Heard Island.

The early morning of the day of the transit was fine at Kerguelen, but the sky gradually clouded, which interfered considerably with the observations. At Station I. Venus was observed until after bisection at ingress; a small cloud then covered the sun completely until after internal contact. Some thirteen photographs, and a few measures of least distance of limbs, and of diameters, were obtained, and internal and external contact, as well as bisection, were fairly observed at egress with the 6- and 4-inch equatorials. At Station II. the sun was well seen at ingress by both observers, but it was cloudy at egress. These results, with the success at Station III., were very satisfactory, considering the reputation for cloud, mist, and high wind enjoyed by this Land of Desolation.

The excellent state of the sky, the good definition of the telescope, and the care taken in the observation, make the time of internal contact at ingress, obtained by Lieutenant Corbet, especially reliable.

Our orders about longitude determinations were very stringent, and our stay on the island was consequently extended to five months, although this necessitated an arrangement about half rations. We obtained for our fundamental longitude nineteen transits of the moon, ninety double altitudes or azimuths, and one occultation. In the meantime the different stations in Kerguelen were connected together by aid of eight special chronometers, and satisfactory gunpowder signals were made for longitude connections on a central island in Royal Sound. These, with a run from the Cape to Kerguelen with all our chronometers, complete the work done to secure satisfactory longitudes.

During our stay the naturalist, the Rev. A. E. Eaton, was indefatigable in collecting the natural products of the island, and at our principal station we carried on, simultaneously with the astronomical work, a very complete series of observations of the elements of the earth's magnetic force, and readings were taken at all the even hours, day and night, of the chief meteorological instruments. The men of the Royal Engineers deserve great credit for the zeal they showed in willingly undertaking this laborious portion of our routine duties. On the advice of Captain Nares a number of rabbits and goats were taken from the Cape and left on the island to propagate. They were doing well when we left.

Two hours after the last observed passage of the moon across the meridian, our two vessels were already under steam; H.M.S. 'Supply' bound for the Cape of Good Hope, and H.M.S. 'Volage' for Ceylon and Aden, *en route* for England. Five months on the Island of Desolation was a sore trial to many, but the kindness and assistance we invariably met with from both Captain Fairfax and Captain Inglis, and from the officers and men, lightened our task very much. Before reaching the equator we passed within 22 miles of the centre of a cyclone. The 'Volage' rolled frequently more than 45°, and the sea poured freely over her hammock nettings, but fortunately we escaped after two days without loss of life. On our homeward journey we availed ourselves of the opportunities afforded of collecting data for the formation of magnetic charts of the declination, dip, and intensity.

With the sole exception of New Zealand, where fine weather was almost a certainty at that season of the year, the observers at the remaining British stations were favoured with excellent weather, and the harvest is proportionately abundant. India was equally successful; and the official astronomers at Melbourne and Sydney can show much that will lessen our disappointment at the failure in New Zealand. But England did not depend solely on her Government expeditions, or on those of her colonies for her share in the work of the late transit. Many private observers, as Mr. Tebbutt and Mr. Hennessy in Australia and India, Admiral Ommanney and Colonel Campbell in Egypt, and others, added their valuable results to those accumulated by official astronomers; but no expedition was more perfectly equipped, or more ably manned, than that which rounded the Cape in the yacht 'Venus,' and we have very great reason to regret that sickness, caused by this journey to Mauritius, has prevented Lord Lindsay from giving us himself a full account of the noble part he took in the transit of 1874.

Unfortunately the very limited time at our disposal will only allow us to do scanty justice to the triumphs of other nations, but we must at least cast a rapid glance at their successes. Russia, with her thirty-two stations, succeeded perfectly at five, partially at eight, and most unfortunately failed wholly at nineteen. France had admirable good fortune at S. Paul's, Nangasaki, and Pekin, but envious clouds interfered wholly at Campbell Island, and partially at Caledonia Island. America was less fortunate in the south than in the north; her best results are chiefly photographic. Germany obtained excellent observations with heliometer and photoheliograph; and Italy, Holland, and Austria added their quota to the total results.

Preparations conducted by such men as Airy and Struve, Puiseux, Auwers, and Newcombe, could not fail of securing the happiest results, when favoured by the fine weather which has fortunately prevailed. But it will be only when the longitudes have been computed, the photographs measured, and the contacts discussed, that we can with any confidence compare the results obtained by Halley and Delisle, by photograph and daguerreotype, by fixed and by movable telescope, by spectroscope and telescope, by indirect and by direct methods; or be able to say to what degree of approximation we have now attained.

But there are many things that we have already learnt from the late observations that are both interesting in themselves, and that will lead to more accurate results in the approaching transit of 1882. The aqueous vapour in the ring round Venus, discovered by Tacchini; the difference between photographic and eye contact, proved by Janssen; the possibility of seeing Venus on the chromosphere, before external contact, without the aid of the spectroscope, also verified by the same astronomer; the absence of ellipticity in the planet; are all useful and instructive additions to our previous knowledge. The difficulty about the ligament and the deformation of Venus, which so much troubled observers of the last century, has almost vanished, and we now find this messenger of darkness replaced by a messenger of light, a bright ring, which threatens to be nearly as troublesome as its predecessor. But difficulties are rarely invincible when we see clearly what we have to meet, and we may confidently expect that the observations of 1874 will not only furnish a close approximation to the solar parallax, but will also, by the lessons it affords, lead to a still closer approximation in 1882.

In this necessarily rapid sketch of what is already known of the observations made on the 8th of December, 1874, many names have doubtless been omitted which might have been mentioned, and full weight could scarcely have been given to many important results touched upon so cursorily. It is hoped, however, that no one will be led to conclude that this arises from any want of a due appreciation of the value of his labours in this field of science.

[S. J. P.]

Friday, February 14, 1879.

C. William Siemens, Esq. D.C.L. F.R.S. Vice-President, in the Chair.

G. Johnstone Stoney, Esq. M.A. F.R.S.

The Story of the November Meteors.

[As some readers may wish to consult the original investigations referred to in this lecture, a list of them is given in a postscript at the end.]

METEORS AS THEY APPEAR IN THE EARTH'S ATMOSPHERE.

When observers band together to watch every quarter of the sky, and to keep on the look-out through the whole night, the number of meteors that present themselves is very great. In this way it has been ascertained that upwards of thirty on the average, which are conspicuous enough to be seen without instruments, come within the view of the observers stationed at one locality. And it is computed that telescopic meteors must be about forty or fifty times as numerous as those visible to the naked eye.

These results may be obtained from observations made at one station; but when concerted observations are carried on at different stations, several other facts of interest come to light. By simultaneous observations at distant stations, it has been discovered that the height of meteors above the surface of the earth usually ranges from 120 down to twenty miles, the average height being about sixty miles; that the direction of their flight is towards the earth, either in a vertical or in a sloping direction; and that their speed in most cases lies between thirty and fifty miles a second.

We thus arrive at the conclusion that *visible* meteors are phenomena of our own atmosphere; and as the atmosphere reaches a height, at most, of 150 miles, and is, therefore, but a thin film over so vast a globe as the earth, it is obvious that the spectators at any one place can see only a very small portion of the meteors which dart about through all parts of this envelope. After making allowance for this, we are forced to conclude that no fewer than 300 millions of these bodies pass daily into the earth's atmosphere, of which about seven millions and a half are large enough to be seen with the naked eye on a clear night, and in the absence of the moon.

From the direction and swiftness of their flight, it is manifest that meteors are visitors from without. They plunge into our atmosphere, and the resistance to which they become then suddenly exposed must raise them to a temperature which exceeds that of the most intense furnace. The heat is enough first to melt and then to dissipate in vapour the most refractory substances, and it only now and then happens that even a part of a meteor escapes this fate, and reaches the ground. They are for the most part lost in vapour ere they get within several miles of us. The difficulty, indeed, is not to account for their incandescence, but to see why they do not emit a greater flood of light where the heat must be so intense. And, in fact, they cannot be other than very small bodies, or they would be much brighter. The average weight of those visible to the unassisted eye appears to be under an ounce, and the telescopic ones, of course, are much lighter.

SPORADIC METEORS, AND METEORIC SHOWERS.

Meteors may be distributed into two very obvious classes—casual meteors, which dart irregularly through the sky, and meteoric showers, which stream into our atmosphere in one definite direction, and at stated intervals of time. We are concerned at present with the meteoric showers. Many such are known to exist, of which the principal are the August shower, through which the earth passes every year upon the 9th, 10th, and 11th of August; and the great November shower, which is discharged upon the earth three times in a century. The November meteors are those about which most is known, and it was of these, therefore, that the lecture chiefly treated.

THE REGIONS FROM WHICH METEORS COME.

To make their history intelligible, it was necessary to explore, in some degree, the regions from which they come. For this purpose a great diagram was exhibited on a scale rather more than thirty times the scale of the accompanying woodcut. Yet, though the diagram was so large, every hundredth of an inch upon it represented a distance in nature equal to the interval between the earth and the moon. The distance from the earth to the sun on this diagram was a decimeter, that is, four inches; and, on the same scale, the nearest fixed star would have to be placed at a distance of twenty kilometers, or upwards of twelve miles.

ORBIT OF THE GREAT NOVEMBER SWARM.

In these vast celestial spaces, there are no rails over the roughnesses of which the train must be made to rattle, if it is to move at all; there are no wheels to be worn out; there is no air in which a wind must be produced, or through which noise will be propagated. The music of the spheres is not a sound audible to the ear, and an impediment to motion: it is harmless, it is altogether good, it is the pleasure of the human mind when it understands the great works of nature. There is no thundering along through the heavens. All is silence and peace round the planets as they swiftly glide. Bodies which sweep in this way without obstruction through the depths of space, are ready to yield at once the due amount of obedience to the attraction of the sun. Accordingly each meteor which traverses the elliptic orbit represented in the diagram, mends its pace so long as it is gliding along that half of its course in which it is approaching the sun, because here the sun is drawing it forwards as well as sideways; and the forward attraction increases its velocity, while the sideward attraction bends its path into the oval form. The meteor takes upwards of sixteen years to traverse this part of its orbit, and all this time its velocity is on the increase. It has attained its greatest speed when it reaches the point of its orbit which is closest to the sun, near to which is the place where it crosses the earth's path. As it passes this point its velocity is twenty-seven miles a second. The earth moves at the rate of nineteen miles a second in very nearly the opposite direction, so that if the meteor happen to strike the earth, the velocity of its approach is the sum of these two numbers, or forty-six miles a second; and it is at this enormous speed that it plunges into our atmosphere. But if it escape the earth, and continue its course along its orbit, it loses speed for the next sixteen years, until it passes the farthest part of its orbit at its slowest pace, which is about a mile and a third per second. In each revolution its velocity oscillates between these extremes. Its orbit is so vast that it takes thirty-three years and a quarter to get round it.

Such is a good picture of the course pursued by each member of the great November swarm. There are countless myriads of meteors in this mighty group, each one moving independently of the rest, each one fulfilling its own destiny. They form, together, an enormous stream of meteors, the dense part of which appears to be about 100,000 miles in width, and of immense length. The orbit along which they travel was represented on the diagram by an ellipse of 207 centimeters, or close upon seven feet, long—i. e. by an oval about as long and broad as the hall-door of a house; and the length, breadth, position, and motion of the swarm in 1865, before it reached the earth, would be represented on the same scale by a thread of the finest sewing silk, about a foot and a half or two feet long, creeping inwards along the orbit, the rear of the column having been between the orbits of Jupiter and Saturn, and the front of it nearly as far in as the earth's orbit. The actual train which is thus represented was so amazingly long that even moving at the rate of twenty-seven miles a second, it took upwards of two years to pass the point where its path crosses the earth's orbit. The earth passes this point on the morning of the 14th of November in every year. The head of the dense part of the stream seems to have reached the same point early in the year 1866. The earth was then in a distant part of its orbit, but on the following 14th of November we came round to the place where the great stream of

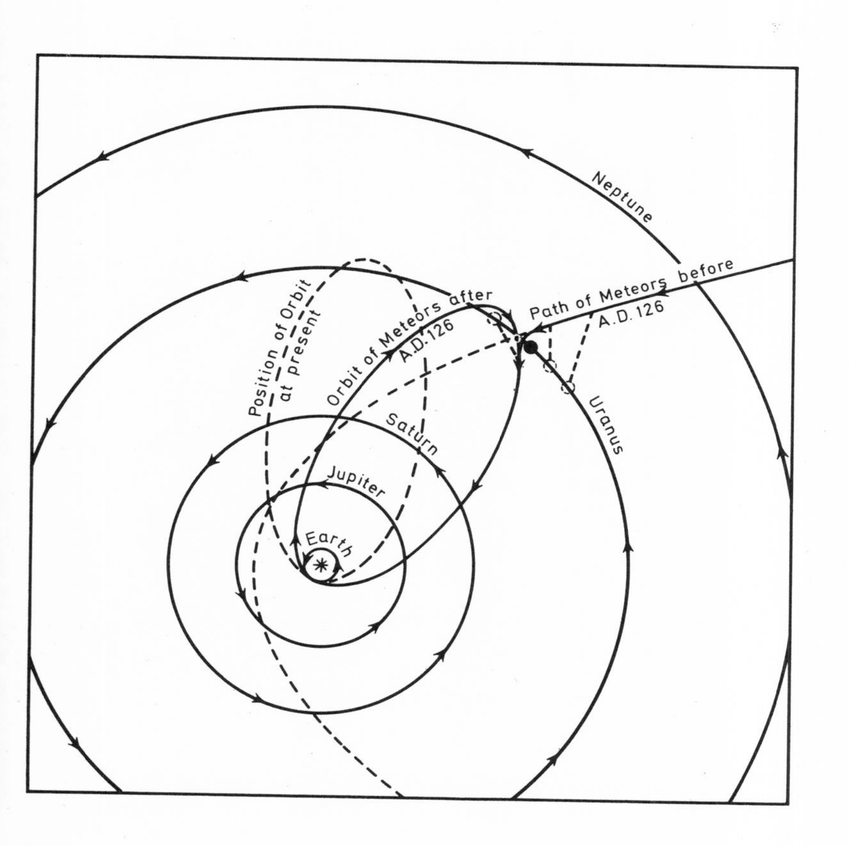

Neptune
Path of Meteors before
A.D. 126
Position of Orbit
at present
Orbit of Meteors after
A.D. 126
Saturn
Uranus
Jupiter
Earth

meteors was pouring across our path. The earth then passed through the swarm, just as you might imagine a speck, too small to be seen by the eye, to be carried on the point of a fine needle in a sloping direction through the thread which represents the meteors. The earth took about five hours to pass through the stream; and it was Europe, Asia, and Africa, which happened at the time to be moving forwards. Accordingly it was upon this side of the earth on that occasion, that the meteors were poured, and they produced the gorgeous display in our atmosphere which many here must remember. In 1867, when we came round again to the same place, the stream of meteors was still there. America, this time, chanced to be the part of the globe which was turned in the right position to receive the shower. In 1868, the mighty swarm had not passed, and in subsequent years, when we came round to the proper place, we still found ourselves among outlying stragglers of the great procession.

[The lecturer next attempted to give an outline of the successive steps by which the path over which the meteors travel had been determined, and in doing so had an opportunity of adding other particulars to the marvellous history of these bodies.]

HUMBOLDT.

In 1799 Humboldt was travelling in South America, and on the morning of the 12th of November in that year the November shower was poured out over the New World. Humboldt's description of this shower seems first to have fixed the attention of scientific men upon the subject. But he contributed still more to the advance of our knowledge by the success with which he insisted that nearly all such phenomena are periodic, and that therefore there is reason to hope that the causes of them are discoverable. Shortly after, the periodic character of the August meteors was established; and when the next return of the November meteors to the earth took place, when there was a magnificent display of them exhibited to Europe in 1832, and a still more impressive spectacle seen in America in the following year, the attention of scientific men was thoroughly aroused.

PROFESSOR H. A. NEWTON.

In England, meteors began to be systematically observed, and in this way all that knowledge about them has been acquired which was referred to in the beginning of the lecture. In France, the records of antiquity and the annals of distant nations were ransacked; and by this most useful antiquarian search, no less than ten visits of the November swarm, previous to the shower observed by Humboldt in 1799, have been brought to light. But the first great step towards gaining a knowledge of their orbit was made by Professor H. Newton, of New Haven in America, who published in 1864 two memoirs, in which he discussed all the accounts that had been collected, extending back to the year A.D. 902. He found by comparing the dates of the old observations with the modern ones, that the phenomenon is one which recurs three times in a century, or more exactly, that the middle of the swarm crosses the earth's path at intervals of $33\frac{1}{4}$ years. He further showed that meteors which thus visit the earth three times in a century must be moving in one or other of five orbits which he described; and that therefore if means could be found for deciding between these five orbits, the problem would be solved. The five possible orbits are —the great oval orbit which we now know the meteors actually do traverse every 33 years and a quarter; a nearly circular orbit, very little larger than the earth's orbit, which they would move round in a few days more than a year; another similar orbit in which their periodic time would be a few days short of a year; and two other small oval orbits lying within the earth's orbit. But we owe even more to Professor Newton. He also pointed out how it was *possible* to ascertain which of these orbits is the true one, although the test he indicated was one so difficult of application that there was at the time little hope that any astronomer would attempt it. Fortunately our own Professor Adams, of Cambridge, was found able to grapple with the difficulties of the problem, and willing to encounter its immense labour, and to him we owe the completion of this great discovery.

PROFESSOR ADAMS.

A comparison of the dates of the successive showers which have been recorded shows that the point where the path of the meteors crosses the earth's orbit is not fixed, but that every time the meteors come round they strike the earth's orbit at a point which is twenty-nine minutes (i. e. nearly half a degree) farther on in the direction in which the earth is travelling. In other words, the meteors do not describe exactly the same orbit over and over again: their path in one revolution is not exactly the same as their path in the next revolution, although very close to it. Thus, their path in A.D. 126 was that which is represented by the strong oval line in the diagram, but in the seventeen centuries which had since elapsed, it has gradually shifted round into the position represented by the dotted ellipse. This kind of motion is well known to astronomers, and its cause is well known. It would not happen if the sun were the only body attracting the meteors, but arises because the planets also draw them in other directions; and although the attraction of the planets is very weak compared with the immense power of the sun, still they are able to drag the meteors a little out of their course round the sun, and in this way occasion that shifting round of the orbit of which we are speaking. Now, in the case of meteors which are really travelling in the large orbit, this shifting of the orbit must be due to the attraction of the planets Jupiter, Saturn, Uranus, and the Earth, while, if they had travelled in any of the four smaller orbits, the planets that would be near enough and large enough to act sensibly upon them would be the

Earth, Venus, and Jupiter. Accordingly, if anyone could be found able to calculate how much effect would be produced in each of the five cases, the calculated amount of shifting of the orbit could be compared with the observed amount, which is 29′ in 33¼ years, and this would at once tell which of the five possible orbits is the true one.

These papers of Professor Newton's were published in 1864. Before the computations which he had indicated could be attempted, it was necessary that the direction in which the meteors enter the earth's atmosphere should be known much more accurately than it then was, in order to enable astronomers to compute the *exact* forms and positions of the five possible orbits. This observation then was of the greatest importance in 1866, and it was on this account that all the astronomers on that occasion devoted nearly all their efforts to determining with the utmost precision the exact point of the constellation Leo from which the meteors seemed to radiate. This important direction was ascertained during the great meteoric shower on the morning of the 14th of November, 1866, and immediately after Professor Adams and his two assistants in the Cambridge Observatory set to work at their arduous task. This great calculation required the solution of a problem in mechanics which had never before been attempted, and involved an immense amount of tedious labour. Amidst all these difficulties Professor Adams triumphed; and after months of toil he was able to announce in the following March that if the meteors are moving in the large orbit, Jupiter would produce a shifting of the orbit in each revolution amounting to 20′, the attraction of Saturn would add to this 7′, Uranus would add 1′; the effect of the earth and the other planets would be insensible. Adding these numbers together, the whole effect, according to Mr. Adams's computation, is 28′, almost exactly the same as the observed amount which had been determined by Professor Newton, which was 29′. But if the meteors were in any of the other four possible orbits, the total amount would never exceed 12′. Here, then, we have reached the final result: *the long orbit is the orbit of the meteors.* This great discovery was published in March, 1867.

PROFESSOR SCHIAPPARELLI.

Meanwhile Signor Schiapparelli, of Milan, was labouring in another direction. It was evident from the observations that the meteors were drawn out into a long stream. What was the cause of this? Signor Schiapparelli pointed out that if a cloud of meteors were started under conditions which were not quite the same, each meteor would pursue its own orbit, which would differ from the others. If they were treated almost exactly, although not quite, alike at starting, their various orbits would lie excessively close to one another, and would be undistinguishable in most respects. But if there be any effect which goes on accumulating from revolution to revolution, such an effect would in the end become very sensible. And such an effect there is. The periodic times differ a little in these different orbits. At the end of the first revolution those meteors which have the longest periodic times are the last to get back to the starting point, and have therefore already fallen a little into the rear of the group, while those with the shortest periodic time have gone a little ahead. At the end of the second revolution the separation is doubled, and in each successive revolution the column is still more lengthened out. After a sufficient number of revolutions it will be spread out over the whole length of the orbit, and form a complete oval ring. This has not yet happened to the November meteors, and we are thus assured that it cannot be any enormous period, speaking cosmically, since the time when they first started on their present path. On the other hand the August meteors, which have returned punctually *every* year since they were first observed, are probably a complete ring, and are at all events of far greater antiquity than the November meteors. But they are also, as might be expected, more scattered, so that the sprinkling of meteors they discharge upon the earth as it passes through them has nothing like the splendour of the great November shower. Signor Schiapparelli also pointed out that there is a comet moving in the track of the August meteors, and another in the track of the November meteors. We shall presently see the significance of this observation.

M. LE VERRIER.

The next great step was made by M. Le Verrier, the late Director of the Paris Observatory. Acting on the suggestion made by Sig. Schiapparelli, M. Le Verrier pointed out that the orbit of the meteors intersects the orbit of Uranus, as represented in the diagram. From its inclined position it does not intersect the path of any of the intermediate planets Saturn, Jupiter, and Mars. M. Le Verrier also calculated back the epochs at which the planet and the meteors were at the point of intersection, and found that early in the year A.D. 126 they were both at that spot, but that this has not happened since. Taking this in conjunction with what Sig. Schiapparelli pointed out, we seem to have a clue to a truly wonderful past history. All would be explained if we may suppose that before the year 126, the meteors had been moving beyond the solar system; and that in that year they chanced to cross the path of the planet Uranus, travelling along some such path as that represented in the diagram. Had it not been for the planet they would have kept on the course marked out with a dotted line, and after having passed the sun, would have withdrawn on the other side into the depths of space, to the same measureless distance from which they had originally come. But their stumbling on the planet changed their whole destiny. Even so great a planet would not sensibly affect them until they got within a distance, which would look very short indeed upon our diagram. But they seem to have almost grazed his surface, and while they were very close to such a planet, he would be able to drag them quite out of their former course. This the planet Uranus seems to have done, and when, pursuing his own course, he again got too far off to influence them sensibly, they found themselves moving slowly backwards, and slowly inwards; and accordingly began the new orbit round the sun, which corresponds to the situation into which they had been brought, and the direction and moderate speed of their new motion.

They seem to have passed Uranus while they were still a small compact cluster. Nevertheless those members of the group which happened to be next the planet as they swept past, would be attracted with somewhat more force than the rest, the farthest members of the group with the least. The result of this must inevitably have been that when the group were soon after abandoned to themselves, they did not find themselves so closely compacted as before, nor moving with an absolutely identical motion, but with motions which differed, although perhaps very little, from one another. These are conditions which would have started them in those slightly differing orbits round the sun, which, as we have seen, would cause them, as time wore on, to be drawn out into the long stream in which we now, after seventeen centuries, find them.

What is here certain is, that there was a definite time when the meteors entered upon the path they are now pursuing—that this time was the end of February or beginning of March in the year 126 is still a matter of probability only. It is, however, *highly* probable, because it explains all the phenomena at present known; but astronomers are not yet in a position to assert that it is ascertained, since one link in the complete chain of proof is wanting. We who live now should be in possession of this link if our ancestors had made sufficiently full observations; and our posterity will have it when they compare the observations they can make with those which we are now carefully placing on record for their use. They will then know whether the rate at which the stream is lengthening out is such as to indicate that A.D. 126 was the year in which this process began. If so, Le Verrier's hypothesis will be fully proved.

MR. STONEY.

Another episode in the eventful history of these meteors is also known with considerable probability. It has been already mentioned that a comet is travelling along the same path as the meteors. It is moving a very little slower than they, and is at present just at the head of the procession which they make through space. Another comet is similarly moving in the track of the great elliptic ring of August meteors. In 1867, the lecturer ventured to suggest an important function which these comets seem to have discharged. Picture to yourselves a mass of gas before it became connected with the solar system, travelling through space at a distance from the sun or any other star. Meteors would now and then pass in various directions, and with various velocities through its substance. For the most part they would go entirely through and pass out again; but in every such case the meteor would leave the comet with less velocity than it had when approaching it. And in some cases this reduced velocity would be such that the future path of the meteor would be an ellipse round the comet. Whenever this was once brought to pass, the meteor would inevitably return again and again to the comet, each time passing through some part of its substance, and at every passage losing speed. After each loss of speed the ellipse it would next proceed to describe would be smaller than the one before, until at last the meteor would sink entirely into the gas and be engulfed by it. In this way meteor after meteor would settle down through the comet, and, in the end, just such a cluster would be formed as came across the planet Uranus in the year 126, or, if such a cluster existed originally within the mass of gas, it would in this way be augmented. As the comet swept past the planet, its outlying parts would seem to have grazed his surface, and in this way the gas was probably somewhat more retarded than the meteors; and in the centuries which have since elapsed the meteors have gone so much ahead of the comet that they are now treading on his heels and on the point of overtaking him, while probably the gas has again brought together a smaller cluster of the meteors.

PROFESSOR GRAHAM.

The question now arises, How the deserts of space which extend from star to star come to be tenanted here and there by a patch of gas or an occasional meteorite? Light has been thrown on this inquiry by discoveries made with the spectroscope in modern times and by observations during eclipses. These have revealed to us the fact that violent outbursts occur upon the sun, and doubtless on other stars, so swift that the up-rush must sometimes carry matter clear away into outer space. Imagine such a mass consisting in part of fixed gas and in part of condensable vapours ejected from some star. As it travels forward the vapours cool into meteorites, while the fixed gas spreads abroad like a great net, to entangle other meteors. In some cases both might travel together; in others the gaseous portion would be retarded before it passed beyond the neighbourhood of the star, and the denser meteors would get ahead. But even so in the lapse of ages other meteors would be caught, so that in any event a cluster would at length be formed. Now, the reasonable suspicion that this is the real origin of meteors has received striking confirmation from the discovery of the late Professor Graham, that meteoric iron contains so much hydrogen occluded within it as indicates that the iron had cooled from a high temperature in a dense atmosphere of hydrogen—precisely the conditions under which the vapour of iron would cool down while escaping from a large class of stars, including our sun.

RECAPITULATION.

We have now traced an outline of the marvellous history of these Arabs of the sky. We have met with outbursts upon stars sometimes of sufficient violence to shoot off part of their substance. We have found the gaseous portion sweeping through space like a net, and the vapours that accompanied it condensed into spatters that have consolidated into meteorites. We have seen this system travelling through boundless space, with nothing near it except an occasional solitary meteor, and we have seen it in the long lapse of ages slowly augmenting its cluster of these little strangers. As it wandered on it passed within the far-spreading reach of the sun's attraction, and perhaps has since been millions of years in descending towards him. Its natural course would have been to have glided round him in a curve, and to have then withdrawn to the same vast abyss from which it had come; but in attempting this, it became entangled with one of the planets, which dragged it out of its course and then flung it aside. Immediately, it entered upon the new course assigned to it, which it has been pursuing ever since. After passing the planet the different members of the group found themselves in paths very close to one another, but not absolutely the same. These orbits differed from one another very slightly in all respects, and amongst others in the time which a body takes to travel round them. Those meteors which got round soonest found themselves, after the first revolution, at the head of the group; those which moved slowest fell into the rear, and the comet was the last of all. Each succeeding revolution lengthened out the column, and the comet soon separated from the rest. Fifty-two revolutions have now taken place, and the little cloud has crept out into an extended stream, stretching a long way round the orbit, while the comet has fallen the greater part of a revolution behind. We can look forward too, and see that in seventeen centuries more the train will have doubled its length, and that ultimately it will form a complete ring round the whole orbit. When this takes place, a shower of these meteors will fall every year upon the earth, but the swarm will be then so scattered that the display will be far less imposing than it now is.

Such is the history of one of the many meteoric streams which cross the path of the earth. There are several of these streams, and no doubt the story of every one of them is quite as strange. And if there are several streams of meteors, which come across that little line in space which constitutes the earth's orbit, what untold multitudes of them must be within the whole length and breadth of the solar system! Perhaps it may even turn out that the mysterious zodiacal light which attends the sun, is due to countless hordes of these little bodies flying in all directions through the space that lies within the earth's orbit.

POSTSCRIPT.

Professor Newton's Memoirs will be found in 'Silliman's Journal' for 1864, vol. xxxvii. p. 377; and vol. xxxviii. p. 53.

The result of Professor Adams's investigations was announced in the 'Comptes Rendus' of the Academy of Sciences of Paris, of the 25th March, 1867, p. 651; and a fuller account of it will be found in the 'Monthly Notices of the Astronomical Society' for April, 1867, p. 247.

An account of Sig. Schiapparelli's contributions will be found in 'Les Mondes' for December, 1866, and the first quarter of 1867. An outline of them in English, from the pen of Professor Newton, will be found in the 'Philosophical Magazine' for July, 1867, p. 34.

M. Le Verrier's communication was made to the French Academy of Sciences, and is published in the 'Comptes Rendus' of the 21st of January, 1867, p. 94.

Mr. Stoney's paper will be found in the 'Monthly Notices of the Astronomical Society' for June, 1867, p. 271; and in the 'Philosophical Magazine' for September, 1867, p. 188.

Professor Graham's experiments are described in the 'Proceedings of the Royal Society' for May, 1867, vol. xv. p. 502, and in the 'Comptes Rendus' of the Academy of Sciences of Paris of the 27th May, 1867, vol. lxiv. p. 1067.

Professor Newton has recently delivered in America an interesting lecture on "The Relation of Meteorites to Comets," which travels over part of the same ground as the present lecture. A report of it is given in the numbers of 'Nature' for the 6th and 13th of February, 1879.

Friday, February 6, 1880.

WARREN DE LA RUE, ESQ., M.A. D.C.L. F.R.S. Secretary and Vice-President, in the Chair.

WILLIAM HUGGINS, D.C.L. LL.D. F.R.S. *M.R.I.*

The Photographic Spectra of the Stars.

IN the year 1863 my friend Dr. William Allen Miller exhibited on the screen in this room a photograph of the spectrum of the star Sirius, which we had taken the evening before in my observatory. The images of stars in the telescope had already been photographed as points, but this was the first time that their rays after dispersion by a prism had recorded themselves upon a photographic plate. For certain instrumental reasons, the photographs which we then took did not possess sufficient purity of the spectrum to give them a scientific value.

Several researches in other directions to which I subsequently devoted myself, prevented me for some years from resuming this inquiry, until a few years ago, when I took up the subject again. I purpose this evening to give an account of this recent work, and of the results which have come out of it.

Our common notion of light is limited not by the actual extent of range of the radiations of a luminous body, but by the power of our eyes to see them. Of the long range of radiations which comes from highly heated matter, the sun for example, only a small portion falls within the power of the eye. Beyond the extreme violet, where visibility ends, a great range of shorter vibrations beats upon the eye, and we know it not. So on the other side below the red all consciousness of light fails us; but here another sense, that of the feeling of heat and warmth, enables us still to know that a radiated influence from the hot body is coming upon us. These two invisibles, the ultra-violet and the ultra-red, though they cannot stimulate our eyes directly, can make themselves known to us mediately, through certain actions on other bodies.

One of these is the disturbing influence they exert on delicately balanced salts of silver, which we call their photographic power. This action was regarded as so exclusively the property of the ultra-violet portion of the spectrum, that these rays have been distinguished by the names, "chemical rays," "photographic rays." Quite recently, however, Captain Abney, by the discovery of a new molecular condition of silver bromide, has brought the whole of the other end of the spectrum, the ultra-red, within the power of the photographic plate. He has, I believe, taken the photograph of a kettle of boiling water in the dark by means of its own radiation.

This evening we shall have to do exclusively with the ultra-violet portion of the spectrum.

In the years 1865 and 1869 I had the honour to bring before this Institution the results of the observations of Dr. Miller and myself on the visible spectra of some of the stars. These eye observations embraced a range of vibrations extending from a little below C in the red to about G in the blue. The recent researches, to which I now at once proceed, begin where the eye observations ended, about G, and carry our knowledge of the stellar spectra beyond O, and in some cases beyond S, in the ultra-violet.

We shall, perhaps, underrate the importance of a knowledge of the ultra-violet spectra of stars, if we regard these photographs as simply adding so much in length to the visible spectrum, for there are reasons why a knowledge of this part of the spectrum may be of exceptional value to us.

I shall describe first, in some little detail, the instrumental methods by which the very great difficulties which present themselves in so delicate an inquiry were successfully overcome. The two principal difficulties with which the inquirer is at once brought face to face, are the feebleness of the star's light after dispersion by a prism, and the circumstance that the stars are in apparent motion, arising from the earth's rotation.

It was therefore necessary to do two things, first, to obtain a sufficiently pure and detailed spectrum with the least possible loss of light, and secondly, to devise some method by which the star's image could be kept absolutely invariable in position within a very narrow slit.

After passing the limit of the visible spectrum, the transparency of glass diminishes rapidly, until at length it becomes opaque to the rays of very high refrangibility; for this reason it was necessary to avoid altogether the use of this substance. A telescope of the reflecting form, in which the light is received upon a metallic speculum, was employed. This instrument has a speculum of 18 inches diameter. The spectrum apparatus must also contain no glass. There were two substances available, Iceland spar and quartz, both of which are very transparent to this part of the spectrum. Quartz is harder and takes a higher polish and was used for the lenses, but its dispersive power is so small that more than one prism would have been needed, introducing loss of light and other drawbacks, if this substance had been employed. Iceland spar possesses a much higher dispersive power; it is, indeed, about equal to moderately dense flint glass. One prism of this substance of 60°, which was beautifully cut for me by Mr. Hilger, was found to be sufficient for the purpose.

The apparatus is represented in this diagram (Fig. 1). It is mounted on a base plate *a* with bevelled edges, which enables it to be accurately adjusted at the end of the telescope. The prism is at *c*. The image of the star is brought upon the slit *b*. The light is rendered parallel by lens *d*; it passes through the prism, and is then, by a second lens of quartz, made to converge and form an image on the photographic plate *f*, which is inclined so as to bring a considerable part of the spectrum to focus upon the plate.

FIG. 1.

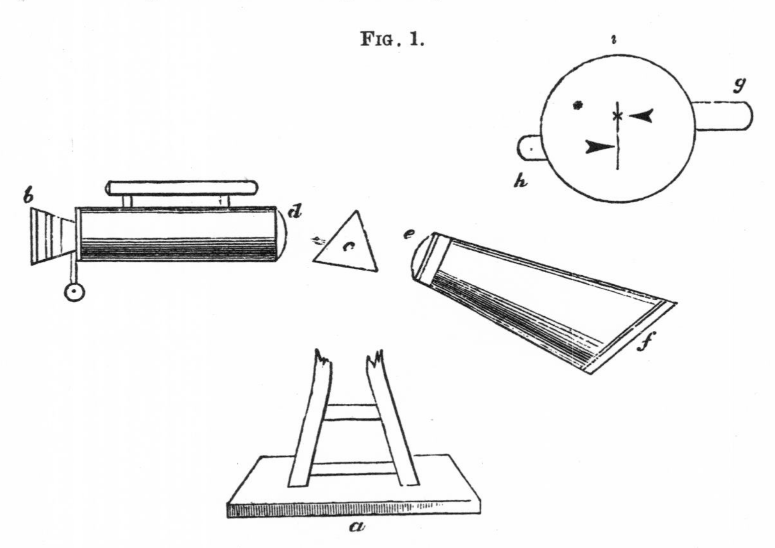

This apparatus was found to meet very satisfactorily the one primary condition of diminishing the star's light to the least possible extent compatible with obtaining a spectrum full of fine details and well defined. The photographs taken with this instrument measure not more than half an inch from G to O, and yet under suitable magnifying power seven lines can be counted between H and K.

The second important difficulty was to find a ready means of bringing the luminous point, into which the star's light is gathered up by the mirror, accurately upon any part of the very narrow chink, the $\frac{1}{350}$ part of an inch, through which the light has to enter the spectrum apparatus, and further to maintain the star's image precisely within the same part of this chink during the whole time of exposure of the photographic plate, which might be as long as one hour or even more.

The telescope was, of course, mounted upon an equatorial stand—that is, one in which the axis of motion is placed parallel to the earth's axis of rotation,—so that the telescope when kept in suitable motion by clockwork will remain invariably pointed to any star, notwithstanding that the rapidly rotating earth is carrying the telescope and the observer round with it. This clock motion is one of exceptional excellence, due to the inventive skill of Mr. Grubb, being furnished with a secondary control by means of a pendulum in electrical connection with a standard clock. But even these instrumental arrangements, although of exceptional excellence, were not delicate enough. It was found necessary to supplement them with a method of continuous supervision and control by hand.

In this diagram (Fig. 2) you have represented a portion of the reflecting telescope, which is of the Cassegrain form. The small mirror was removed, and the spectrum apparatus accurately adjusted

Fig. 2.

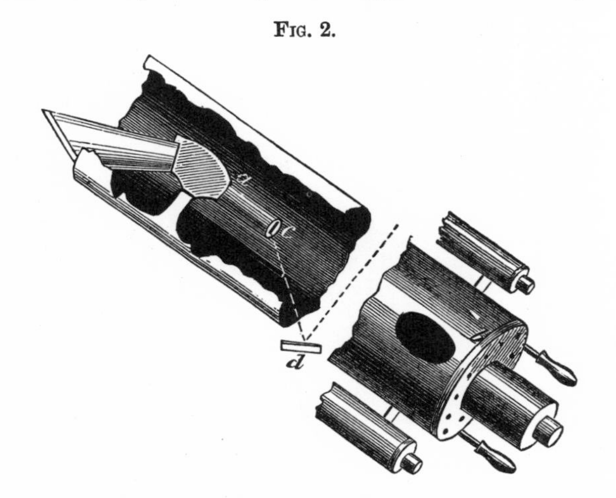

by its sliding base plate, so that the slit was brought precisely to the principal focus of the large speculum. Now over this slit is placed a highly polished silver plate *c*, with a narrow opening rather larger than the slit.

The next point was to fix on the side of the telescope a small mirror *d*, by which artificial yellow light could be thrown upon the plate. One point further. The great speculum has a central hole; now behind this, in place of the usual eyepiece, is fixed a small Galilean telescope or opera-glass.

Now if the observer directs the telescope to a star, and then looks into this small telescope, he sees before him the silver plate and the slit within the opening by means of the artificial illumination, and also at the same time the star's image as a bright point somewhere on the plate. It is then easily within the observer's power to bring the star's image exactly upon any desired part of the slit. In the figure at *i*, Fig. 1, you have represented what the observer sees. The star's image being rather larger than the width of the slit, its place, even when upon the slit, can be seen. If, therefore, the observer keeps his eye fixed upon the star's image during the whole time of exposure, half an hour, one hour, or it may be two hours, he can instantly correct by hand any small irregularities of the motion of the telescope, and so maintain the star's image invariably fixed upon the slit.

Further, it was necessary to obtain the photographs under such conditions that it should be possible afterwards to determine with accuracy the value in wave lengths of the positions in the spectrum of the stellar lines.

For this purpose the slit was provided with two small shutters, as represented at *h* and *g*, Fig. 1. One of these only remains open while the photograph of the star is taken.

When the exposure is finished this shutter is closed. The other can then be opened, and a second spectrum upon the same plate for the purpose of comparison taken. It may be the solar spectrum reflected from the moon, or the spectrum of a known star, or a terrestrial spectrum, or the apparatus may remain until the following day, and then the solar spectrum be taken upon the plate directly.

Afterwards, from these comparison spectra, by the aid of a suitable measuring apparatus attached to a microscope, the wave lengths of the stellar lines were determined. And for this purpose use was made of the excellent map of M. Cornu of the ultra-violet, and of his determinations, and those of Mascart, of the wave lengths of the lines of cadmium, aluminium, and zinc. Various photographic methods were tried, but the great sensitiveness which may be given to gelatine plates, as well as the great advantage of employing plates in a dry state, led to the exclusive use of this method of photography.

I was about to complain of how few nights sufficiently fine for this work present themselves during a whole year—they may be counted upon the fingers—but I forbear when I remember that, notwithstanding the terrible drawbacks of our climate, no country contributes more largely than our own to the advance of astronomy.

Before proceeding to the results of my work, I will endeavour to make visible to you some portion of the ultra-violet part of the spectrum.

Besides their photographic power, there is another mode of action by which the ultra-violet rays may make themselves visible to us. There are some substances which absorb these very rapid vibrations, and then give back the energy they have received, in the form of vibrations which are sufficiently long to come within the power of the eye. They transform the invisible energy into visible light. This property of fluorescence is possessed in a high degree by sulphate of quinine, and by æsculin, a substance which exists in the bark of the horse-chestnut. I have a small screen which has been brushed over with a solution of this substance.

Professor Dewar has kindly placed at my disposal one of his electric-arc crucibles. I cannot forbear congratulating Professor

Dewar on having inaugurated so fruitful a method of spectroscopic investigation. Instead of the usual optical arrangement of glass, I have substituted a lens of quartz, and a prism of Iceland spar, similar to that which I have used in my star work.

I will now ask Mr. Cottrell to throw first upon the usual screen the visible spectrum. Even now, when no glass is used, you see how brilliant are the blue and violet parts of the spectrum. The part of the spectrum we shall have to do with in the stars lies for the most part beyond. Now, if this prepared screen be held beyond, you see that the invisible energy is translated for us into characters which the eye can read. In the crucible we have the vapours of calcium and aluminium, and we now see, not merely the ultra-violet light, but the bright lines of these substances in this part of the spectrum.

I now proceed to the results which have come out of this work.

In 1865 I exhibited on the screen several coloured drawings of spectra taken from the observations of Dr. Miller and myself in illustration of the different kinds of spectra which the stars present. It is desirable that I pass three or four of them in review before I exhibit the photographic spectra corresponding to them.

The first diagram represents the spectrum of Sirius. The spectrum of this star may be taken as typical of the stars which shine with white light. Most of the photographs belong to this class of star. Very early Dr. Miller and myself called attention to the distinctive characteristics of the spectra of stars of this class. The great distinguishing feature of their spectra consists of three or four very broad and intense lines. By a method of direct comparison we found three of these lines to coincide with lines of hydrogen. The first corresponds to C of the solar spectrum, the second with solar F, and the third with a line of hydrogen near G. This last line near G appears as the first strong line of the photographic spectrum. There are, indeed, numerous very fine lines also present, but these are so delicate as to be seen, fitfully only, except under the most favourable conditions. We satisfied ourselves of the double line of sodium at D, the least refrangible of the magnesium group, and a line at E—a line of iron—and some others. This class includes the largest number of bright stars. The spectra of the different stars of this class are chiefly distinguished from each other by the greater or less breadth and diffuseness of these lines of hydrogen, and also by various degrees of strength and visibility of the finer lines.

I will now show you the spectrum of another class of stars of which the light is tinged with yellow. This spectrum is that of the star Aldebaran. This class includes our sun. In this star the lines of hydrogen are reduced to about the proportion they possess in the solar spectrum. The other lines of the spectrum are no longer fine and difficult to see. Here we have in full the triple line of magnesium. I now show the spectra of two stars of a different class indeed, but in both cases the light is orange. I will not stop to describe these spectra, but pass to one more class, the stars in the light of which the predominant colour is less refrangible. These stars are of a full red colour.

Now we return to the class of white stars. [The original photographs were exhibited on the screen.] As this photograph is a negative, of course the black lines are represented by transparent spaces and the continuous spectrum by a dark band. We shall be able, therefore, better to study the peculiarities of the spectrum if we substitute for it a positive taken by direct superposition. Here (Fig. 3) the dark and light are not reversed. The circumstance, which is so marked as to compel us to give it first attention, is the distinctly symmetrical character of this strong group of lines. When the negative is examined under suitable conditions of illumination, twelve lines may be counted. As the refrangibility increases, the lines diminish in breadth and the distance between any two lines is less as the refrangibility of the lines increases. It is also of importance to notice that the spectrum does not end with them. Beyond the last of the group of lines the continuous spectrum runs on far beyond S in the ultra-violet. The point where the group ends is between M and N.

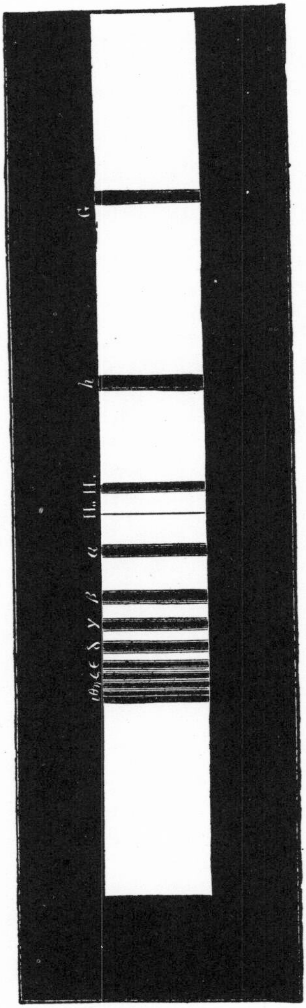

The first in order of refrangibility of these lines is the well-known line of hydrogen near G, which you saw in the visible spectrum of the star. The second of these lines is also a line of hydrogen, coincident with h of the solar spectrum. The next line coincides in position with the strong line H of the solar spectrum. But where is H_2 or K? It is represented by this very thin line, which is barely recognziable. You remember how narrow a slit was used, and that if this were a photograph of the solar spectrum, some seven lines or more would be clearly visible in this space. We shall now be able to

FIG. 3.

study the spectrum better by a reference to this diagram (Fig. 4), in which the lines are put down according to their wave lengths. The two lines H and K, as is well known, coincide with two lines of calcium, and we attribute them usually to the vapour of that substance.

FIG. 5.

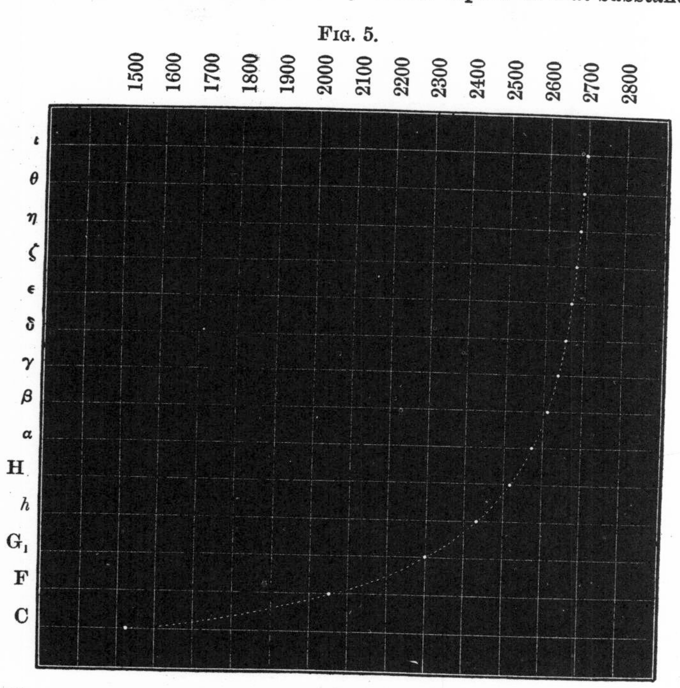

The remarkable behaviour of these lines in the stars was pointed out by me at the end of 1876. A few months previously, Mr. Lockyer suggested that photographs of the brighter stars might show a modification of this character, and that if such were the case, it would support his view as to the dissociation of the vapour of calcium in the hottest stars. In a subsequent paper to the Royal Society Mr. Lockyer explained in more detail his views of the dissociation of the terrestrial elements, and also the bearing of his views in connection with the different kinds of visible star spectra. I wish also here to acknowledge the kindness of Professor Dewar, who permitted me to witness the experiments conducted by himself and Professor Liveing. I saw the lines of calcium corresponding to H and K in the emission spectrum of that substance vary in relative brightness until for a moment the line corresponding to H alone remained.

Are this thick line of the star spectrum and the thin line at K really due to calcium?

Now, beyond these lines there is another pair of strong lines in the calcium spectrum more refrangible. The question arises, are these represented in the star's spectrum by any strong lines? They are not. Two strong lines are near them, but do not coincide with them.

Again, in the photographic spectrum of hydrogen there is a line at the position of H. It is seen in Mr. Capron's photographic spectra. It is also present in my own photographs. Mr. Lockyer called attention to the coincidence of this line with H in December, 1879.

In this diagram the photographic spectrum is completed by the addition of the strong lines in the visible one, and laid down according to the scale of wave lengths. Of these the first four are certainly lines of hydrogen; the fifth, H, coincides both with calcium and hydrogen. I suspect in some of my photographs of hydrogen fine lines at the positions of several of the more refrangible lines of the stellar group.*

That all the lines of this remarkable group are members of a common physical system becomes very highly probable indeed, if we convert the wave lengths into their reciprocals, the wave frequencies, and then plot them down, as is done in this diagram (Fig. 5). It then becomes evident that they lie on, or very near, a definite curve, a state of things which we could not suppose to happen by chance. Mr. Johnstone Stoney, in a letter to me, remarks on this point:—

"The question whether they lie actually or only near a definite curve is, if I mistake not, of great significance in the theory. If they lie on a curve, obeying any exact mathematical law, their connection must, I think, be attributed to their corresponding to the consecutive overtures of some vibrating system. If, on the other hand, they lie near but not on the curve, this circumstance would support the hypothesis (which seems to accord with other facts) that the visible lines are members of a long series of harmonics, most of the members of which are invisible, those which are seen being those whose positions chance nearly to fulfil a definite state of things which I have

* Since this discourse was given, Dr. H. W. Vogel has called my attention to a paper on the "Spectrum of Hydrogen" in the 'Monatsbericht der Königl. Acad. der Wissenschaften zu Berlin,' July 10, 1879. In this paper Vogel pointed out coincidence of a line of hydrogen with H; and among the lines given by him are three others which agree with stellar lines. Vogel's wave lengths for these lines are—

3968 H	3834
3887	3795

The wave lengths of the twelve typical lines are—

H near G	4340	δ	3767·5
h	4101	ε	3745·5
H	3968	*s*	3730
α	3887·5	η	3717·5
β	3834	θ	3707·5
γ	3795	ι	3699

February 28, 1879.

shown to exist in some acoustic arrangements, and which where it exists exalts the intensity of the harmonics whose positions nearly fulfil the requisite condition. I converted the wave lengths into wave frequencies. . . . I think it must be accepted that the lines do not lie on, but near a definite curve. This appears to be corroborated by finding that H_1 and G_1 (hydrogen line near G) are connected harmonially, these rays being exactly the 35th and 32nd harmonies of a vibration whose fundamental is $\frac{\tau}{72 \cdot 003}$ (τ being the time in which light travels a millimetre in air) . . . ."

Under these circumstances one is led to regard the whole series of lines as due to hydrogen. In this connection it may be stated that Messrs. Dewar and Liveing find that the line of calcium K is more easily reversed than the line at the position of H.

This spectrum of Vega may be taken conveniently as typical of the whole class of white stars, so that in our consideration of the other stars of this class we shall consider the distinctive features peculiar to each, as modifications, or departures, from this common typical form. To facilitate these comparisons I have distinguished the typical lines by the letters of the Greek alphabet, beginning with the line more refrangible than H.

In this map (Fig. 4) I have arranged the spectra of five other stars of the white group in their order of change, approximately at least, from the spectrum of Vega.

I will point out some of the directions in which these changes show themselves, and I will then exhibit upon the screen the photographs themselves of these stars.

There are principally three directions in which the changes take place:—

1. In the breadth and greater or less marginal diffuseness of the typical lines.

2. In the presence or absence of K, and, if present, in its breadth and intensity relatively to H.

3. In the number and distinctness of other lines in the spectrum.

Now in these stars we see modifications in these three directions, a successive diminution of breadth of the typical lines, and of the nebulosity at the edges; the lines become at the same time narrower and defined at the edges.

In Sirius the lines are about the same thickness as in *α* Lyræ, and the line corresponding to K of about the same fineness.

In the next star, *α* Ursæ Majoris, we have the same typical group, but the lines are less broad and rather more defined at the edges. There is no fine line at the position of K, but some other lines make their appearance.

The star next in order is *α* Virginis. Here the typical lines are still narrower and more defined. K is stronger relatively to H, and numerous lines are visible beyond the last of the typical group.

In the spectrum of *α* Cygni the typical lines are still narrower

Fig. 4.

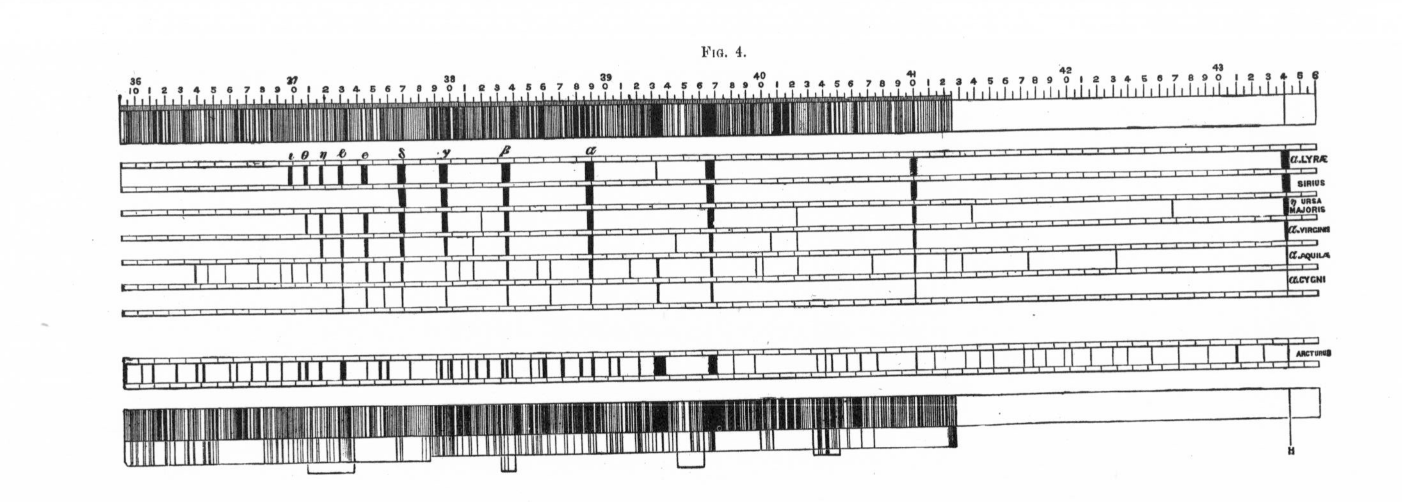

36
37
38
39
40
41
42
43
ι θ η ζ ε δ γ β α
α LYRÆ
SIRIUS
η URSA MAJORIS
α VIRGINIS
α AQUILÆ
α CYGNI
ARCTURUS
H

and more defined. The line at K is nearly as broad as H, and there are other lines present.

In the last spectrum of the map, that of Arcturus, we come to that of a star of another order, which includes the solar type, but this star appears to be further removed than the sun is, in the order of change from the typical form, as we meet with it in Vega and Sirius. Here the typical lines are no longer present as a strong group. The line at K is stronger relatively to H than it is in the solar spectrum. The spectrum is crowded with fine lines, and in the visible part resembles the solar spectrum, but beyond H the lines are more intense and differently grouped.

We cannot resist the feeling that we have here to do with a star which has departed farther from the condition in which Vega now is than our sun has yet done.

The question presents itself—Have we before us stars of permanently different orders, or have we to do with some of the life-changes through which all stars pass?

Does the sun's position somewhere before Arcturus in the order of change indicate also his relative age?

On these points we know nothing certainly. If I may give some play to the scientific use of the imagination, I would ask you to imagine an inhabitant from some remote part of the universe seeing for the first time an old man with white hair and wrinkled brow, to ask, Was he born thus? The answer would be, No; in this child, this youth, this man of mature age, you see some of the life-changes through which the old man has passed. So, giving play to the scientific imagination, there may have been a time when a photograph of the solar spectrum would have presented the typical lines only which are still in Vega. At a subsequent period these would have been narrower and more defined, and other lines would have made their appearance. And if we allow this scientific imagination to project these Friday evenings into the far future, the lecturer, clad it may be in the skin of a white bear, may have to describe how the spectrum of the then feeble sun has already passed into the class of spectra which now distinguishes the stars which shine with red light.

There remain only two other points. In 1865 I described the method of observing the spectrum of a planet compared directly with the solar spectrum under *similar conditions* of terrestrial atmosphere. The planet is observed in the early evening, when the light from the sky is bright enough to give a spectrum. With a long slit one sees a broad spectrum of the sky, and then upon it the brighter spectrum of the planet. Making use of this method, spectra were taken of the planets Venus, Mars, and Jupiter.

I will now exhibit upon the screen the spectrum of Venus. This broad spectrum is that of the light from the sky. The narrow stronger spectrum is that of the planet Venus. You see line corresponds to line, and that there are no modifications or additions which indicate a planetary atmosphere.

The same is true of the planets Mars and Jupiter. These two last-named planets do show indications of atmospheric absorption in the visible part of the spectrum. Similar photographs taken of different small areas of the moon under different conditions of illumination are negative as to any lunar atmosphere. It must not be supposed that such observations are necessarily antagonistic to the existence of a lunar atmosphere. They simply tell us nothing as to its existence.

There are many other directions in which the photographic arrangements I have described may be doubtless successfully employed. I hope to photograph any lines that may exist in the ultra-violet part of the spectra of the gaseous nebulæ. The apparatus will give us the spectra of different portions of a sun-spot. It may enable us to determine the difference of velocity in the line of sight of two stars; and also we may record by it the sun's rotation by the alteration in refrangibility of the lines of the spectra of opposite limbs.

One of the great charms of the study of Nature lies in the circumstance that no new advance, however small, is ever final. There are no blind alleys in scientific investigation. Every new fact is the opening of a new path. As the description of a first step in a new and broad highway, I venture to hope the last hour's discourse has not been wholly wanting in interest.

[W. H.]

Friday, February 11, 1881.

WILLIAM BOWMAN, Esq. F.R.S. Vice-President, in the Chair.

ROBERT S. BALL, Esq. LL.D. F.R.S.
ASTRONOMER ROYAL OF IRELAND.

The Distances of the Stars.

EVERY one who is acquainted with the rudiments of astronomy knows that the sun with its attendant planets is merely an island group in the vast realms of space.

An island the size of this room in the middle of the Atlantic would be over a thousand miles from the coasts of Europe and America on either side. Yet that island would not be more remotely apart from the surrounding shores than is our solar system from the bodies which surround it in space. To determine the distance from this solar island to the stars which surround it, is the problem for our consideration to-night.

Recent Researches on 61 *Cygni.*

It is now almost exactly forty years (February 12, 1841) since the gold medal of the Royal Astronomical Society was awarded to Bessel for his discovery of the annual parallax of 61 Cygni. On that occasion Sir John Herschel delivered an address, in which he glanced at the labours of Struve and Henderson as well as Bessel. The discovery of the distances of the stars was alluded to as "the greatest and most glorious triumph which practical astronomy has ever witnessed." From this date the history of our accurate knowledge of the subject may be said to commence. Each succeeding race of astronomers takes occasion to investigate the parallax of 61 Cygni anew, with the view of confirming or of correcting the results arrived at by Bessel.

[The parallactic ellipse which the stars appear to describe, having been briefly explained, the method of deducing the distances of the stars was pointed out.]

The attention of Bessel was directed to 61 Cygni by its proper motion of five seconds per annum. When Bessel was at his labours in 1838, the pair of stars forming the double were in the position indicated on Fig. 1. When O. Struve undertook his labours in 1853 the pair of stars forming 61 Cygni had moved considerably, as is shown on the figure. Finally, when the star was observed at Dunsink in 1878, it had made another advance in the same direction as before. In forty years this object had moved over an arc of the heavens upwards of three minutes in length.

The diagram contains four other stars besides the three positions of 61 Cygni. These are but small telescopic objects, they do not parti-

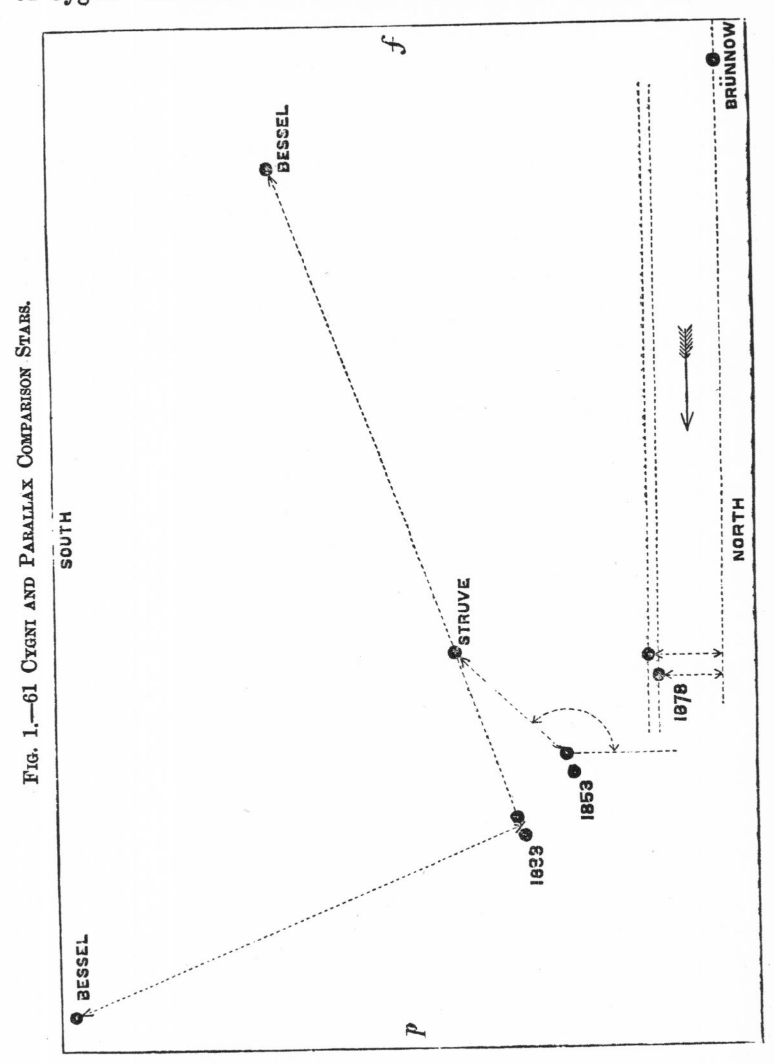

FIG. 1.—61 CYGNI AND PARALLAX COMPARISON STARS.

cipate in the large proper motion of 61 Cygni, and they may be presumed to be much more remote from us. Bessel chose as the comparison stars the two objects marked with his name. He measured the distance from the central point of 61 Cygni to each of the two comparison stars. From a series of such measures he discovered the parallactic ellipse of 61 Cygni. He was led to the same ellipse by each of the two comparison stars.

Fifteen years latter (1853) Struve undertook a new determination. He chose a comparison star different from either of those Bessel had used. Struve's method of observing was also quite different from Bessel's. Struve made a series of measures of the distance and position of the comparison star from 61 (B) Cygni. Struve succeeded also in measuring the parallactic ellipse.

There was, however, an important difference between their results. The distance, according to Bessel, was half as much again as Struve found. Bessel said the distance was sixty billions of miles; Struve said it could not be more than forty billions.

The discrepancy may be due to the comparison stars. If Bessel's comparison stars were only about three times as far as 61 Cygni, while Struve's star was about eight or ten times as far, the difference between Struve's result and Bessel's would be accounted for.

To settle the question, observations were subsequently made by Auwers and others; the latest of these investigations is one which has recently been completed at Dunsink Observatory.

Dr. Brünnow commenced a series of measures of the difference in declination between 61 Cygni and a fourth comparison star. The carrying out of this work devolved on the lecturer, as Dr. Brünnow's successor. Two series of observations have been made, one with each of the components of 61 Cygni. The results agree very nearly with those of Struve.

On a review of the whole question, there seems no doubt that the annual parallax of 61 Cygni is nearer to the half second found by Struve, than to the third of a second found by Bessel.

To exhibit the nature of the evidence which is available for the solution of such a problem, a diagram showing the observations has been prepared. In the accompanying Fig. 2 the abscissæ are the dates of the second series of observations made at Dunsink. The ordinates indicate the observed effect of parallax on the difference of declinations between 61 (B) Cygni and the comparison star. Each dot represents the result of the observations made on the corresponding night. The curve indicates where the observations should have been with a parallax of $0''\cdot47$. The discordances seem in many cases very considerable. They are not, however, intrinsically so great as might perhaps be at first thought. The distance from the top of the curve to the horizontal line represents an angle of four-tenths of a second. This is about the apparent diameter of a penny-piece at the distance of ten miles. The discordance between the observations and the curve is in no case much more than half so great. It therefore appears that the greatest error we have made in these observations amounts to but two or three tenths of a second. This is equivalent to the error of pointing the telescope to the top edge of a penny-piece instead of to the bottom edge when the penny-piece was fifteen or twenty miles off.

FIG. 2.—PARALLAX IN DECLINATION OF 61 (B) CYGNI.

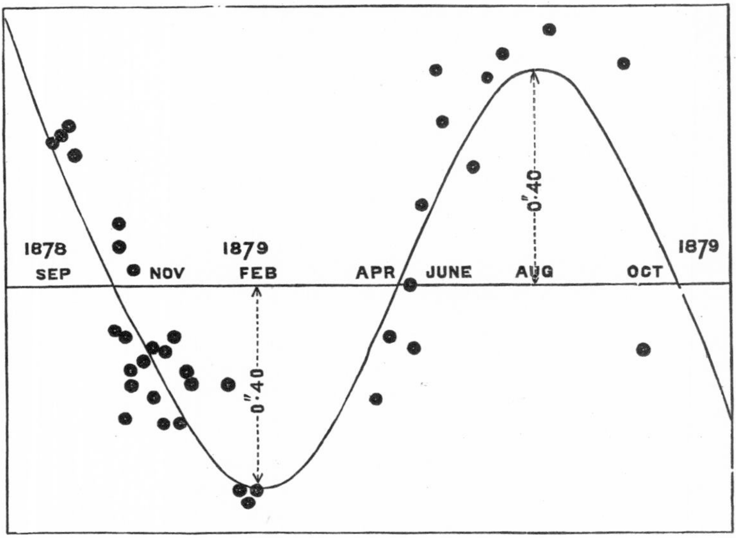

Ordinates indicate parallax.
Dots indicate observations.

Still, however, the entire quantity to be measured is so small that the errors, minute as they are, bear a large proportion to the parallax. In this lies the weakness of such work. By sufficiently multiplying the numbers of the observations, and by discussing them with the aid of the method of least squares, considerable confidence may be attached to the results.

Groombridge 1830.

This star has been the subject of much parallax work. It has a proper motion of seven seconds annually. Mr. Huggins or Mr. Christie could perhaps ascertain by the spectroscope what its motion may be in the line of sight. From the theory of probabilities it may not improbably be nine seconds. We shall, however, take it at seven seconds. The parallax has been determined by Struve and by Brünnow. It is very small, being one-tenth of a second. The actual velocity of 1830 Groombridge must therefore be 70 radii of the earth's orbit per annum, or 200 miles per second.

Newcomb has employed this result to throw light on the question as to whether all our stars form one system. If an isolated body in our system is to remain there for ever, the theory of gravitation imposes the imperative condition, that the velocity of the body must not exceed a certain amount. Assuming that the stars are 100,000,000 in number, and that each star is five times as large as the sun, assuming also that they are spread out in a thin layer of such dimensions that a ray of light takes 30,000 years to pass it, Newcomb shows that the critical velocity is 25 miles per second.

As this is only the eighth-part of the velocity of Groombridge 1830, we are thus led to the dilemma that either the masses of the bodies in our system must be much greater than we have supposed, or Groombridge 1830 is a runaway star, which can never be controlled and brought back.

Search for Stars with Parallax.

The lecturer has been engaged for some years at Dunsink Observatory in a systematic search for stars which have an appreciable parallax. Up to the present about three hundred stars have been examined. In the majority of cases each of these stars has been observed only twice. The dates of the observations have been chosen so as to render the effects of parallax as manifest as possible. It is not of course expected that a small parallax of a few tenths of a second could be detected by this means.

The errors of the observations would mask any parallax of this kind. It seems, however, certain that no parallax could have escaped detection if it at all approached to that of α Centauri.

The stars examined have been chosen on various grounds. It had been supposed that some of the red stars were possibly among the sun's neighbours, and consequently many of the principal red stars were included in our list. No appreciable parallax has, however, been detected in any of the red stars up to the present. Many of the principal double stars are also included in the list. Other stars have been added on very various grounds, among them may be mentioned the Nova, which some time ago burst out in the constellation Cygnus, and dwindled down again to a minute point. The earth's orbit, however, does not appear any larger when seen from Nova Cygni than from any of the other stars on our list.

Groombridge 1618.

We have, however, found one star which seems to have some claim to attention as one of the sun's neighbours. The star in question is Groombridge 1618. It lies in the constellation Leo, and is 6·5 magnitude. Groombridge 1618 has a proper motion of $1''\cdot 4$ annually. From a series of measurements of its distance made on fifty-five nights from a suitable comparison star, the parallax of Groombridge 1618 appeared to be about one-third of a second. As this seemed to be a result of considerable interest, measures were renewed for a second series of forty nights. The result of the second series confirms the first. Measurements of the position angle were also made at the same time. Some difficulties not yet fully explained have arisen, but on the whole the measurements of the positive angle seem to confirm the supposition that the parallax of Groombridge 1618 is about one-third of a second. No doubt this is but a small quantity. The orbit of the earth viewed from Groombridge 1618 is about the same size as a penny-piece at the distance of seven miles.

Proper Motions of the Stars.

Geologists have made us acquainted with the enormous intervals of time which have elapsed since the earth first became the abode of living animals. Regarding a period of 50,000,000 of years as comparable with geologic time, some considerations were adduced as to the effect of proper motions during such an interval. It was pointed out that in all probability none of the stars now visible to the unaided eye can have then been visible from the earth.

The Nature of Space.

The possible connection of parallax work with the problems of the nature of space was then alluded to. It was shown that if space be hyperbolic the observed parallax is smaller than the true parallax, while the converse must be the case if space be elliptic. The largest triangle accessible to our measurements has for base a diameter of the earth's orbit, and for vertex a star. If the defect of such a triangle be in any case a measurable quantity, it would seem that it can only be elicited by observations of the same kind as those which are made use of in parallax investigations.

[R. S. B.]

Friday, January 20, 1882.

GEORGE BUSK, Esq. F.R.S. Treasurer and Vice-President, in the Chair.

WILLIAM HUGGINS, Esq. D.C.L. LL.D. F.R.S.

On Comets.

IN the olden time comets were looked upon as the portents of all kinds of woe. In the words of Du Bartas, as rendered by Sylvester* :—

> "There with long bloudy haire, a Blazing Star
> Threatens the World with Famin, Plague and War:
> To Princes, death: to Kingdoms, many crosses:
> To all Estates, ineuitable Losses:
> To Heard-men, Rot: to Plough-men, hap-lesse Seasons:
> To Saylers, Storms: to Cities, ciuill Treasons."

In the past year, including telescopic comets, no fewer than seven of these blazing stars have threatened us, and certain contemporary events might seem, indeed, to justify this view of their malign influence. But though comets are no longer a terror to us, they are still, in some respects, a great mystery. When we attempt to explain the marvellous phenomena they present, by the rigid application of the laws of physics, we find ourselves confronted by prodigious difficulties. We are almost led to think that in the heavens, at least, there is more "than is dreamt of in our philosophy"—some profound and still unknown mystery of nature. Not to mention the many absurd theories which are heard on all sides when a great comet appears, on no phenomenon of nature have we so many guesses at truth by the masters of science on different and even opposing principles of explanation. At the present moment there is no consensus of opinion as to the nature of comets.

But within the last few years, from two opposite directions—from the use of the spectroscope and from mathematical investigation applied to the periodical displays of shooting stars—much knowledge has been gained of their nature, though there are still many points on which we can only speculate.

It is my purpose this evening to give chief prominence to the new knowledge which these two methods of research have placed within our reach, and to distinguish as clearly as may be possible between what we know about comets and what is not more than speculation.

To carry out this purpose we must first study carefully the phe-

* Du Bartas, translated by J. Sylvester, fol. 1621, p. 33.

nomena which have to be explained, namely, the essential appearances and changes which comets present during their approach to the sun, at the time that they are visible to us.

It is not necessary here to describe in detail the more purely astronomical side of the subject. It will be sufficient to say, in two words, that some comets have become permanent members of our system, while others probably visit us once only, never to return. It depends upon a comet's velocity whether its course shall be a hyperbola, a parabola, or an ellipse. In the latter case only can it become a permanently attached member of our system. If the velocity of the comet when at the earth's distance from the sun exceeds 26 miles a second, the comet must go off into space, never to come back to us. If the comet is moving less swiftly its path will return into itself, and it will visit us periodically after longer or shorter wanderings. In the case of many comets, including the brightest comet of last year, their velocity is so near the parabolic limit that it is scarcely possible, from observations made in the small part of their orbit near the sun, to be quite sure whether they will return to us or not. A number of comets, chiefly small ones, are certainly periodic, and of some comets several returns, true to the calculated time, have been observed.

The small portion of the comet's life during which we are able to study it is quite unlike its ordinary humdrum existence. It consists of the short period of extreme excitement into which it is thrown by a more or less near approach to the sun—a state of things which is accompanied by rapid and marvellous changes, often on a stupendous scale.

The appearances which comets put on under the sun's influence differ widely from each other. A few of these forms, passing from an almost invisible nebulosity up to a brilliant comet of the grand type, are represented on these diagrams. In nearly all these forms three essentially distinctive parts may be seen.

1. *The nucleus.* With the aid of a telescope, in the heads of most comets a minute bright point may be found. This apparently insignificant speck is truly the heart and kernel of the whole thing—potentially it is the comet. It is this small part alone which conforms rigorously to the laws of gravitation, and moves strictly in its orbit. If we could see a great comet during its distant wanderings when it has put off the gala trappings of perihelion, it would be a very sober object, and consist of little more than nucleus alone. It is only this part of the comet which can have any claim to solidity, or even appreciable weight. Though many of the telescopic comets are of extremely small mass, nucleus included—so small, indeed, that they are unable to perturb such small bodies as Jupiter's satellites—yet in some large comets the nucleus may be a few hundred miles in diameter, and may consist of solid matter. I need not say that the collision of a cometary nucleus of this order with the earth would be fraught with danger on a very wide scale.

2. *The coma.* This appears usually as a luminous fog surrounding the nucleus, and gradually shading off from it. The nucleus and the coma form together the head of the comet.

3. *The tail.* The tail may be considered as a continuation, in a direction opposite to that of the sun, of the luminous fog of the coma. This appendage may be scarcely distinguishable as a slight elongation of the coma, or it may extend half across the heavens, and be many millions of miles in length. The tail may be single, or composed of several branches.

We must now study more closely the cometary appearances as they may be seen when a large telescope is directed to a brilliant comet. I have selected for this purpose the Great Comet of 1858, and I shall exhibit on the screen a series of views of this comet, taken at intervals of a few days. The first set shows the growth, position, and forms of the tail, as a whole. The second group represents the more detailed structure and changes of form of the head of the comet, as viewed in a large telescope. These views are, of course, from sketches made at the telescope. Last year several attempts were made to photograph the comet which appeared in June. Mr. Janssen has kindly sent me a positive taken from the original negative. It is now upon the screen. Mr. Janssen purposely sacrificed detail in the head of the comet, for the sake of obtaining the structure and form of the tail, exposing the plate for thirty minutes. From a careful examination of several similar negatives, Janssen made a drawing of the comet. A photograph of this drawing is now upon the screen. Mr. Common, at Ealing, with a fine three-foot reflector of his own construction, also photographed the comet, but his object, different from that of Janssen, was to get the form of the nucleus. For this purpose he gave an exposure of only ten minutes—far too short to obtain an impression of the tail. The comet was also photographed by Dr. Draper, of New York. My own work was confined to the comet's spectrum, of which I shall speak presently.

We may now advance to the consideration of two primary questions:—

1. Does a comet shine wholly by reflected solar light, or has it also light of its own?

2. Of what materials is a comet composed?

The spectroscope has furnished us with information on both these points. The first successful application of the spectroscope to a comet was in 1864, when Donati discovered in its light three bright bands. In 1866 I was able to distinguish two kinds of light from a telescopic comet—the one kind giving a continuous spectrum and presumably solar light, and the other a spectrum of three bright bands, similar to those which had been seen by Donati. But in 1868 a great advance was made. The close agreement of measures I took of the bands of the comet *b* of that year with those I had previously taken of the spectrum of certain compounds of carbon led me to compare, directly, in conjunction with my friend Dr. W. Allen Miller, the spectrum of the induction spark in olefiant gas with the comet's spectrum, in the manner shown upon the screen.

The next diagram shows the result of this direct comparison.* There could be no longer any doubt of the oneness of chemical nature of the cometary stuff with the gas we were using, in fact, that carbon, in some form or in some state of combination, existed in the cometary matter. From that time some twenty comets have been examined by different observers. The general close agreement, notwithstanding some small divergencies, of the positions of the three bands with those seen in the flame spectrum of hydrocarbons, leaves no doubt whatever that the original light of comets is really due to matter containing carbon in combination with hydrogen.

At first, indeed, for certain reasons, I was led to consider this spectrum to be that of carbon itself in the form of gas, a view still held by some physicists; but subsequent researches by several experimentalists on this point appear to me to be strongly in favour of carbon combined with hydrogen.†

Last year another advance was made. For the first time since the spectroscope has been in the hands of the astronomer the coming of a bright comet made it possible to extend this mode of research into the more refrangible region of the spectrum. Making use of the apparatus and arrangements which I employed for photographing the spectra of stars,‡ I succeeded in obtaining a photograph of the spectrum of the head of comet *b*.

A copy of this spectrum is now upon the screen (see plate). There is a continuous spectrum which can be traced from about G to beyond K, in which are seen distinctly several of the Fraunhofer lines, G, *h*, H, K, and many others. The presence of these lines was crucial, and made it certain that this continuous spectrum was really due to reflected solar light.§

But there was also present a second spectrum consisting chiefly of two groups of *bright* lines. These evidently were due to the same light which is resolved, in the visible region, into the three bright groups.

I regarded them with intense interest, for there was certainly hidden within these hieroglyphics some new information for us. Measures of their position in the spectrum, taken under the microscope, brought out that these groups were undoubtedly the same

* These hydrocarbon groups may be seen with a pocket spectroscope in the blue base of a candle flame or in the flame of a Bunsen burner.

† Among many papers, I may refer to 'Ueber die Spectra der Cometen,' Dr. Hasselberg, Mem. Acad. des Sciences, St. Pétersbourg, vii. ser. tome xxviii. No. 2; and papers by Professors Liveing and Dewar and by Mr. Lockyer in recent volumes of the 'Proceedings of the Royal Society.'

‡ 'Trans. R. S.' 1880, part 2, p. 671. 'Proceedings R. Instit.' vol. ix. part 3, p. 285.

§ See observations of the visible spectrum of this comet by Professor Young, the Astronomer Royal, Professor Vogel, Professor Wright, Dr. Von Konkoly, Dr. Hasselberg, and others.

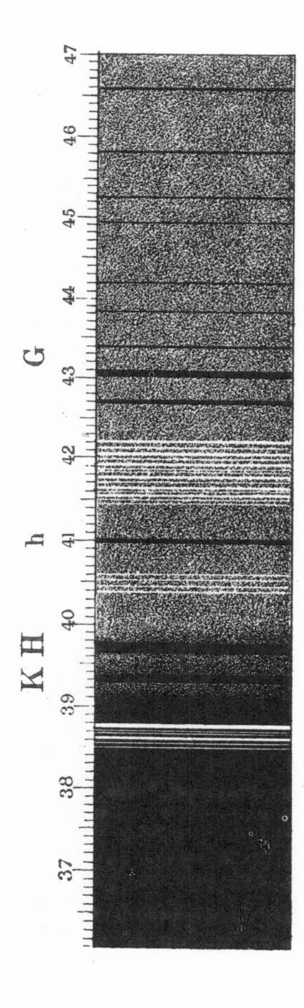

Spottiswoode & C° Lith. London.

which appear in certain compounds of carbon. Professors Liveing and Dewar had recently shown that these groups indicate a *nitrogen* compound of carbon, namely, cyanogen. On this view there must be in the cometary matter, besides carbon and hydrogen, the element *nitrogen*.

A few days after my photograph was taken, Dr. Draper succeeded in obtaining a photograph of the comet's spectrum, which appears to confirm mine so far as the bright lines, but does not give the Fraunhofer lines.

About the same time that the observations were made on the comet, Professor Dewar succeeded in confirming his results, by the reversal of the groups, employing either titanic cyanide or boron nitride.

The positions and characters of these bands, together with those in the visible spectrum, leave no doubt that the substances, carbon, hydrogen, and nitrogen, and probably oxygen, are present in the cometary matter, and that this light-emitting stuff appears to be essentially of the same chemical nature for all the comets, some twenty, which have been observed up to the present time. Certain minor modifications of the common type of spectrum are often present, and show, as was to be expected, that the conditions prevailing in different comets, and indeed in any one comet from day to day, are not rigidly uniform.

The temperature, the state of tenuity, the more or less copious supply from the nucleus of the gaseous matter, must be subject to continual variation. At times it is probable that the hydrocarbon spectrum is complicated by traces of the spectrum of the oxygen compounds of carbon. These and other possible variations betray themselves to us in the spectrum, by the length of range of refrangibility through which each group can be traced, by an alteration in the position of maximum brightness in the groups, by the relative brightness of the groups, by a more or less breaking up of the shaded light of the bands and the visibility or otherwise of bright lines, by a more or less distinctness of the violet group, and, lastly, by the visibility in the brightest comet of last year of a less refrangible band of the hydrocarbon spectrum which occurs between C and D of the spectrum.*

We must now consider the information about the nature of comets which has come to us from a wholly different source.

On almost any fine night, after a short watch of the heavens, we shall see the well-known appearances of "shooting stars." At ordinary times, these are small, and appear indifferently in all parts of the heavens, but on certain nights they show themselves in great numbers, and of such brilliancy as to present a spectacle of much magnificence. On such occasions one remarkable feature presents

* For these reasons measures of these bands should be considered as strictly applicable to the particular comet at the time of observation only, and not necessarily as applicable to other comets.

itself, which is well marked in the diagram on the screen. The meteors all shoot forth from one spot, which is called the radiant point. A little consideration will show that this appearance is really due to perspective, and represents the vanishing point of the parallel courses in which the meteors are moving. Hence we learn that they all belong to an enormous swarm of these bodies which the earth is meeting, and further, we may find the direction in which the swarm is moving relatively to the earth. Now the researches of Olbers, H. A. Newton, and Adams showed that the November shower is really a planetary swarm, revolving round the sun in about $33\frac{1}{4}$ years. Further investigations of Schiaparelli, Leverrier, and Oppolzer brought out the astonishing result that the path of the November meteors is really identical with that of a comet discovered by Tempel in 1865. Schiaparelli showed further, that another independent group of meteors which appears in August, has an orbit identical with the third comet of 1862. We are thus led to see the close physical connection, and oneness of origin, if not indeed identity of nature, of comets and of these meteors. Now the meteors on these occasions are too minute to pass through the ordeal of ignition by our atmosphere, they are burnt up before they reach the earth, but at other times small celestial masses come down to us, which, there can be little doubt, are of the same order of bodies, and similar in chemical nature. The meteorites we have in our hands, contain matter of the same kind probably as that which gives rise to cometary phenomena. These two small meteorites, which fell at Estherville, were kindly sent to me by Professor Newton, as probably good examples of the sort of stuff of which the nuclei of comets are composed. The question arises, are the revelations of the spectroscope about comets in harmony with what we know of the chemical nature of these celestial waifs and strays?

Meteorites may be arranged in a long series, passing from metallic iron alloyed with nickel at one extremity, to those of a stony nature, chiefly silicates, at the other. In meteorites more than twenty of the elementary bodies have been found, including hydrogen, carbon, and nitrogen, which the spectroscope has shown to be in comets. It may be, however, that in the sun's action on comets, we have to do not with the decomposition of the cometary matter, but with the setting free of gases occluded within the meteoric matter, forming the comet's nucleus. If the meteoric matter were decomposed, we should expect a more complicated spectrum.

In the year 1867 Professor Odling, lecturing on Professor Graham's researches, lighted up this room with the gas brought by a meteorite from celestial space. This meteorite, of the iron type, yielded nearly three times its volume of gas, of which 85 per cent. was hydrogen, 5 per cent. was carbonic oxide, and 10 per cent. nitrogen. Since that time Professor A. W. Wright has experimented with a meteorite of the stony type, containing, however, numerous very small grains of metallic iron and sulphide of iron scattered through the mass. This meteorite gave off about two and a half times the volume of the meteorite as a whole, or twenty times that of the iron scattered within it. The same gases came off, but in a different proportion; there being a larger proportion of the oxide of carbon, at a low temperature carbon dioxide was chiefly given off.* Now in all these cases, a spectrum similar to that of comets would be given by these gases under suitable conditions.

Some years ago, in conjunction with my friend Professor Maskelyne, I examined the spectra of certain meteorites, and obtained in several cases a spectrum similar to that of comets. Some meteorites like that of Bokkveldt, contain a large percentage of hydrocarbons. Professor Vogel has recently experimented in the same direction, and finds that the gas which comes off from the meteorite he used gives a hydrocarbon spectrum mixed with that of carbonic oxide, and under certain conditions the spectrum of hydrocarbon predominates and becomes almost exactly similar to that of comet *b* 1881.† We are at a disadvantage in one particular, for we cannot get at meteorites as they exist in celestial space, but only after superficial ignition in passing through the air.

The experiments hitherto made throw but little light on the question, whether cyanogen ready formed is present in combination or otherwise in the comet, or whether it is formed at the time by the interaction of carbonaceous and nitrogenous matter. In the latter case we should have to admit a high temperature, which would be in favour of the view of an electric origin of the comet's light. Professor A. Herschel and Dr. Von Konkoly have pointed out that the spectra of the periodic meteors are different for different groups. I may also mention that Captain Abney considers that he has evidence of hydrocarbons in the outer portion of the sun's atmosphere.

---

* 'American Journal of Science and Arts,' vol. x. July 1875.

† 'Publicationen des Astrophysikalischen Observatoriums,' Band ii. p. 182.

Since this Discourse was given, Dr. Flight has presented to the Royal Society a paper on the Meteorite of Cranbourne, Australia, and the Rowton Meteoric Iron. In the case of the former, the occluded gases amounted to 3·59 the volume of the iron, and consisted of—

Carbonic acid	0·12
Carbonic oxide	31·88
Hydrogen	45·79
Marsh gas	4·55
Nitrogen	17·66
	100·00

The Rowton Iron gave 6·38 times its bulk of gas, as follows—

Carbonic acid	5·155
Hydrogen	77·778
Carbonic oxide	7·345
Nitrogen	9·722
	100·000

We have now advanced to the extreme boundary of the solid ground of our knowledge of comets. Before us lies the enchanted region of speculation. Without being too venturesome, we may well consider a few points which may explain more in detail some of the phenomena of comets. Of whatever nature we may regard the tremendous changes which take place in them to be, we must certainly look for the primary disturbing cause to the sun. Is the solar heat sufficient to account directly for the self-light of comets, or does it act the part of a trigger, setting free chemical or electrical forces? On this point of the sufficiency of the solar radiation we must not look to the few cases of exceptionally close approach to the sun, but to the more average distance of comets at perihelion. Professor Stokes has suggested that some results obtained by Mr. Crookes may throw light upon this question. He concluded from his experiments that in such vacua as exist in planetary space the loss of heat, which in such cases would take place only by radiation, would be exceedingly small.* In this way the heat received from the sun by the comet would accumulate, and we should get a much higher temperature than would otherwise be possible. In this connection may be mentioned the remarkable persistence of the bright trains of meteors in the cold upper air, which sometimes remain visible for three-quarters of an hour before the light fades out by the gradual dissipation of the energy.

I need hardly say that the enormous tails of bright comets, many millions of miles in length, cannot be considered as one and the same material object, brandished round like a great flaming sword, as the comet moves about the sun. It is but little less difficult to suppose that the cometary mass is of so large an extent as to include all the space successively occupied by the sweep of the tail at perihelion. On the material theory we seem to be shut up to the view that the tail is constantly renewed and reformed, either by matter streaming from the nucleus or in some other way. But this view involves velocities far greater than the force of gravitation can account for. Let us consider the order of the phenomena. Under the sun's influence, luminous jets issue from the matter of the nucleus on the side exposed to the sun's heat. These are almost immediately arrested in their motion sunwards, and form a luminous cap; the matter of this cap then appears to stream out into the tail, as if by a violent wind setting against it. Now, one hypothesis supposes these appearances to correspond to the real state of things in the comet, and that there exists a repulsive force of some kind acting between the sun and the gaseous matter, after it has been emitted by the nucleus. On this hypothesis the forms of the tails of comets, which are usually curved, and denser on the convex side, admit of explanation. Each particle of matter of the tail must be moving in a curved course, under the influence of the motion it originally possessed, combined with that of this hypothetical repulsive force. But in the form which the tail assumes for us we

* 'Proceedings R. S.' 1880, p. 243.

have not only to consider the effect of perspective, but also that the comet itself is advancing, so that the visible tail is due to the portion of space which at the time contains all the repelled matter, each particle describing its own independent orbit, and reflecting to the eye the solar light or giving out its own light, as the case may be.* The value of the repulsive force which would be necessary on this

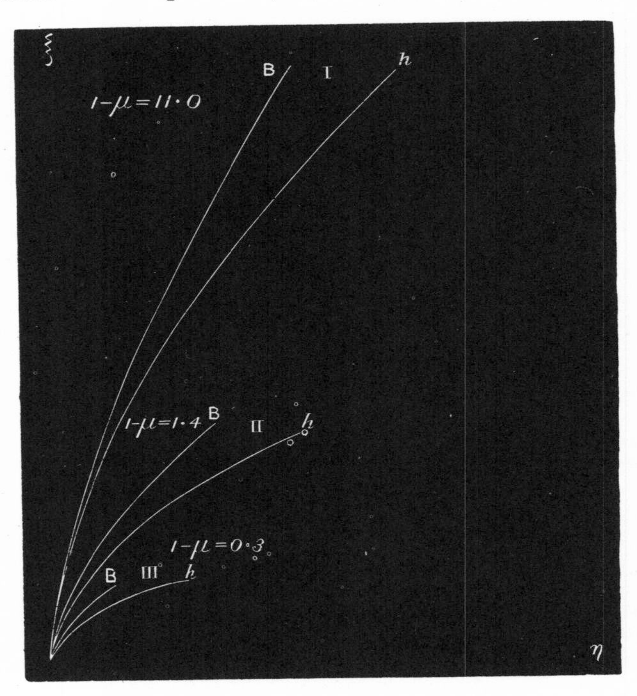

theory has been investigated by Bessel, Peirce, and others.† Recently Bredichin ‡ has investigated the curvatures of the tails of a number of comets. According to him, they fall into three classes, which are represented in this diagram, each type of curve depending upon a different assumed value of the repulsive force. This leads to another point, namely, the secondary tails which are often present. Some of these appear to be darted off with an energy of repulsion so enor-

* As a rule, the tails of comets appear to be luminous by reflected solar light, but at times the stuff which emits the light giving a spectrum of bright bands is carried into the tail to a greater or less distance from the head.

† See numerous papers by Faye in the 'Comptes Rendus.'

‡ 'Annales de l'Observatoire de Moscou,' vol. v. liv. 2, p. 30; and 'Astr. Nachr.' No. 2411.

mously great that the original motion of the nucleus tells for very little, and hence the secondary tail is but slightly curved, or even is sensibly straight. Again, if we take the hypothesis that this repulsive force, of whatever character it may be, varies as the surface, and not, like gravity, as the mass, substances of different specific gravity would be differently affected and separated from each other, and these secondary straight, or nearly straight tails would, on this view, consist of the lightest matter.

On this hypothesis a comet would suffer of course a large waste of material at each return to perihelion, as the nucleus would be unable to gather up again to itself the scattered matter of the tail; and this view is in accordance with the fact that no comet of short period has a tail of any considerable magnitude.

A theory, based on chemical decomposition, has been proposed by Professor Tyndall,* but as this view has been illustrated here by the eloquent author himself, I will not now enter upon it.

A different view of the whole matter has been suggested by Professor Tait.† He supposes, not the nucleus only, but the whole comet, to consist of a swarm, of enormous dimensions, of minute meteoroids, which become self-luminous at and about the nucleus, in consequence of the impacts of the various meteoric masses against each other, giving rise to incandescence, melting, the development of glowing gas, and the crushing and breaking up of the bodies into fragments of different sizes, and endowed with a great variety of velocities. The tail he conceives to be a portion of the less dense part of the train illuminated by sunlight, and visible or invisible to us, according not only to circumstances of density, illumination, and nearness, but also of tactic arrangement, as of a flock of birds under different conditions of perspective, or the edge of a cloud of tobacco smoke.

On this hypothesis we should expect to find a more complicated spectrum, and the spectra of comets to differ greatly from each other.

There seems to be a rapidly-growing feeling among physicists that both the self-light of comets and the phenomena of their tails belong to the order of electrical phenomena. One of the most distinguished of the American astronomers wrote to me recently: "As to the American views of the self-light of comets I cannot speak with authority for any one but myself, still I think the prevailing impression amongst us is that the light is due to an electric, or, if I may coin the word electric-oid action of some kind." Here I confess I tread most cautiously, for we have no longer any stepping-stones of fact on which to place our feet. I am ready to admit that the spectroscopic evidence, especially that furnished by the photographs of last year, favours, though it does not necessarily demand, the view that the self-light of comets is due to electric discharges. I do not attach

* Phil. Soc. Cambridge, and 'Phil. Mag.' April 1869.

† 'Proceedings R. Society Edinburgh,' vol. vi. p. 553.

much importance to the fact that the bright groups in the visible spectrum of comet *b* agreed with those of the so-called "flame spectrum," for the reason that the same spectrum may be obtained from the induction spark, when suitable arrangements are used to make the discharge one of comparatively low temperature.*

As we are now fairly on the wide ocean of speculation, I need not say that the precise modes of application of the principle of electricity which have been suggested are many. Broadly, they group themselves about the common idea that great electrical disturbances are set up by the sun's action in connection with the vaporization of some of the matter of the nucleus, and that the tail is matter carried away, possibly in connection with electric discharges, in consequence of the repulsive influence of the sun, which is supposed to be in a state of constant high electrical potential of the same name. Further, it is supposed that the luminous jets and streams and caps and envelopes belong to the same order of phenomena as the aurora, the electrical brush, and the stratified discharges of exhausted tubes. Views resting more or less on this basis have been suggested by several physicists, and, in particular, have been elaborated at great length by Zöllner, who endeavours to show that on certain assumed data, which appear to him to be highly probable, the known laws of electricity are fully adequate to the explanation of the phenomena of comets.†

All the theories we have considered assume that the bright lines seen in the spectra of comets indicate heated luminous gas. An alternative hypothesis has been suggested by Professor Wright,‡ and especially by Mr. Johnstone Stoney,§ who considers that the compound of carbon vapour is opaque in reference to the particular rays which appear as bright lines, and they appear as bright lines in consequence of sending back to us the sun's rays falling upon the vapour. Further, he considers the phenomenon to be of the order of phosphorescent bodies, and he states that the conditions existing in the cometary gas are such as will eminently promote phosphorescence, and therefore visibility, in presence of a luminary.‖

Here I must stop. May I venture to hope that the experience of the past hour has not been such as to confirm in your minds the old view to which I referred at the beginning of the lecture, that the influence of comets is always a malign and woeful one. [W. H.]

* See Professor Piazzi Smyth, 'Nature,' vol. xxiv. p. 430.

† 'Astr. Nachr.' Nos. 2057–2060, 2082–2086, and 'Ueber die Natur der Cometen,' Leipzig, 1872.

‡ 'American Journ. S. and A.' vol. x. July 1875.

§ British Association Report, 1879, p. 251.

‖ Respighi ('Comptes Rendus,' 5 Sept. 1882) has sought indeed to explain the occurrence of bright bands by supposing them to be simply the remaining portions of the continuous spectrum of reflected sunlight after absorption through the enormous depth of the comet's atmosphere. This view appears to me for many reasons improbable, especially if we take into account the extreme relative brilliancy of the most refrangible group in the photographic spectrum of comets.

Friday, May 5, 1882.

WARREN DE LA RUE, Esq. M.A. D.C.L. F.R.S. Vice-President, in the Chair.

PROFESSOR R. GRANT, M.A. LL.D. F.R.S.

The Proper Motions of the Stars.

THE spectacle presented by the stellar heavens as viewed by ordinary observers is characterised by two remarkable features, the absence of uniformity in the brightness, and the absence of uniformity in the distribution of the stars. Certain of the stars soon came to be recognisable by their superior lustre, and certain groups of stars became familiarly known as so many landmarks in the stellar firmament. The way was thus prepared for an important discovery. It was ascertained respecting a limited number of the stars that their places in the heavens relatively to the general multitude of the stars were continually changing. They consequently received the appellation of planets, or wandering stars, while, on the other hand, the stars in general, in consequence of their always maintaining the same relative position, were denominated *fixed* stars. Ptolemy, in his great work upon the astronomy of the ancients, places the earth in the centre of the universe, and assumes the sun, moon, and planets to be revolving in orbits around it, while beyond all was the sphere of the fixed stars, which revolved with a uniform motion around the earth, effecting a complete revolution once in every twenty-four hours. No opinion is expressed respecting the nature of the stars, nor is any allusion made to the possibility of the stars being endued with a proper motion.

When Copernicus propounded the true system of the universe, he made the earth a planet revolving like the other planets round the sun, and he explained the phenomenon of the diurnal revolution of the starry sphere by the revolution of the earth upon a fixed axis in the opposite direction. No opinion was expressed by him respecting the physical nature of the celestial bodies, or their having any probable community with the earth in this respect. Indeed, it could hardly be said that any new light was thrown upon the physics of astronomy by the theory of Copernicus. As a mathematical exposition of the movements of the celestial bodies it was eminently successful. Indeed, it wanted only the discoveries of Kepler respecting the elliptical movements of the planets to make it perfect in this respect. But it must be acknowledged that in the system propounded by Copernicus the earth was regarded as the body of paramount importance in the universe.

It was the invention of the telescope and its application to the purposes of astronomical observation which first revealed to the human mind the marvellous extent of the physical universe, and suggested the idea that the earth might be a mere atom in comparison with the vastness of the material system beyond. When it was discovered that the planets are round dark bodies like the earth, shining only by the reflected light of the sun, and that they presented apparent diameters of sensible magnitude when viewed through the telescope, no doubt was henceforward entertained that the planets are bodies comparable with the earth in magnitude, and that the earth is merely one of a family of similar bodies, which revolve in orbits of different magnitudes around the sun. It is worthy of remark that Galileo, to whom is due the telescopic discoveries which first disclosed the vast extent of the material universe, has nowhere expressed any opinion respecting the nature of the stars. His mind was probably too much occupied with the more immediate consequences of his discoveries to indulge in speculations leading to more remote conclusions; and a similar remark is generally applicable to his successors in the field of telescopic exploration, who flourished during the seventeenth century. It was reserved for Huyghens to propound the doctrine that the stars are suns. This he did in a work on Cosmical Astronomy, which was published in 1699, shortly after his death. Henceforward the stars have been regarded by astronomers as self-luminous bodies, comparable in magnitude and splendour with the sun.

While more correct ideas were being formed respecting the nature of the stars, the method for ascertaining the exact position of an object in the celestial sphere underwent at the same time a complete revolution. The telescope in its original form was not suited for aiding the observer in fixing the precise position of a star in the heavens, but the subsequent form of the telescope, consisting of a combination of two convex lenses, suggested the admirable invention of telescopic sights, which may be said to constitute the foundation of all exact astronomy. The places of the stars were now determined with a vastly greater degree of precision, and the way was thus prepared for the consideration of the important question whether the epithet *fixed* is strictly applicable to those bodies, or whether they might be rather endued with a movement, so extremely slow as to have hitherto eluded detection.

To Halley is due the discovery of the important fact that some of the stars have a proper motion. In 1717 he communicated a paper to the Royal Society, in which he showed that a comparison of the places of Sirius, Arcturus, and Aldebaran, as determined by Hipparchus about the year 130 B.C., with corresponding observations of the same stars made by himself, clearly indicated that during the intermediate interval the stars had sensibly moved southwards with respect to the ecliptic, and he obtained a further confirmation of this result by examining the account of an occultation of Aldebaran by the moon, observed at Athens in the year 509 A.D.

A few years after Halley announced this important fact, Bradley made his famous discovery of the aberration of light, and its effect upon the apparent place of a star; and subsequently the same astronomer discovered the apparent sidereal movement depending on the nutation of the earth's axis. The astronomer could now ascertain the true place of a star in the heavens with a precision to which the results of previous efforts could offer no comparison, and it seemed probable that ere long the great problem of the proper motions of the stars might be attacked with some hope of success.

To ascertain the proper motion of a star it is necessary to have two well determined places of the star separated from each other by a sufficiently great interval of time. Down to the middle of the last century no such materials may be said to have existed, if we except a few isolated cases such as those referred to by Halley, for the probable errors in the observed places of a star far exceed in magnitude the minute quantity which was the object of inquiry. To Bradley is due a great work of observational astronomy which has constituted the basis of the more extensive investigations of the present day relating to the proper motions of the stars. This consisted in a series of star observations executed by that astronomer at the Royal Observatory, Greenwich, from 1750 to 1762, but which it was reserved for Bessel, the great German astronomer, to reduce, and finally to publish in the year 1818. A comparison of those star places with the corresponding results obtained at the Greenwich Observatory in the present century by Sir George Airy, the late Astronomer Royal, has conducted astronomers to important conclusions respecting the proper motions of the stars. Materials tending to elucidate the same great question have also been derived from the star observations of several other astronomers of the present century.

[The lecturer here exhibited a diagram containing the following illustrations of the proper motions of the stars:—

Star.	Magnitude.	Proper Motion in a Thousand Years. ″
Sirius	1	1360
Procyon	1	1210
Arcturus	1	2230
α Centauri	1	3710
Capella	1	250
Rigel	1	20
Antares	1	30
Groombridge, 1830	7	7106
61 Cygni	6	3200
O² Eridani	4	4100
Lalande, 27,744	6	1681
Lalande, 30,044	7	1607
Lalande, 30,694	6	1789
Weisse's Bessel XVII., 322	7	1476]

The last four proper motions have been recently detected at the Glasgow Observatory, where a system of star observing has been prosecuted since the year 1860.

It must strike every one who inspects the foregoing list that the proper motion of a star has no relation whatever to the apparent magnitude of the star. Thus Rigel, one of the most brilliant stars in the heavens, has a proper motion of only 20″ in a thousand years. On the other hand, the star 1830 Groombridge, which has a proper motion of 7106″ in a thousand years, is a star of only the seventh magnitude. The same remark obviously applies to the other stars in the list. And yet one would have thought that the brighter stars, being presumably nearer to us than the fainter stars, would for that reason have a larger proper motion. With respect to α Centauri and 61 Cygni, which we know, from the researches of astronomers on their parallax, to be the two nearest stars, it turns out conformably to what one might expect, that they have also large proper motions; but what are we to think of 1830 Groombridge, which, although a star of only the seventh magnitude, and one which hardly indicates any sensible parallax, exhibits notwithstanding the largest proper motion of any star in the heavens? These anomalies are doubtless attributable to differences in the absolute magnitude and intrinsic splendour of the stars, and furthermore to the fact that the proper motions as revealed by the telescope are only the motions which are resolved at right angles to the line of sight.

Heretofore the proper motion of a star has been found to take place constantly in the same direction, and as the angular amount of proper motion is in all cases exceedingly small, the same result will probably continue to manifest itself for ages to come. The mean apparent diameter of the sun amounts to 1944″, consequently Arcturus would require nearly a thousand years to describe, in virtue of his proper motion, an arc of a great circle of the celestial sphere, equal to the mean apparent diameter of the sun.

The lecturer next adverted to the interesting spectroscopic researches of Huggins, and Christie the present Astronomer Royal, on the proper motions of the stars in the direction of the line of sight, and he concluded with some remarks on the great problem of the motion of the solar system in space.

[R. G.]

Friday, April 27, 1883.

George Busk, Esq. F.R.S. Treasurer and Vice-President, in the Chair.

Sir William Siemens, D.C.L. LL.D. F.R.S. *M.R.I.*

Some of the Questions involved in Solar Physics.

The lecturer introduced his subject by drawing attention to the circumstance that the idea of the sun being an exceedingly hot body was of very modern date; that both ancient and modern writers up to the early portion of the present century attributed to him a glorious and supernatural faculty of endowing us with light and heat of the degree necessary for our well-being; whilst even Sir William Herschel had attempted to find an explanation in justification of the time-honoured conception that the body of the sun might be at a low temperature and inhabitable by beings similar to ourselves, which he did in surrounding the inhabitable surface by a non-conducting atmosphere—the penumbra—to separate it from the scorching influence of the exterior photosphere.

It was not till the views of Kant, the philosopher, had been developed by La Place, the astronomer, in his famous 'Mécanique Céleste,' that the opinion gained ground that our central orb was a mass of matter in a state of incandescence, representing such an enormous aggregate as to enable it to continue radiation into space for an almost indefinite period of time.

The lecturer illustrated by means of a diagram the fact that of all the heat radiated away from the sun, only $\frac{1}{2250000000}$ part could fall upon the surface of our earth, vegetation and force of every kind being attributable to this radiation; whilst all but this fractional proportion apparently went to waste.

Recent developments of scientific research had enabled us to know much more of the constitution of the sun and other heavenly bodies than had formerly been possible. Comte says in his 'Positive Philosophy' (Martineau's translation of 1853) that "amongst the things impossible for us ever to know was that of telling what were the materials of which the sun was composed;" but within only seven years of that time Messrs. Bunsen and Kirchhoff published their famous research, showing that by connecting the dark Fraunhofer lines of the solar spectrum with the bright lines observed in the spectra of various metals, it was possible to prove the existence of those substances in the solar photosphere, thus laying the foundation of spectrum analysis, the greatest achievement of modern science.

Dr. Huggins and others, applying this mode of research to other heavenly bodies, including the distant nebulæ, had extended our chemical knowledge of them in a measure truly marvellous.

Solar observation had thus led to an analytical method by which chemistry had been revolutionised; and it would be, in the lecturer's opinion, through solar observation that we should attain to a much more perfect conception of the nature and effect of radiant energy in its three forms of heat, light, and actinism, than we could as yet boast of. The imperfection of our knowledge in this respect was proved by the circumstance that whereas some astronomers and physicists, including Waterston, Secchi, and Ericsson, had, in following Sir Isaac Newton's hypothesis, attributed to the sun a temperature of several millions of degrees Centigrade, others, including Pouillet and Vicaire, in following Dulong and Petit, had fixed it below 1500° C. Between these two extremes, other determinations, based upon different assumptions, had fixed the solar temperature at between 60,000° and 9000°.

The lecturer having conceived a process by which solar energy may be thought to a certain extent self-sustaining, had felt much interested for some years in the question of solar temperature. If the temperature of the solar photosphere should exceed 3000° C., combustion of hydrogen would be prevented by the law of dissociation, as enunciated by Bunsen and Sainte Claire Deville; and his speculative views regarding thermal maintenance must fall to the ground. To test the question, he in the first place mounted a parabolic reflector on a heliostat with a view of concentrating solar rays within its focus, which, barring comparatively small losses by absorption in the atmosphere and in the metallic substance of the reflector, should reproduce approximately the solar temperature. By introducing a rod of carbon through a hole at the apex of the reflector until it reached the focus, its tip became vividly luminous, producing a light comparable to electric light. When a gas burner was arranged in such a way that the gas flame played across the focal area, combustion appeared to be retarded, but was not arrested, showing that the utmost temperature attained in the focus did not exceed materially that producible in a Deville oxy-hydrogen furnace, or in the lecturer's regenerative gas furnace, in which the limit of dissociation is also reached.

Having thus far satisfied himself, his next step was to ascertain whether terrestrial sources of radiant energy were capable of imitating solar action in effecting the decomposition of carbonic acid and aqueous vapour in the leaf-cells of plants, which led him to undertake a series of researches on electro-horticulture, extending over three years, a subject he had brought before the Royal Society and the Royal Institution two years ago. By these researches he had proved that the electric arc possessed not only all the rays necessary to plant-life, but that a portion of its rays (the ultra-violet) exceeded in intensity the effective limit, and had to be absorbed by filtration through clear glass, which, as Professor Stokes had shown, produced this effect without interference with the yellow and other luminous and intense heat rays. He next endeavoured to estimate the solar temperature by instituting a comparison between the spectra due to different known luminous intensities. Starting with the researches of Professor Tyndall on radiant energy, supplementing them by experiments of his own on electric arcs of great power, and calling to his aid Professor Langley, of the Alleghany Observatory, to produce for him a complete spectrum of an Argand burner, he concluded that with the temperature of a radiant source, the proportion of luminous rays increased in a certain ratio; whereas in an Argand gas burner only $2\frac{1}{2}$ per cent. of the rays emitted were luminous and mostly red and yellow, the most brilliant portion of a gas flame emitted 4 per cent., as shown by Tyndall, the carbon thread of an incandescent electric light between 5 and 6 per cent., a small electric arc 10 per cent., and in a powerful 5000-candle electric arc as much as 25 per cent. of the total radiation was of the luminous kind. Professor Langley, in taking his photometer and bolometer up the Whitley mountains, 18,000 feet high, had proved that of the solar energy not more than 25 per cent. was luminous, and that the loss of solar energy sustained between our atmosphere and the sun was chiefly of the ultra-violet kind. These rays, if they penetrated our atmosphere, would render vegetation impossible, as proved by the lecturer's own experiments above referred to. It was thus shown that the temperature of the solar photosphere could not materially exceed that of a powerful electric arc, or, indeed, of the furnaces previously alluded to, leading him to the conclusion already foreshadowed by Sainte Claire Deville, and accepted by Sir William Thomson, that the solar temperature could not exceed 3000° C. The energy emitted from a source much exceeding this limit would no longer be luminous, but consist mainly of ultra-violet rays, rendering the sun invisible, but scorching and destructive of all life. The accompanying diagram (Fig. 1) of the spectra alluded to shows clearly the gradual advance of the luminous band, as marked by the letters A to H.

Not satisfied with these inferential proofs, the lecturer had endeavoured to establish a definite ratio between temperature and radiation, which formed the subject of a very recent communication to the Royal Society.* The experiment consisted in heating, by means of an electric current, a platinum or iridio-platinum wire, a metre long, and suspended between binding screws, as shown in the accompanying sketch (Fig. 2); the energy of the current was measured by two instruments—an electro-dynamometer, giving it in ampères, and a galvanometer of high resistance giving the electro-motive force between the same points in volts. The product of the two readings gave the volt-ampères, or Watts of energy communicated to the wire, and dispersed from it by radiation and convection. A reference to the

* 'Proc. of the Royal Society,' vol. xxxv. p. 166.

Fig. 1.

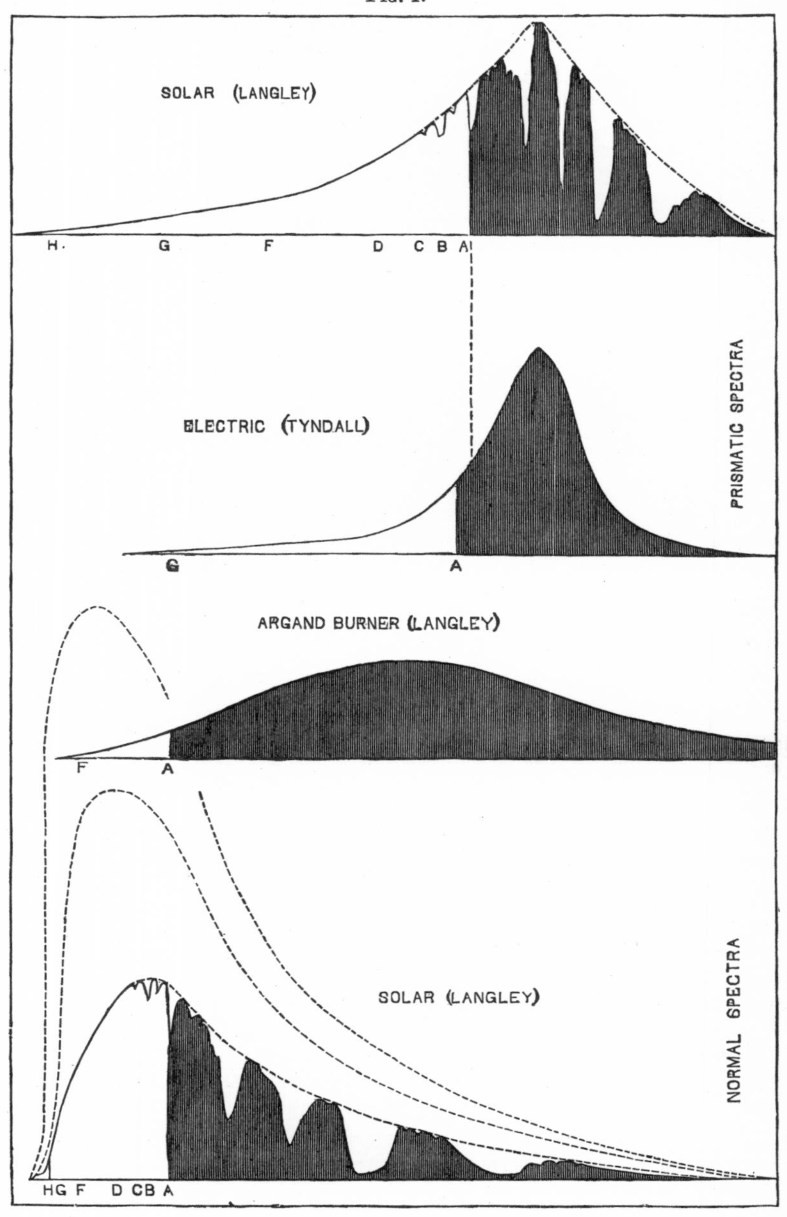
SOLAR (LANGLEY)
H. G F D C B A
ELECTRIC (TYNDALL)
G A
PRISMATIC SPECTRA
ARGAND BURNER (LANGLEY)
F A
SOLAR (LANGLEY)
NORMAL SPECTRA
HG F D CB A

Fig. 2.

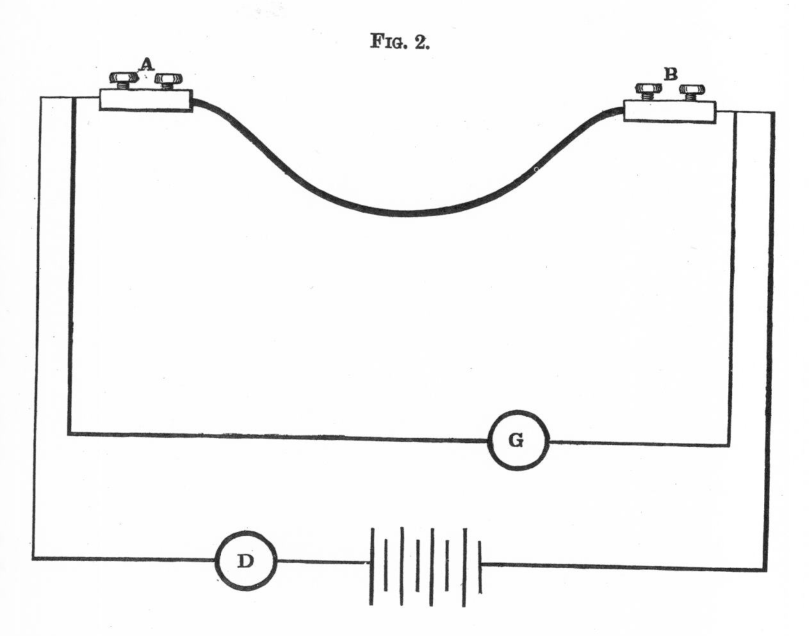
A
B
G
D

Fig. 3.

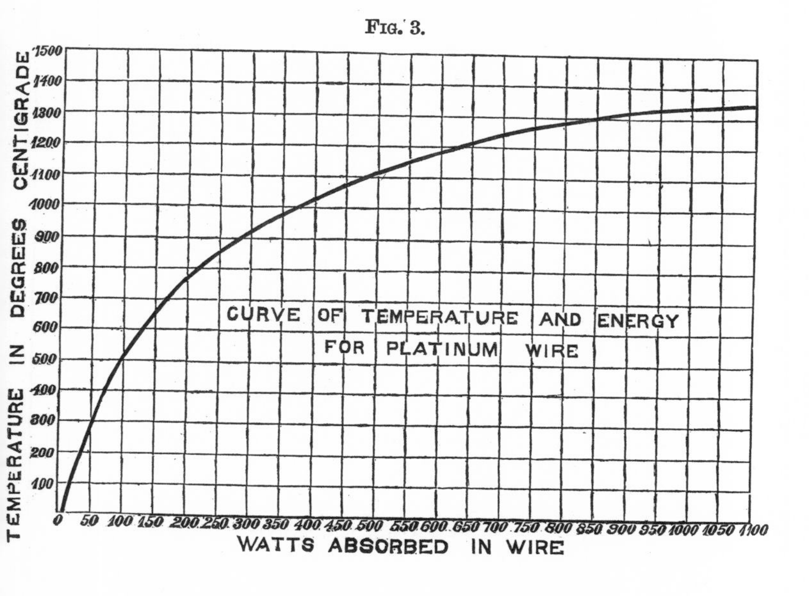
CURVE OF TEMPERATURE AND ENERGY FOR PLATINUM WIRE
TEMPERATURE IN DEGREES CENTIGRADE
100
200
300
400
500
600
700
800
900
1000
1100
1200
1300
1400
1500
0
50
100
150
200
250
300
350
400
450
500
550
600
650
700
750
800
850
900
950
1000
1050
1100
WATTS ABSORBED IN WIRE

lecturer's paper on the Electrical Resistance Thermometer, which formed the Bakerian Lecture of the Royal Society in 1871, would show that the varying electro-motive force in volts observed on the galvanometer was a true index of the temperature of the wire while being heated by the passage of the current. By combining his former experiments on the dependence of resistance upon temperature, with his recent one, a law of increase of radiation with temperature was established experimentally up to the melting-point of platinum; this, when laid down in the form of a diagram, gave very consistent results expressible by the simple formula

$$\text{Rad}^{\text{tn}} = M t^2 + \phi t,$$

M being a coefficient due to substance radiating; an expression represented in the accompanying diagram (Fig. 3), in which the abscissæ represent energy dispersed and the ordinates the corresponding temperatures.

Sir William Thomson had lately shown that the total radiating energy from a unit of surface of the carbon of the incandescent lamp amounted to $\frac{1}{67}$th part of the energy emitted from the same area of the solar photosphere, and taking the temperature of the incandescent carbon at 1800° C. (the melting-point of platinum, which can just be heated to the same point), it follows in applying Sir William Thomson's deductions to the lecturer's formula that the solar photosphere does not exceed 2700° C., or, adding for absorption of energy between us and the sun, about 2800° C., a temperature already arrived at by the lecturer by a different method. The character of the curve was that of a parabola slightly tipped forward, and if the ratio given by that curve held good absolutely beyond the melting-point of platinum, it would lead to the conclusion that at a point exceeding 3000° C. radiation would become, as it were, explosive in its character, rendering a surface temperature beyond that limit physically difficult to conceive.

Clausius had proved that the temperature obtainable in a focus could never exceed that of the radiating surface, and Sainte Claire Deville that the point of dissociation of compound vapours rises with the density of the vapour atmosphere. Supposing interstellar space to be filled with a highly attenuated compound vapour, it would clearly be possible to effect its dissociation at any point where, by the concentration of solar rays, a sufficient focal temperature could be established; but it was argued that the higher temperature observable in a focal sphere was the result only of a greater abundance of those solar vibrations called rays within a limited area, the intensity of each vibration being the outcome of the source whence it emanated: thus, in the focal field of a large reflector the end of a poker could be heated to the welding-point, whereas in that of a small reflector the end of a very thin piece of wire only could be raised to the same temperature. If, however, a single molecule of vapour not associated or pressed upon by other molecules could be sent through the one focus or the other, dissociation in obedience to Deville's law must take place irrespective of the focal area; but, inasmuch as the single solar ray represented the same potential of energy or period of vibration as numerous rays associated in a focus, it seemed reasonable that it should be as capable of dealing with the isolated molecule as a mere accumulation of the same within a limited space, and must therefore possess the same dissociating influence. Proceeding on these premises, the lecturer had procured tubes filled with highly attenuated vapours, and had observed that an exposure of the tubes to the direct solar rays or to the arc of a powerful electric light effected its partial or entire dissociation; the quantity of matter contained within such a tube was too slight to be amenable to direct chemical test, but the change operated by the light could be clearly demonstrated by passing an electric discharge through two similar tubes, one of which had, and the other had not, been exposed to the radiant energy from a source of high potential. If space could be thought filled with such vapour, of which there was much evidence in proof, solar rotation would necessarily have the effect of emitting such vapour equatorially by an action of circulation which might be likened to that of a blowing fan. When reaching the solar photosphere, by virtue of solar gravitation this dissociated vapour would, owing to its increased density, flash into flame, and could thus be made to account in great measure for the maintenance of solar radiation, whilst its continual dissociation in space would account for the continuance of solar radiation into space without producing any measurable calorific effect.

Time did not permit him to enter more fully into these subjects, which formed part of his solar hypothesis, his main object on this occasion having been to elucidate the point of cardinal importance to that hypothesis, that of the solar temperature.

[W. S.]

Friday, May 23, 1884.

The Earl of Rosse, D.C.L. LL.D. F.R.S. Manager and Vice-President, in the Chair.

David Gill, Esq. LL.D. F.R.S.

Her Majesty's Astronomer at the Cape of Good Hope.

Recent Researches on the Distances of the Fixed Stars, and some Future Problems in Sidereal Astronomy.

There has ever been a desire to burst aside the constraints imposed upon our research by the distances of space; to pass from the study of the planets of our solar system to that of the suns and galaxies that surround us; to determine the position and relative importance of our own system in the scheme of the universe, and the whence we have come and the whither we are drifting through the realms of space.

The galaxy or Milky Way—what is it? Is our sun one of its members? What is the shape of that galaxy? What are its dimensions? What is the position of our sun in it?

The star-clusters—what are they? Are these clusters galaxies? Have these suns real dimensions comparable with those of our sun, and is it distance alone that renders their light and dimension so insignificant to the naked eye? Or are the real dimensions of the clusters small as compared with our galaxy? Are their component suns but the fragments of some great sun that has been shattered by forces unknown to us, or have they originated from chaotic matter, which, instead of forming one great whirlpool and condensing by vortex action into one great sun, has been thrown into numerous minor vortices, and so become rolled up into numerous small suns?

The nebulæ—what are they? Are they, too, condensing into clusters or stars, or will their ghost-like forms remain for ever unchanged amongst the stars? or do they play some part in the scheme of nature of which we have as yet no conception?

These and many others are the questions which press on the ardent mind that contemplates the subject; and there arises the intense desire to answer such questions, and, where facts are wanting, to supply facts by fancy. The history of deep and profound thought in some of these subjects goes back through 2000 years, but the history of real progress is but as of yesterday. The foundation of sidereal astronomy may be said to have begun with the art of accurate observation. Bradley's meridian observations at Greenwich about 1750, his previous discovery of the aberration of light in 1727, and Herschel's discovery of the binary nature of double stars, his surveys of the heavens, and his catalogues of double stars—these are solid facts, facts that have contributed more to the advancement of sidereal astronomy than all the speculations of preceding centuries. They point to us the lesson that "art is long and life is short," that human knowledge, in the slow developing phenomena of sidereal astronomy, must be content to progress by the accumulating labours of successive generations of men, that progress will be measured for generations yet to come more by the amount of honest, well-directed and systematically-discussed observation than by the most brilliant speculation, and that in observation concentrated systematic effort on a special thoughtfully-selected problem will be of more avail than the most brilliant but disconnected work.

I hope that no one present thinks from what I have said that I undervalue the imaginative fervid mind that longs for the truth, and whose fancy delights to speculate on these great subjects. On the contrary, I think and I believe that without that fervid mind, without that longing for the truth, no man is fitted for the work required of him in such a field—for it is such a mind and such desires that alone can sweeten the long watches of the night, and transform such work from drudgery into a noble labour of love.

It is for like reasons that I ask you to leave with me the captivating realms of fancy this evening, and to enter the more substantial realms of fact.

We suppose ourselves, then, face to face with all the problems of sidereal astronomy to which I have hastily referred—the human mind is lost in speculation, and we are anxious to establish a solid groundwork of fact.

Now what in such circumstances would be the instinct of the scientific mind?

The answer is unquestionable—viz. to measure—and no sooner were astronomical instruments made of reasonable exactness than astronomers did begin to measure, and to ask, are the distances of the fixed stars measurable?

I should like to have given a short history of the early attempts of astronomers to measure the distance of a fixed star. But I must come at once to the time when the long baffled labours of astronomers began to be crowned with success.

Before I begin, it will save both time and circumlocution if I define a word that we must frequently use—viz. the word "parallax."

It may be defined as the change in the apparent place of a star produced by viewing it from a point other than that of reference. [The lecturer here gave some practical illustrations of parallax.] Our point of reference for stars is the sun, and as we view the stars now from one side of the sun, and six months afterwards from a point on the opposite side of the sun—that is, from two points 186 millions of miles apart—we might expect to find a considerable change in their apparent places.

But previous to 1832 astronomers could not discover with any certainty that such changes were sensible—or, putting it another way, the stars were so distant that the diameter of the earth's orbit viewed from the nearest star subtended a smaller angle than their instruments could measure. Bradley felt sure that if the star γ Draconis were so near that its parallax amounted to $1''$ of arc he would have detected it—that is, if the earth's orbit viewed from γ Draconis measured $2''$ in diameter (or as big as a globe 1 foot in diameter would look if viewed at 40 miles distant) he would have detected it. But the real distances of the stars were greater than that.

The time at last arrived when the two great masters of modern practical astronomy, Bessel and Struve, were preparing by elaborate experiment and study for the researches which led to ultimate success. After vain attempts to obtain conclusive results by endeavours to determine the apparent changes in the absolute direction of a star at different seasons of the year, both astronomers had recourse to a method which, originally proposed by Galileo in 1632, was carried out first on a large scale by Sir William Herschel. I shall refer in the first place to the researches of the great Russian astronomer Struve.

Astronomers had sufficiently demonstrated that the distances of the stars were very great, and it was reasonable to argue that as a rule the brighter stars would be those nearest to us. If, therefore, two stars are apparently near each other—the one bright, the other faint—the chances are that in reality they are far apart, though accidentally nearly in a line.

If two such stars are represented by S *s* in Diagram I., they would

DIAGRAM I.

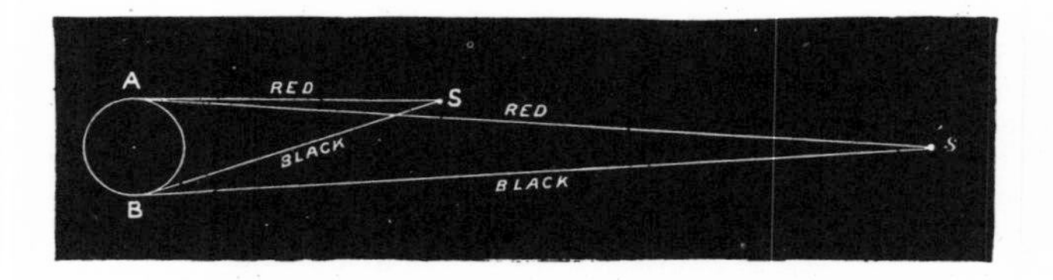

appear near each other viewed from one side of the earth's orbit at A, but not so near each other viewed from B—the opposite side of the earth's orbit, the red lines obviously indicating the apparent angle between the stars when they are viewed from A, and the black lines the apparent angle when they are viewed from B. Struve selected for the star S the bright star Vega (α Lyræ). From its brilliancy he considered it probably one of our nearest neighbours amongst the stars, and a faint star apparently near it seemed to afford a suitable representative of the really distant star *s*. Struve was careful to ascertain that this comparison star was not physically connected with α Lyræ, and he was able to prove this from the fact that whilst α Lyræ has a small annual motion relative to all neighbouring stars, this motion is not shared by the faint comparison star. Struve was provided with a telescope driven by clockwork to follow the diurnal motion of a star, and thus the hands of the observer were free to make the necessary measures. These were accomplished by an instrument, such as I hold in my hands, applied to the telescope. This micrometer contains two parallel spider-webs each attached to a slide, one slide being moved by one screw, the other by the other screw. The screws are provided with drum-heads divided into 100 parts. One web was placed on the image of α Lyræ, the other upon that of the faint comparison star, and the angle between the stars was thus read off in terms of the number of revolutions and decimals of a revolution of the screws. A number of such observations was made on each night, and the result for each night depended on the mean of the numerous observations made each night.

By observations on ninety-six nights between November 1835 and August 1838, he showed that the distance between α Lyræ and the faint comparison star changed systematically with a regular annual period, and that the maxima and minima of those distances corresponded with the times of the year at which these maxima and minima should occur if the brighter star were really much nearer than the fainter one.

Assuming that the fainter star is at a practically immeasurable distance, Struve showed that α Lyræ had a parallax that amounted to about a quarter of a second of arc, which is equivalent to the statement that a globe whose diameter is equal to that of the earth's orbit —that is, to 186 millions of miles—would at the distance of α Lyræ present an apparent diameter of half a second of arc. If you wish to realise this angle, place a globe 1 foot in diameter at a distance of 80 miles, or look at a coin half the diameter of a silver threepenny-piece at a distance of 1 mile from the eye, and try to measure it.

The great German astronomer, Bessel, was simultaneously engaged in like work at Königsberg. He selected as the object of his researches a very remarkable double star—61 Cygni.

This star had already been the subject of similar researches on his part with much inferior means. He now attacked the problem with the splendid heliometer which had been made for him by Frauenhofer for the purpose. The principle of this instrument I shall presently explain. His reasons for choosing 61 Cygni were that the two components of this star, though not remarkable for brightness—they are just visible to the naked eye—yet have this peculiarity, that they have a remarkably large proper motion, the largest then known, though now known to be surpassed by that of two other stars which I shall afterwards mention. The components of 61 Cygni have an apparent angular motion relative to other stars of more than five seconds of arc per annum.

Struve had argued that if the stars were on the average of similar brightness, those stars which were brightest would probably be those nearest to us, and Bessel, in like manner, argued that if the absolute motions of the stars were similar on the average, those motions which appeared the largest belonged to stars which on the average were nearest to us—just as the motion of a snail could be easily watched at the distance of two or three feet from the eye, but could not be detected except after a long interval if the animal were a good many yards distant.

Bessel employed two faint comparison stars at right angles to each other with respect to 61 Cygni, and he made two separate series of observations, the first extending from August 1837 to October 1838, the second from October 1838 to March 1840.

Both series confirm each other, and the results deduced separately from the measures of the two comparison stars also agree within very narrow limits. From all the observations combined Bessel found the parallax of 61 Cygni to be 35/100 of a second—a quantity which has been shown by the modern researches of Prof. Auwers and Dr. Ball to be more nearly half a second of arc. Thus at 61 Cygni the diameter of the earth's orbit round the sun would appear of the same size as a globe a foot in diameter viewed at 40 miles distance, or of a silver threepenny-piece a mile off. But whilst these great masters of astronomy—Struve and Bessel—had been exhausting the resources of their skill in observation, and that of the astronomical workshops of Europe in supplying them with the most refined instruments, a quiet and earnest man had been at work at the Cape of Good Hope, and, without knowing it at the time, had *really made the* FIRST *observations* which afforded strong presumptive evidence of the existence of the parallax of any fixed star.

Henderson occupied the post of Her Majesty's Astronomer at the Cape of Good Hope in 1832 and 1833, and during his brief and brilliant tenure of office there, he made, amongst many others, a fine series of meridian observations of α Centauri—a bright and otherwise remarkable double star. When, after his return to England, Henderson reduced these observations, and compared them with the earlier observations of other astronomers, he found that α Centauri had a large proper motion; he was therefore led to examine and see whether his observations gave any indication of an annual parallax. He found that they did so, and not of a small parallax but of one amounting to nearly a second of arc. But it was not till this was confirmed, not only by the observations with the mural circle but by those of the transit instrument also, not only by his own observations but by those of Lieut. Meadows, his assistant, that Henderson ventured to publish his remarkable result.

In the year 1842 it was felt by the astronomical world at large that the problem which hitherto had baffled astronomers had begun to yield, that some approximation to the truth had at last been arrived at with regard to the distance of a fixed star, and it was fit and proper that the Royal Astronomical Society of London should acknowledge the labours of him who had most effectually contributed to this end.

Henderson's results seemed sufficiently convincing, but they depended upon determinations of the absolute place of α Centauri. The experiences of the skilful astronomer Brinkley at Dublin were still fresh in the minds of astronomers. He had arrived by similar, though less perfect, means at results like those of Henderson; but his results had been proved to be fallacious, though the causes of their being so still remain somewhat inexplicable. In the case of Struve's observations the weight of evidence which he produced and the excellence of his method were admitted, but men were not prepared by experience for accepting as accurate the minute changes of angle which Struve had to measure; nor was the proof afforded by his series of observations so entirely convincing as that afforded by the series of Bessel. Therefore, to Bessel the well-earned medal was given, but the labours of Struve and Henderson received high and honourable mention. I quote from the speech of Sir John Herschel in awarding that medal. He says of Henderson's researches on α Centauri:—

"Should a different eye, and a different circle continue to give the same result, we must of course acquiesce in the conclusion; and the distinct and entire merit of the *first* discovery of the parallax of *a* fixed star will rest indisputably with Mr. Henderson. At present, however, we should not be justified in anticipating a decision which time alone can stamp with the seal of absolute authority."

So much for Sir John Herschel's officially expressed opinion. I can state now, and as Henderson's successor I do so with pride and pleasure, that a different eye (that of his able and sympathetic successor, Sir Thomas Maclear) fully confirmed Henderson's result with another circle; and further, that Henderson's result has been still further confirmed by additional researches of which I shall presently speak.

I must now pass over briefly the history of succeeding researches, and indeed it has been so admirably and so recently told within these walls by Dr. Ball, that it is quite unnecessary I should enter upon it in detail. The most reliable values arrived at for the parallaxes of the stars of the northern hemisphere are given in the following table, and to these results I shall afterwards refer:—

TABLE I.—PARALLAXES OF STARS WHICH HAVE BEEN DETERMINED IN THE NORTHERN HEAVENS WITH CONSIDERABLE ACCURACY.

	Magnitude.	Proper Motion.	Parallax.
		"	"
61 Cygni	6	5·14	0·50
Lalande 21185..	7¼	4·75	0·50
α Tauri	1	0·19	0·52
34 Groombridge	8	2·81	0·29
Lalande 21258..	8½	4·40	0·26
O. Mg. 17415	9	1·27	0·25
σ Draconis	5½	1·87	0·25
α Lyræ	1	0·31	0·20
p Ophiuchi	4½	1·0	0·17
α Bootis	1	2·43	0·13?
Groombridge 1830	7	7·05	0·09
Bradley 3077	6	2·09	0·07
85 Pegasi	6	1·38	0·05

The recent researches referred to in the title of this evening's lecture are some investigations which, in conjunction with a young American friend, Dr. Elkin, who was my guest for two years, I have recently carried out at the Cape of Good Hope.

The instrument employed was a heliometer—my own property—the good qualities of which I had previously tested at Mauritius in 1874, and at the Island of Ascension in 1877.

[The lecturer here described the heliometer, and illustrated the method of its use.]

I have said that the angle between the stars is measured in terms of the scale of the heliometer, but the scale-value, in seconds of arc, may change by the effects of temperature and from other causes.

Bessel in his researches on the parallax of 61 Cygni, determined by independent means the effect of temperature on his scale-value, and applied corresponding corrections to his observations. But he also took the precaution to employ two stars of comparison situated at right angles to each other with respect to the principal star, so that the effect of parallax would be at a maximum for one comparison star at the season of the year when it was at zero for the other, and *vice versâ*.

But in the course of previous researches I found that there were sources of error other than mere change of the temperature of the air, viz. differences of temperature in different parts of the instrument, and changes in the normal focus of the observer's eye, which exercised a very sensible influence on the results. It was necessary to devise some method by which these should also be eliminated.

There is a very simple means of doing this. Instead of taking two comparison stars at right angles, take two comparison stars situated nearly symmetrically on opposite sides of the star whose parallax is to be determined—such, for example, as the stars α and β in Diagram II. Now observe these distances in the order α, β, β, α, on each night of observation; so that on each night the observations at both distances are practically made at the same instant. Then, whatever causes have combined to create a systematic error in the measurement of one of these distances, precisely the same causes must create precisely similar systematic error in the measurement of the other distance. Thus if, by the regular or irregular effects of temperature or by changes in the normal condition of the observer's eye, we measure the distance α too great, so for the simultaneous observations of the distance β we shall, from precisely the same causes, measure that distance too great also.

But the *difference* of the distances will be entirely free from all errors of the kind; and if the distances are not quite equal, it is very easy to apply a correction on the assumption that the sum of the distances is a constant.

In Diagram II. the circle represents a radius of 2° surrounding the star α Centauri. The distance of the component stars α_1 and α_2 Centauri in the diagram is enormously exaggerated for the sake of clearness. Guided by the principles just explained, search was made for comparison stars in pairs symmetrically situated with respect to α Centauri, and otherwise favourably situated for measurement of parallax.

DIAGRAM II.

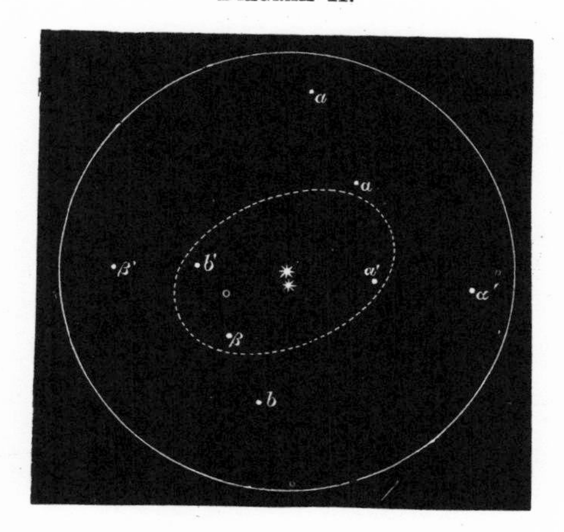

Showing comparison stars employed in determining the parallax of α Centauri.

You will remember that from the effects of parallax all stars appear to describe small ellipses about a mean position; stars near the pole of the ecliptic describing nearly circles, and those near the ecliptic very elongated ellipses. Obviously, then, those pairs of stars are most favourable—other conditions being equal—which lie near the major axis of the parallactic ellipse. The dotted ellipse in Diagram II. represents the *form* of the parallactic ellipse; that is to say, the form of the apparent path which α Centauri must describe if it is affected by parallax. Of course the *size* of the ellipse is exaggerated—in fact in the diagram nearly 5000 times—therefore, remember that the diagram represents only that which we can compute before we have observed, viz. the *shape* of the ellipse, or the relations of the lengths of the two axes; the *absolute* size has to be determined from the observations.

The most favourable couple of comparison stars in our drawing is that marked α and β—they are nearest to the major axis of the parallactic ellipse, and they are very symmetrically situated with respect to α Centauri.

Now turn to Diagram III. Here is exhibited the results of my measures on a very large scale—in a manner similar to that in which the height of the barometer for different hours of the day, or the comparative price of wheat at different seasons of the year or in different

DIAGRAM III.

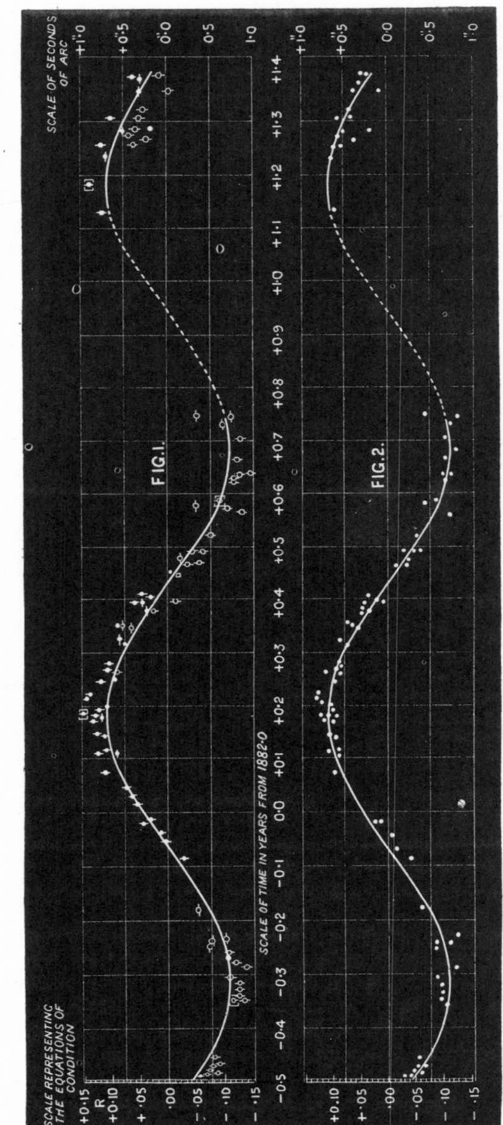

Curves showing the results of the observations of **α** Centauri relative to the comparison stars **α** and **β**.

years, is now exhibited in the daily papers. Imagine the star α about a mile immediately below any point of that curve, and the star β rather over three-quarters of a mile immediately above the same point, and you would then have a diagram to scale.* The middle horizontal line represents the mean difference of these two distances, and each dot or mark on Fig. 1 of the diagram represents the variation of that distance according to each successive observation. The different kinds of dot represent measures made at different hour angles, or when the relation of the direction of measurement to the line joining the observer's eye is different. These different kinds of personal errors were separately investigated, and they were then allowed for and the observations were corrected accordingly.

The observations so corrected are represented in Fig. 2, where each black dot expresses the result of the observations of a single night, and the curve is the commuted curve resulting from the mathematical discussion of the observations.

You must be careful to understand that this is not simply the kind of curve which best represents the observations. The curve is limited by purely geometrical conditions to have its maximum on March 7, and its minimum on September 10, and to follow a precise form of curve according to a simple law. The observations only determine the range from maximum to minimum, and yet you see how perfectly the maximum of the observations agrees with the maximum of the curve, and the minimum of the observations with the minimum of the curve, and how closely the law is followed throughout.

The result was that from these observations the parallax of α Centauri was $0''\cdot747$, or practically three-quarters of a second of arc.

But I was not content with this result alone. I wished further confirmation, and selected another pair of stars, α' and β', shown in Diagram II.

From similar observations with these comparison stars I obtained for the parallax of α Centauri $0''\cdot760$, a result which is identical with the last within the limits of the probable error of either.

My friend Dr. Elkin selected the stars $a\,b$ and $a'\,b'$ as his comparison stars, and in a precisely similar way he obtained as the mean of his results a parallax of $0''\cdot752$, a result identical with my own, so that we may conclude as one of the most certainly established facts of astronomy that the parallax of α Centauri relative to an average star of the seventh or eighth magnitude is three-quarters of a second of arc.

It is therefore beyond all doubt that Henderson's discovery was a real one. Herschel's verdict must therefore be confirmed, and the palm for first breaking down the barriers that separated us from any knowledge of the distances of the fixed stars be accorded to the memory of the Cape Astronomer, Henderson.

---

* In the wall diagram one second of arc was represented by about fifteen inches.

So far as all existing researches go, α Centauri is the nearest of the fixed stars. Regarding the faint comparison stars as practically infinitely distant, let us try to realise how near or how far distant α Centauri really is.

If we wish to deal with distance so immense, we must adopt a more convenient unit of measure.

The most convenient unit for our purpose is the number of years that light would take to reach us. Light takes almost exactly 500 seconds of time to come from the sun; this is a figure easy to remember, and is probably exact to a single unit. The sun is 93 millions of miles distant, and this figure I believe to be exact within 200,000 miles.

Quite recently the accuracy of these figures has been confirmed in a very remarkable way by different kinds of investigations by different observers; otherwise I should not have quoted them with so much confidence.

The parallax of α Centauri is three-quarters of a second of arc; therefore its distance is 275,000 times the distance of the earth from the sun, and therefore light, which travels to the earth from the sun in 500 seconds (i. e. in $8\frac{1}{3}$ minutes) would take 4·36, or a little more than $4\frac{1}{3}$ years to come from α Centauri.

You will find in the accompanying table a specific account of the other results which were arrived at by Dr. Elkin and myself by precisely similar means, and you will find on the wall diagrams representing my own detailed observations in the case of Sirius and ε Indi. (See Diagrams IV. and V.)

TABLE II.—RESULTS OF RECENT RESEARCHES ON THE PARALLAX OF STARS IN THE SOUTHERN HEMISPHERE.

Name of Star.	Observer.	Star's magnitude.	Annual proper motion in arc.	Parallax.	Star's distance in light units, or number of years in which light from star would reach the earth.	Velocity of star's motion in miles per second at right angles to line of sight.
α Centauri ..	G. & E.	1	3″·67	0″·75	4·36	14·4
Sirius ..	G. & E.	1	1·24	0·38	8·6	9·6
Lacaille 9352	G.	$7\frac{1}{2}$	6·95	0·28	11·6	73
ε Indi ..	G. & E.	$5\frac{1}{4}$	4·68	0·22	15	63
o_2 Eridani ..	G.	$4\frac{1}{2}$	4·10	0·17	19	69
e Eridani ..	E.	$4\frac{1}{2}$	3·03	0·14	23	64
ζ Tucanæ ..	E.	6	2·05	0·06	54	101
Canopus ..	E.	1	0·00	Insensible	—	—
β Centauri	G.	1	—	Insensible	—	—

Time does not permit me to go into more detail as to each of these separate results, full of interest though they are, and each of them representing months of labour.

My object now is to generalise, to put out the conclusions that must be drawn from these two tables of parallax (Tables I. and II.), and to see what are the broad lessons that they teach us.

A glance is sufficient to show that neither apparent magnitude nor apparent proper motion can afford a definite criterion of the distance of any fixed star—that different stars really differ greatly in absolute brightness and in absolute motion.

And now what is the work before us in the future?

The great cosmical problem that we have to solve is not so much what is the parallax of this or that particular star, but we have to solve the much broader questions—

1. What are the average parallaxes of stars of the *first*, *second*, *third*, and *fourth* magnitudes, compared with those of fainter magnitude?

2. What connection does there subsist between the parallax of a star and the amount and direction of its proper motion, or can it be proved that there is no such relation or connection?

With any approximate answer to these questions we should probably be able to determine the law of absorption of star-light in space, and be provided with the data at present wanting for determining with more precision the constant of precession and the amount and direction of the solar motion in space. And who can predict what hitherto unknown cosmical laws might reveal themselves in the course of such an investigation?

It is important to consider whether such a scheme of research is one that can be realised in the immediate future, or one that can only be carried to completion by the accumulated labours of successive astronomers.

I have very carefully considered this question from a practical point of view, and I have prepared a scheme, founded on the results of my past experience. I have submitted that scheme for the opinion of the most competent judges, and in their opinion, as well as my own, the work can be done, with honest hard work for one hemisphere, within ten years. I have offered to do that work for the southern hemisphere with my own hands, and a proposal for the necessary instruments and appliances is now under the consideration of my Lords Commissioners of the Admiralty. I need hardly add that in this matter I look confidently for that complete consideration and that efficient support which I have never failed to receive at their hands since I have had the honour to serve them.

The like work will be undertaken for the northern hemisphere by my friend Dr. Elkin, who is now in charge of the heliometer at Yale College in America. It is at present the finest instrument of the kind in the world, and a photograph of it you have already seen upon the screen.

I most earnestly trust that we may be granted health and strength for this work, and that no unforeseen circumstances will prevent its complete accomplishment.

DIAGRAM IV.

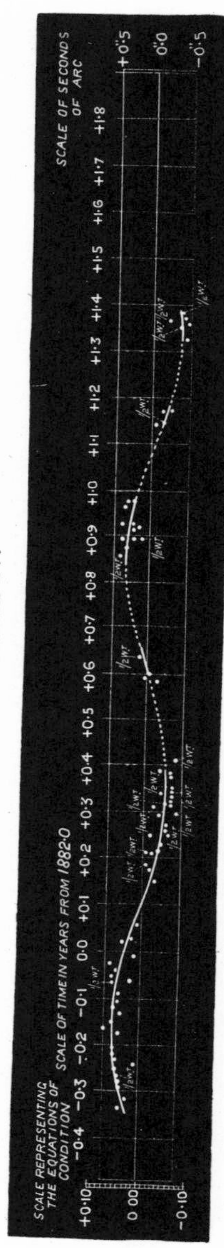

Curve showing results of observations of Sirius for parallax.

DIAGRAM V.

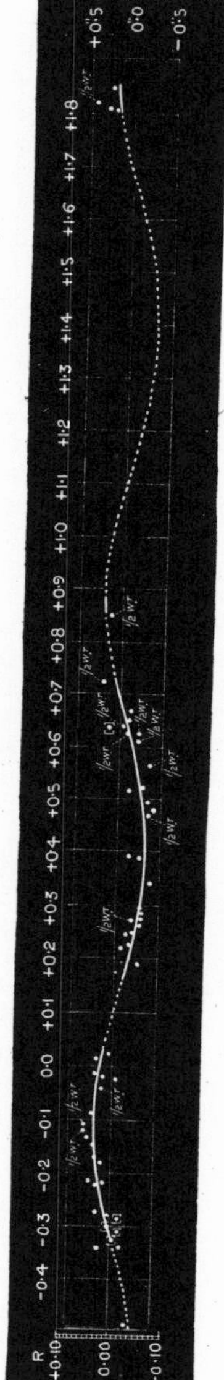

Curve showing results of observations of ε Indi for parallax.

Before closing this lecture I wish briefly to allude to another engine of research in sidereal astronomy which quite recently has received an enormous development, and whose application appears to offer a rich harvest of results. I refer to the application of photography to astronomical observation.

Your respected member, Mr. De la Rue, is the father of this method. Time does not permit me to dwell on his early endeavours and his successful results, but they are well known to you all. He opened up the field, and he cleared the way for his successors.

The recent strides in the chemistry of photography and the production of dry plates of extreme sensibility, have permitted the application of the method to objects that formerly could not be photographed. Here, on the screen, are the spectra of stars photographed directly from the stars by Dr. Huggins, the lines which tell of the chemical constitution and temperature of the star's atmosphere being sharply defined.

Here are photographs of the great comet of 1882, which, with the co-operation of Mr. Allis of Mowbray, I obtained at the Cape, by attaching his ordinary camera to an equatorially mounted telescope, and with its aid following the comet exactly for more than two hours. Each one of the thousands of points of light that you see is the picture of a fixed star. The photograph suggests the desirability of producing star maps by direct photography from the sky.

Here on the screen is a photograph of the great nebula of Orion, or rather a series of photographs of it, made by Mr. Common of Ealing. You will note the gradual development of detail by increase of exposure, and the wonderful amount of detail at last arrived at. Here are photographs from drawings of the same, and you will note the discrepancies between them. And here is a photograph of a star cluster, also by Mr. Common.

No hand of man has tampered with these pictures. They have a value on this account which gives them a distinct and separate claim to confidence above any work in which the hand of fallible man has had a part.

The standpoint of science is so different from that of art. A picture which is a mere copy of nature, in which we do not recognise somewhat of the soul of the artist, is nothing in an artistic point of view; but in a scientific point of view the more absolutely that the individuality of the artist is suppressed, and the more absolutely a rigid representation of nature is obtained, the better.

Here is a volume compiled by one of the most energetic and able of American astronomers—Prof. Holden. It contains faithful reproductions of all the available drawings that have been made by astronomers of this wonderful nebula of Orion from the year 1656 to recent times.

If now we were to suppose one hundred years to elapse, and no further observation of the nebula of Orion to be made in the interval; if in some extraordinary way all previous observations were lost, but that astronomers were offered the choice of recovering this photograph of Mr. Common's, or of losing it and preserving all the previous observations of the nebula recorded in Prof. Holden's book—how would the choice lie? I venture to say that the decision would be —Give us Mr. Common's photograph.

Is it not therefore now our duty to commence a systematic photographic record of the present aspect of the heavens? Will not coming generations expect this of us? Does not photography offer the only means by which, so far as we know, man will be able to trace out and follow some of the more slowly developing phenomena of sidereal astronomy?

Huggins has shown how the stars may be made to trace in the significant cipher of their spectra the secrets of their constitution and the story of their history. Common has shown us how the nebulæ and clusters may be separately photographed, and it is not difficult to see how that process may be applied, not only to special objects, but piece by piece to the whole sky, till we possess a photographic library of each square half-degree of the heavens. But such a work can only be accomplished by consummate instruments, and with a persistent systematic continuity which the unaided amateur is unable to procure and to employ. It is a work that must be taken up and dealt with on a national scale, on lines which Huggins and Common have so well indicated, and which has already been put in a practical form by a proposal of Norman Lockyer's at a recent meeting of the Royal Astronomical Society.

I would that I had the power to urge with due force our duty as a nation in this matter, but my powers are inadequate to the task.

I employ rather the words of Sir John Herschel, because no words of mine can equal those of him who was the prose-poet of our science, whose glowing language was always as just as it was beautiful, and whose judgment in such matters has never been excelled. They were spoken in the early days of exact sidereal astronomy, when the strongholds of space were but beginning to yield the secret of their dimensions to the untiring labour and skill of Bessel, of Struve, and of Henderson. Think what they would have been *now* when they might have told how Huggins' spectroscope had determined the kinship of the stars with our sun, how it had so far solved the mysteries of the constitution of the nebulæ, and pointed out the means of determining the absolute velocity of the celestial motions in the line of sight. Think what Herschel would have said of those photographs by Common that we have seen to-night of that nebula that Herschel himself had so laboriously studied, and whose mysterious convolutions he had in vain endeavoured adequately to portray; and think of the lessons of opportunity and of duty that he would have drawn from such discoveries, as you listen to his words spoken forty-two years ago:—

"Such results are among the fairest flowers of civilisation. They justify the vast expenditure of time and talent which have led up to them; they justify the language which men of science hold, or ought to hold, when they appeal to the Governments of their respective countries for the liberal devotion of the national means in furtherance of the great objects they propose to accomplish. They enable them not only to hold out but to redeem their promises, when they profess themselves productive labourers in a higher and richer field than that of mere material and physical advantages.

"It is then, when they become (if I may venture on such a figure without irreverence) the messengers from heaven to earth of such stupendous announcements as must strike every one who hears them with almost awful admiration, that they may claim to be listened to when they repeat in every variety of urgent instance that these are not the last of such announcements which they shall have to communicate, that there are yet behind, to search out and to declare, not only secrets of nature which shall increase the wealth or power of man, but TRUTHS which shall ennoble the age and country in which they are divulged, and, by dilating the intellect, react on the moral character of mankind. Such truths are things quite as worthy of struggles and sacrifices as many of the objects for which nations contend and exhaust their physical and moral energies and resources. They are gems of real and durable glory in the diadems of princes, and conquests which, while they leave no tears behind them, continue for ever unalienable."

[D. G.]

Friday, February 20, 1885.

SIR FREDERICK BRAMWELL, F.R.S. Manager and Vice-President, in the Chair.

WILLIAM HUGGINS, Esq. D.C.L. LL.D. F.R.S. *M.R.I.*

On the Solar Corona.

IF it were usual to prefix a motto to these evening discourses, I might have selected such words as "Seeing the Invisible," for I have to describe a method of investigation by which what is usually unseeable may become revealed. We live at the bottom of a deep ocean of air, and therefore every object outside the earth can be seen by us only as it looks when viewed through this great depth of air. Professor Langley has shown recently that the air mars, colours, distorts, and therefore misleads and cheats us to an extent much greater than was supposed. Langley considers that the light and heat absorbed and scattered by the air and the particles of matter floating in it amount to no less than 40 per cent. of the light falling upon it. In consequence of this want of transparency and of the presence of finely divided matter always more or less suspended in it, the air, when the sun shines upon it, becomes itself a source of light. This illuminated aerial ocean necessarily conceals from us by overpowering them any sources of light less brilliant than itself which are in the heavens beyond. From this cause the stars are invisible at midday. This illuminated air also conceals from us certain surroundings and appendages of the sun, which become visible on the very rare occasions when the moon coming between us and the sun cuts off the sun's light from the air where the eclipse is total, and so allows the observer to see the surroundings of the sun through the cone of unilluminated air which is in shadow. It is only when the aerial curtain of light is thus withdrawn that we can become spectators of what is taking place on the stage beyond. The magnificent scene never lasts more than a few minutes, for the moon passes and the curtain of light is again before us. On an average, once in two years this curtain of light is lifted for from three to six minutes. I need not say how difficult it is from these glimpses at long intervals even to guess at the plot of the drama which is being played out about the sun.

The purpose of this discourse is to describe a method by which it is possible to overcome the barrier presented to our view by the bright screen of air, and so watch from day to day the changing scenes taking place behind it in the sun's surroundings.

The object of our quest is to be found in the glory of radiant beams and bright streamers intersected by darker rifts which appears about the sun at a total solar eclipse. The corona possesses a structure of great complexity, which is the more puzzling in its intricate arrangement because though we seem to have a flat surface before us, it exists really in three dimensions. If we were dwellers in Flatland and the corona were a sort of glorified catherine-wheel, the task of interpretation would seem less difficult. But as we are looking at an object having thickness as well as extension, the forms seen in the corona must appear to us more or less modified by the effect of perspective. This consideration tells us also that the intrinsic brightness of the corona towards the sun's limb is much less than its apparent brightness as seen by us, of which no inconsiderable part must be due to the greater extent of corona in the line of sight as the sun is approached. The corona undergoes great and probably continual change, as the same coronal forms are not present at different eclipses.

The attempts which have been made from time to time to see the corona without an eclipse have been based mainly upon the hope that if the eye were protected from the intense direct light of the sun, and from all light other than that from the sky immediately about the sun, then the eye might become sufficiently sensitive to perceive the corona. These attempts have failed because it was not possible to place the artificial screen where the moon comes, outside our atmosphere, and so keep in shadow the part of the air through which the observer looks. The latest attempts have been made by Professor Langley at Mount Whitney, and Dr. Copeland, assistant to Lord Crawford, on the Andes. Professor Langley says, "I have tried visual methods under the most favourable circumstances, but with entire non-success." Dr. Copeland observed at Puno, at a height of 12,040 feet. He says : "It ought to be mentioned that the appearances produced by the illuminated atmosphere were often of the most tantalising description, giving again and again the impression that my efforts were about to be crowned with success."

There are occasions on which the existence of the brighter part of the corona near the sun's limb can be detected without an eclipse. The brightness of the sky near the sun's limb is due to two distinct factors, the air-glare and the corona behind it, which M. Janssen considers to be brighter than the full moon. When Venus comes between us and the sun, it is obvious that the planet as it approaches the sun, comes in before the corona, and shuts off the light which is due to it. To the observer the sky at the place where the planet is appears darker than the adjoining parts, that is to say, the withdrawal of the coronal light from behind has made a sensible diminution in the brightness of the sky. It follows that the part of the sky behind which the corona is situated must be brighter in a small degree than the adjoining parts, and it would perhaps not be too much to say that the corona would always be visible when the sky is clear, if our eyes were more sensitive to small differences of illumination of adjacent areas. My friend Mr. John Brett, A.R.A. tells me that he is able to see the corona in a telescope of low power.

The spectroscopic method by which the prominences can be seen fails because a part only of the coronal light is resolved by the prism into bright lines, and of these lines no one is sufficiently bright, and co-extensive with the corona, to enable us to see the corona by its light, as the prominences may be seen by the red, the blue, or the green line of hydrogen.

The corona sends to us light of three kinds. (1) Light which the prism resolves into bright lines, which has been emitted by luminous gas. (2) Light which gives a continuous spectrum, which has come from incandescent liquid or solid matter. (3) Reflected sunlight, which M. Janssen considers to form the fundamental part of the coronal light.

The problem to be solved was how to disentangle the coronal light from the air-glare mixed up with it, or in other words how to give such an advantage to the coronal light that it might hold its own sufficiently for our eyes to distinguish the corona from the bright sky.

When the report reached this country in the summer of 1882 that photographs of the spectrum of the corona taken during the eclipse in Egypt showed that the coronal light seen from the earth, as a whole is strong in the violet region, it seemed to me probable that if by some method of selective absorption this kind of light were isolated, then when viewed by this kind of light alone the corona might be at a sufficient advantage relatively to the air-glare to become visible. Though this light falls within the range of vision, the eye is less sensitive to small differences of illumination near this limit of its power. This consideration and some others led me to look to photography for aid, for it is possible by certain technical methods to accentuate the extreme sensitiveness of a photographic plate for minute differences of illumination. [A cardboard on which a corona had been painted by so thin a wash of Chinese white that it was invisible to the audience, had been photographed. The photograph thrown upon the screen showed the corona plainly.] This cardboard represents the state of things in the sky about the sun. The painted corona is brighter than the cardboard, but our eyes are too dull to see it. In like manner the part of the sky near the sun where there is a background of corona, is brighter than the adjoining parts where there is no corona behind, but not in a degree sufficiently great for our eyes to detect the difference.

A photographic plate possesses another and enormous advantage over the eye, in that it is able to furnish a permanent record of the most complex forms from an instantaneous exposure.

In my earlier experiments the necessary isolation of violet light was obtained by interposing a screen of coloured glass or a cell containing potassic permanganate. The possible coming of false light upon the sensitive plate from the glass sides of the cell, as well as from precipitation due to the decomposition of the potassic permanganate under the sun's light, led me to seek to obtain the necessary light-selection in the film itself. Captain Abney had shown that argentic bromide, iodide, and chloride, differ greatly in the kind of light to which they are most sensitive. The chloride is most strongly affected by violet light from *h* to a little beyond K. It was found possible by making use of this selective action of argentic chloride to do away with an absorptive medium. To prevent reflected light, the back of the plate was covered with asphaltum varnish, and frequently a small metal disc a little larger than the sun's image was interposed in front of the plate to cut off the sun's direct light.

The next consideration was as to the optical means by which an image of the sun, as free as possible from imperfections of any kind, could be formed upon the plate. For several obvious reasons the use of lenses was given up, and I turned to reflection from a mirror of speculum metal. My first experiments were made with a Newtonian telescope by Short. With this instrument, during the summer of 1882 about twenty plates were taken on different days, in all of which coronal forms are to be seen about the sun's image. After a very critical examination of these plates, in which I was greatly helped by the kind assistance of Professor Stokes and Captain Abney, there seemed to be good ground to hope that the corona had really been obtained on the plates. [One of these negatives, obtained in August 1882, was shown upon the screen.]

In the spring of the following year, 1883, the attack upon the corona was carried on with a more suitable apparatus. The Misses Lassell were kind enough to lend me a seven-foot Newtonian telescope made by Mr. Lassell which possesses great perfection of figure and retains still its fine polish. For the purpose of avoiding the disadvantage of a second reflection from the small mirror, and also of reducing the aperture to 3½ inches, which gives a more manageable amount of light, I adopted the arrangement of the instrument which is shown in the following woodcut.

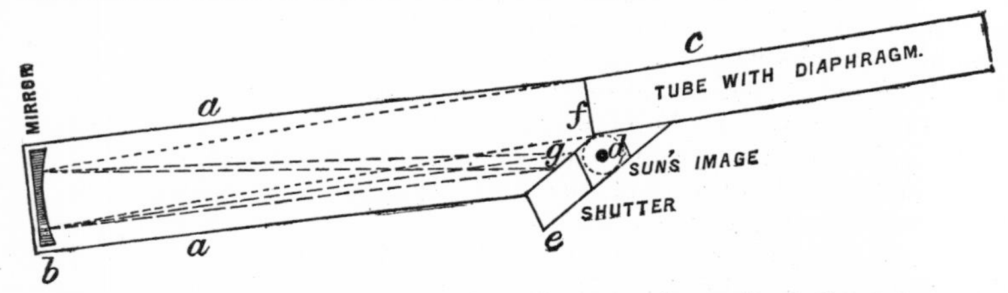

The speculum *b* remains in its place at the end of the tube *a*, *a*, by which the mechanical inconvenience of tilting the speculum within in the tube as in the ordinary form of the Herschelian telescope is avoided.

The small plane speculum and the arm carrying it were removed. The open end of the tube is fitted with a mahogany cover. In this cover at one side is a circular hole *f*, $3\frac{1}{4}$ inches diameter, for the light to enter; below is a similar hole over which is fitted a framework to receive the "backs" containing the photographic plates, and also to receive a frame with fine-ground glass for putting the apparatus into position. Immediately below, towards the speculum, is fixed a shutter with an opening of adjustable width, which can be made to pass across more or less rapidly by the use of indiarubber bands of different degrees of strength. In front of the opening *f* is fixed a tube *c*, six feet long, fitted with diaphragms, to restrict as far as possible the light which enters the telescope to that which comes from the sun and the sky immediately around it. The telescope-tube *a, a*, is also fitted with diaphragms, which are not shown in the diagram, to keep from the plate all light, except that coming directly from the speculum. It is obvious that, when the sun's light entering the tube at *f* falls upon the central part of the speculum, the image of the sun will be formed in the middle of the second opening at *d*, about two inches from the position it would take if the tube were directed axially to the sun. The exquisite definition of the photographic images of the sun shows, as was to be expected, that this small deviation from the axial direction, two inches in seven feet, does not affect sensibly the performance of the mirror. The whole apparatus is firmly strapped on to the refractor of the equatorial in my observatory, and carried with it by the clock motion.

The performance of the apparatus is very satisfactory. The photographs show the sun's image sharply defined; even small spots are seen. When the sky is free from clouds, but presents a whity appearance from the large amount of scattered light, the sun's image is well defined upon a uniform background of illuminated sky, without any sudden increase of illumination immediately about it. It is only when the sky becomes clear and blue in colour that coronal appearances present themselves with more or less distinctness. [Several negatives taken during the summer of 1883 were shown on the screen.] In our climate the increased illumination of the sky where there is a background of coronal light is too small to permit the photographs which show this difference to be otherwise than very faint. A small increase of exposure, or of development, causes it to be lost in the strong photographic action of the air-glare. For this reason, the negatives should be examined under carefully arranged illumination. They are not, therefore, well adapted for projection on a screen. [A negative taken with a whity sky, showed a well-defined image of the sun, with a sensibly uniform surrounding of air-glare, but without any indication of the corona. In the case of the other negatives exhibited, which were taken on clearer days, an appearance, very coronal in character, was to be seen about the sun.]

On May 6, the corona was photographed during a total eclipse at Caroline Island by Messrs. Lawrence and Woods. This circumstance furnished a good opportunity of subjecting the new method to a crucial test, namely, by making it possible to compare the photographs taken in England where there was no eclipse, with those taken at Caroline Island of the undoubtedly true corona during the eclipse. On the day of the eclipse the weather was bad in this country, but plates were taken before the eclipse, and others taken later on. These plates were placed in the hands of Mr. Wesley, who had had great experience in making drawings from the photographs taken during former eclipses. Mr. Wesley drew from the plates before he had any information of the results obtained at Caroline Island, and he was therefore wholly without bias in the drawings which he made from them. [Photographs of Mr. Wesley's drawings were projected on the screen, and then a copy of the Caroline Island eclipse photograph. The general resemblance was unmistakable, but the identity of the object photographed in England and at Caroline Island was placed beyond doubt by a remarkably formed rift on the east of the north pole of the sun. This rift, slightly modified in form, was to be seen in a plate taken about a solar rotation period before the eclipse, and also on a plate taken about the same time after the eclipse. The general permanence of this great rift certainly extended over some months, but no information is given as to whether the corona rotates with the sun. For from the times at which the plates were taken, one about a rotation period before and the other a rotation period after the eclipse, it is obvious the rift might have gone round with the sun, but there is no positive evidence on this point.*]

As the comparison of the English plates with those taken at Caroline Island possesses great interest, I think it well to put on record here a letter written by Mr. Lawrence to Professor Stokes, dated September 14, 1883:—

"Dr. Huggins called upon Mr. Woods this morning and showed us the drawings Mr. Wesley has made of his coronas. He told us that he particularly did not wish to see our negatives, but that he would like us to compare his results with ours. We did so, and found that some of the strongly marked details could be made out on his drawings, a rift near the north pole being especially noticeable; this was in a photograph taken on April 3, in which the detail of the northern hemisphere is best shown, while the detail of our southern hemisphere most resembles the photograph taken on June 6; in fact, our negatives seem to hold an intermediate position. Afterwards I went with Dr. Huggins and Mr. Woods to Burlington House to see the negatives. The outline and distribution of light in the inner corona of April 3 is very similar to that on our plate which had the shortest exposure; the outer corona is, however, I think, hidden by atmospheric glare. As a result of the comparison, I should say that Dr. Huggins's coronas are certainly genuine as far as 8′ from the limb."

Though the plates which were obtained during the summer of 1883 appeared to be satisfactory to the extent of showing that there could

* See Plates XI. and XIA, British Association Report, 1883, p. 348.

be little doubt remaining but that the corona had been photographed without an eclipse, and therefore of justifying the hope that a successful method for the continuous investigation of the corona had been placed in the hands of astronomers, yet as the photographs were taken under the specially unfavourable conditions of our climate, they failed to show the details of the structure of the corona.

The next step was obviously to have the method carried out at some place of high elevation, where the large part of the glare which is due to the lower and denser parts of our atmosphere would no longer be present. I ventured to suggest to the Council of the Royal Society that a grant from the fund placed annually by the Government at the disposal of the Royal Society, should be put in the hands of a small committee for this purpose. This suggestion was well received, and a committee was appointed by the Council of the Royal Society. The committee selected the Riffel near Zermatt in Switzerland, a station which has an elevation of 8500 feet, and the further advantages of easy access, and of hotel accommodation. The committee was fortunate in securing the services, as photographer, of Mr. Ray Woods, who as assistant to Professor Schuster had photographed the corona during the eclipse of 1882 in Egypt, and who in 1883, in conjunction with Mr. Lawrence, had photographed the eclipse of that year at Caroline Island.

Mr. Woods arrived at the Riffel in the beginning of July 1884, with an apparatus, similar to one shown in the woodcut on a former page, constructed by Mr. Grubb.

Captain Abney who had made observations on the Riffel in former years, had remarked on the splendid blue-black skies which were seen there whenever the lower air was free from clouds or fog. But unfortunately during the last year or so a veil of finely divided matter of some sort has been put about the earth, of which we have heard so much in the accounts from all parts of the earth of gorgeous sunsets and afterglows. This fine matter was so persistently present in the higher regions of the atmosphere during last summer, that Mr. Woods did not get once a really clear sky. On the contrary, whenever visible cloud was absent, then instead of a blue-black sky there came into view a luminous haze, forming a great aureole about the sun, of a faint red colour, which passed into bluish white near the sun. Mr. Woods found the diameter of the aureole to measure about 44°. This appearance about the sun has been seen all over the world during last summer, but with greatest distinctness at places of high elevation.

The relative position of the colours, blue inside and red outside, shows that the aureole is a diffraction phenomenon due to minute particles of matter of some kind. Mr. Ellery, Captain Abney, and some others consider the matter to be water in the form probably of minute ice spicules; others consider it to consist of particles of volcanic dust projected into the air during the eruption at Krakatoa; but whatever it is, and whencesoever it came, it is most certainly matter in the wrong place so far as astronomical observations are concerned, and in a peculiar degree for success in photographing the corona. We are only beginning to learn that whether in our persons or in our works, it is by minimised matter chiefly that we are undone. So injurious was the effect of this aureole that it was not possible to obtain any photographs of the corona at my observatory near London. This great diffraction aureole went far to defeat the object for which Mr. Woods had gone to the Riffel, but fortunately the great advantage of being free from the effects of the lower 8000 feet of denser air told so strongly, that notwithstanding the ever-present aureole Mr. Woods was able to obtain a number of plates on which the corona shows itself with more or less distinctness. [Three untouched photographic copies of the plates taken at the Riffel were shown upon the screen.] From the presence of the aureole the negatives show less detail than we have every reason to believe would have been the case if the sky had been as blue and clear as in some former years. This circumstance makes great care necessary in the discussion of these plates, and it would be premature to say what information is to be obtained from them.

[As an illustration of the differences of form which the corona has assumed at different eclipses, photographs taken in 1871, 1878, 1882, and 1883 were projected on the screen. Attention was called to the equatorial extension seen in the photograph taken in 1878, and to the suggestion which had been put forward that this peculiar character was connected with the then comparative state of inactivity of the sun's surface, at a period of minimum sun-spot action, especially as an equatorial extension was observed in 1867.]

It is now time that something should be said of the probable nature of the corona.

Six hypotheses have been suggested:—

1. That the corona consists of a gaseous atmosphere resting upon the sun's surface and carried round with it.

2. That the corona is made up, wholly or in part, of gaseous and finely divided matter which has been ejected from the sun, and is in motion about the sun from the forces of ejection, of the sun's rotation, and of gravity,—and possibly of a repulsion of some kind.

3. That the corona resembles the rings of Saturn, and consists of swarms of meteoric particles revolving with sufficient velocity to prevent their falling into the sun.

4. That the corona is the appearance presented to us by the unceasing falling into the sun of meteoric matter and the débris of comets' tails.

5. That the coronal rays and streamers are, at least in part, meteoric streams strongly illuminated by their near approach to the sun, neither revolving about nor falling into the sun, but permanent in position and varying only in richness of meteoric matter, which are parts of eccentric comet orbits. This view has been supported by

Mr. Proctor, on the ground that there must be such streams crowding richly together in the sun's neighbourhood.

6. The view of the corona suggested by Sir William Siemens in his solar theory.

It has been suggested, even, that the corona is so complex a phenomenon that there may be an element of truth in every one of these hypotheses. Any way this enumeration of hypotheses more or less mutually destructive, shows how great is the difficulty of explaining the appearances which present themselves at a total solar eclipse, and how little we really know about the corona.

An American philosopher, Professor Hastings, has revived a prior and altogether revolutionary question: Has the corona an objective existence? Is it anything more than an optical appearance depending upon diffraction? Professor Hastings has based his revival of this long discarded negative theory upon the behaviour of a coronal line which he saw, in his spectroscope, change in length east and west of the sun during the progress of the eclipse at Caroline Island. His view appears to rest on the negative foundation that Fresnel's theory of diffraction may not apply in the case of a total eclipse, and that at such great distances there is a possibility that the interior of the shadow might not be entirely dark, and so to an observer might cause the appearance of a bright fringe around the moon.*

Not to speak of the recent evidence of the reality of the corona from the photographs which have been taken when there is no intervening moon to produce diffraction, there is the adverse evidence afforded by the peculiar spectra of different parts of the corona and by the complicated and distinctly peculiar structure seen in the photographs taken at eclipses. The crucial test of this theory appears to be, that if it be true, then the corona would be much wider on the side where the sun's limb is least deeply covered, that is to say the corona would alter in width on the two sides during the progress of the eclipse. Not to refer to former eclipses where photographs taken at different times and even at different places have been found to agree, the photographs taken during the eclipse at Caroline Island show no such changes. M. Janssen says: "Les formes de la couronne ont été absolument fixes pendant toute la durée de la totalité." The photographs taken by Messrs. Lawrence and Woods also go to show that the corona suffered no such alterations in width or form as would be required by Professor Hastings' theory during the passage of the moon.

We have therefore, I venture to think, a right to believe in an objective reality of some sort about the sun corresponding to the appearance which the corona presents to us. At the same time some very small part of what we see must be due to a scattering of the

---

* Report of the Eclipse Expedition to Caroline Island, May 1883. Memoir of the National Academy of Sciences, Washington.

coronal light itself by our air, but the amount of this scattered light over the corona must be less than what is seen over the dark moon.

That the sun is surrounded by a true gaseous atmosphere of relatively limited extent there can be little doubt, but many considerations forbid us to think of an atmosphere which rises to a height which can afford any explanation of the corona, which streams several hundred thousand miles above the photosphere. For example, a gas at that height, if hundreds or even thousands of times lighter than hydrogen, would have more than metallic density near the sun's surface, a state of things which spectroscopic and other observations show is not the case. The corona does not exhibit the rapid condensation towards the sun's limb which such an atmosphere would present, especially when we take into account the effect of perspective in increasing the apparent brightness of the lower regions of the corona. There is, too, the circumstance that comets have passed through the upper part of the corona without being burnt up or even sensibly losing velocity.

There can scarcely be doubt that matter is present about the sun wherever the corona extends, and further that this matter is in the form of a fog. But there are fogs and fogs. The air we breathe, when apparently pure, stands revealed as a dense swarming of millions of motes if a sunbeam passes through it. Even such a fog is out of the question. If we conceive of a fog so attenuated that there is only one minute liquid or solid particle in every cubic mile, we should still have matter enough, in all probability, to form a corona. That the coronal matter is of the nature of a fog is shown by the three kinds of light which the corona sends to us. Reflected solar light scattered by particles of matter solid or liquid, and secondly light giving a continuous spectrum, which tells us that these solid or liquid particles are incandescent, while the third form of spectrum of bright lines, fainter and varying greatly at different parts of the corona and at different eclipses, shows the presence also of light-emitting gas. This gas existing between the particles need not necessarily form a true solar atmosphere, which the considerations already mentioned make an almost impossible supposition, for we may well regard this thin gas as carried up with the particles, or even to some extent to be furnished by them under the sun's heat.

It will be better to consider first the probable origin of this coronal matter, and by what means it can find itself at such enormous heights above the sun.

There is another celestial phenomenon, very unlike the corona at first sight, which may furnish us possibly with some clue to its true nature. The head of a large comet presents us with luminous streamers and rifts and curved rays, which are not so very unlike, on a small scale, some of the appearances which are peculiarly characteristic of the corona.* We do not know for certain the con-

* See "Comets," Royal Institution Proceedings, vol. x. p. 1.

ditions under which these cometary appearances take place, but the hypothesis which seems on the way to become generally accepted, attributes them to electrical disturbances, and especially to a repulsive force acting from the sun, possibly electrical, which varies as the surface and not like gravity as the mass. A force of this nature in the case of highly attenuated matter can easily master the force of gravity, and as we see in the tails of comets, blow away this thin kind of matter to enormous distances in the very teeth of gravity.

If such a force of repulsion is experienced in comets, it may well be that it is also present in the sun's surroundings. If this force be electrical it can only come into play when the sun and the matter subjected to it have electric potentials of the same kind, otherwise the attraction on one side of a particle would equal the repulsion on the other. On this theory, the coronal matter and the sun's surface must both be in the same electrical state, the repelled matter negative if the sun is negative, positive if the sun is positive.

The grandest terrestrial displays of electrical disturbance, as seen in lightning and the aurora, must be of a small order of magnitude as compared with the electrical changes taking place in connection with the ceaseless and fearful activity of the sun's surface, but we do not know how far these actions, or the majority of them, may be in the same electrical direction, or what other conditions there may be, so as to cause the sun's surface to maintain a high electrical state, whether positive or negative. A permanence of electric potential of the same kind would seem to be required by the phenomena of comets' tails.

If such a state of high electric potential at the photosphere be granted as is required to give rise to the repulsive force which the phenomena of comets appear to indicate, then considering the gaseous irruptions and fiery storms of more than Titanic proportions which are going on without ceasing at the solar surface, it does not go beyond what might well be, to suppose that portions of matter ejected to great heights above the photosphere and often with velocities not fa removed from that which would be necessary to set it free from the sun's attraction, and very probably in the same electric state as the photosphere, might so come under this assumed electric repulsion as to be blown upwards and to take on forms such as those seen in the corona: the greatest distances to which the coronal streamers have been traced are small as compared with the extent of the tails of comets, but then the force of gravity which the electrical repulsion would have to overcome near the sun would be enormously greater.

It is in harmony with this view of things that the positions of greatest coronal extension usually correspond with the spot zones where the solar activity is most fervent; and also that a careful examination of the structure of the corona suggests strongly that the forces to which this complex and varying structure is due have their seat in the sun. Matter repelled upwards would rise with the smaller rotational velocity of the photosphere, and lagging behind would give rise to curved forms; besides, the forces of irruption and subsequent electrical repulsion might well vary in direction and not be always strictly radial, and under such circumstances a structure of the character which the corona presents might well result. The sub-permanency of any great characteristic coronal forms, as, for example, the great rift seen in the photographs of the Caroline Island eclipse and also in those taken in England a month before the eclipse and about a month afterwards, must probably be explained by the maintenance for some time of the conditions upon which the forms depend, and not to an unaltered identity of the coronal matter; the permanency belonging to the form only, and not to the matter, as in the case of a cloud over a mountain top, or of a flame over the mouth of a volcano. If the forces to which the corona is due have their seat in the sun, the corona would probably rotate with it; but if the corona is produced by conditions external to the sun, then the corona might not be carried round with the sun.

We have seen that the corona consists probably of a sort of incandescent fog, which at the same time scatters to us the photospheric light. Now we must bear in mind the very different behaviour of a gas, and of liquid or solid particles in the near neighbourhood of the sun. A gas need not be greatly heated, even when near the sun, by the radiated energy; heated gas from the photosphere would rapidly lose heat; but on the other hand liquid or solid particles, whether originally carried up as such, or subsequently formed by condensation, would absorb the sun's heat, and at coronal distances would soon rise to a temperature not very greatly inferior to that of the photosphere. The gas which the spectroscope shows to exist along with the incandescent particles of the coronal stuff, may therefore have been carried up as gas, or have been in part distilled from the coronal particles under the enormous radiation to which they are exposed. Such a view would not be out of harmony with the very different heights to which different bright lines may be traced at different parts of the corona and at different eclipses. For obvious reasons, gases of different vapour density would be differently acted upon by a repulsive force which varies as the surface and would to some extent be winnowed from each other; the lighter the gas the more completely would it come under the sway of repulsion, and so would be carried to a greater height than the gas more strongly held down by gravity. The relative proportions, at different heights of the corona, of the gases which the spectroscope shows to exist there (and recently Captain Abney and Professor Schuster have shown that in addition to the bright lines already known, the spectrum of the corona of 1882 gave the rhythmical group of the ultra-violet lines of hydrogen which are characteristic of the photographic spectra of the white stars, and some other lines also) would vary from time to time, and depend in part upon the varying state of activity of the photosphere, and so probably establish a connection with the spectra of the prominences. This view of the corona would bring it within the charmed circle of interaction which seems to obtain among the phenomena of sun-spots and terrestrial magnetic disturbances and auroræ.

Many questions remain unconsidered; among others, whether the light emitted by the gaseous part of the corona is due directly to the sun's heat, or to electrical discharges taking place in it of the nature of the aurora. Further, what becomes of the coronal matter on the theory which has been suggested? Is it permanently carried away from the sun, as the matter of the tails of comets is lost to them? Among other considerations it may be mentioned that electric repulsion can maintain its sway only so long as the repelled particle remains in the same electrical state: if through electric discharges it ceases to maintain the electrical potential it possessed, the repulsion has no more power over it, and gravity will be no longer mastered. If, when this takes place, the particle is not moving away with a velocity sufficiently great to carry it from the sun, the particle will return to the sun. Of course, if the effect of any electric discharges or other conditions has been to change the potential of the particle from positive to negative, or the reverse, as the case may be, then the repulsion would be changed into an attraction acting in the same direction as gravity. In Mr. Wesley's drawings of the corona, especially in those of the eclipse of 1871, the longer rays or streamers appear not to end, but to be lost in increasing faintness and diffusion, but certain of the shorter rays are seen to turn round and to descend to the sun.*

It is difficult for us living in dense air to conceive of the state of attenuation probably present in the outer parts of the corona. Mr. Johnstone Stoney has calculated that more than twenty figures are needed to express the number of molecules in a cubic centimetre of ordinary air, and Mr. Crookes shows us in his tubes that matter, even when reduced to one-millionth part of the density of ordinary air, can become luminous under electrical excitement. [A glass bulb about 4 inches in diameter, kindly lent to me by Mr. Crookes, was exhibited, in which a metal ball about half an inch in diameter formed the negative pole. Under a suitable condition of the induction current, this ball was seen to be surrounded by a corona of blueish-grey light which was sufficiently bright to be seen from all parts of the theatre.] Yet it is probable that these tubes must be looked upon as crowded cities of molecules as compared with the sparse molecular population of the great coronal wastes.

I forbear to speculate further, as we may expect more information as to the state of things in the corona from the daily photographs which will be shortly commenced at the Cape of Good Hope by Mr. Ray Woods under the direction of Dr. Gill. [W. H.]

* For a history of opinion of the nature of the corona, see Papers by Prof. Norton, Prof. Young, and Prof. Langley in the 'American Journal of Science'; also 'The Sun,' by Prof. Young; and 'The Sun the Ruler of the Planetary System,' and various essays by Mr. R. A. Proctor.

Friday, February 26, 1886.

WILLIAM HUGGINS, Esq. D.C.L. LL.D. F.R.S. Vice-President, in the Chair.

A. A. COMMON, Esq. F.R.S. *M.R.I.*

Photography as an Aid to Astronomy.

IN many kinds of astronomical work the old method of direct observation seems likely to be superseded by the use of photography; and the astronomer of the near future, instead of examining with eye and telescope the various objects of the heavens, will prefer to deal with the automatic records they leave on the sensitive plate. In some work this state of things already exists, and its extension to all kinds seems but a matter of time. It appears strange that any indirect way of seeing an object can be better than the direct way, but in some cases we shall certainly find it to be so. Between the construction of the eye and of the apparatus of the photographer there are many points of great similarity. Both have optical means of producing an image, a camera, or dark chamber to keep out other light than that going to form this image; and both have a similar screen on which this image is received. It is when we come to consider the action of the screen in dealing with the image that the distinctive difference becomes apparent. In the eye the retina receives this image, and in some occult way the impressions are carried to the brain. With the sensitive plate the varying amount of light that makes up the images produces corresponding changes in the chemical nature of the film which allows the reproduction of this image in a visible form afterwards.

As I understand the action of the eye, the retina acts only as a transmitter of the sensations produced on its delicate structure by the image; and the brain records these sensations in a more or less perfect manner according to its capacity to deal with all those sensations that are transmitted; hence the power of the eye is limited by the power of the brain to record, and this is evidently in many cases less than that of the eye to perceive, whilst the power of the eye itself is limited in more than one important point as regards our subject, the retina becoming insensible to an image however perfect if insufficiently illuminated. With the sensitive plate the measure of the effective light of an image is not, as in the eye, the amount going to form the image, but the total amount that can be accumulated by a sufficiently long exposure. Hence, we have the remarkable fact that I have elsewhere before mentioned, that a certain object such as a star or nebula, that would be just beyond the power of the eye, however long the gazing was continued, could be photographed with a sufficiently long exposure; and this holds good whatever optical power be employed to increase the amount of light brought to the eye; as with the same optical power, the power of the sensitive plate, if allowed sufficient exposure, will always be greater. There is from this another consequence:—The enlargement of an image produced by a telescope is limited in one direction by the light, faint objects becoming too faint to be seen if greatly enlarged, the sensitive plate, however, may still utilise this image. There are other points of difference between the action of the retina and of the sensitive plate. The power of the latter is greater to record rays of light of quicker vibration, and it may be possible to obtain photographs of celestial objects radiating light that the eye is not adapted to receive; and it is also quite possible that plates may be prepared that will be sensitive to the visual rays so that the magnitudes of stars in stellar photographs would agree with magnitudes as obtained by the eye.

Though the image seems to be clearly formed in the eye over a large angular extent, the central parts only are clearly seen, and the image has to be traversed across this central part piece by piece to be properly examined.

With the sensitive plate, the image, no matter how complex, acts equally over its extent and records itself with fidelity.

As was well said many years ago by Dr. de la Rue, who has been rightly called the father of astronomical photography, "the sensitive film is a retina that never forgets."

I will try and show to-night how important the difference between the ordinary eye observations and the work done by photography may become; not only in cases where the ordinary visual observations have been used, but in cases where the use of the micrometer attached to the telescope has been the only means of accurate observations. Looking for a moment at the history of our subject, it appears that the earliest application of photography to a celestial object was made by Professor J. W. Draper in 1840 within one year of the announcement by Daguerre of the details of the photographic process with which his name will always be associated. Professor Draper obtained a picture of the moon in twenty minutes, using a lens and a heliostat.

An extract from the minutes of the New York Lyceum of Natural History was to this effect:—

"March 23, 1840.—Dr. Draper announced that he had succeeded in getting a representation of the moon's surface by the daguerreotype. The time occupied was twenty minutes, and the size of the figure about one inch in diameter. Daguerre had attempted the same thing but did not succeed. This is the first time that anything like a distinct representation of the moon's surface has been obtained.—Signed, Robert H. Brownne, Secretary."

Remembering that this entry was made less than one year after the publication of Daguerre's process, the negative statement that Daguerre had failed where Draper had succeeded, is strange; and the allusion to the distinct representation of the moon's surface rather implies that other representations existed. It is, however, the earliest record I can find, and we may consider it the starting-point.

Beyond experimental work little seems to have been done with the daguerreotype.

Some astronomers, notably G. P. Bond in America, assisted by two skilful photographers, with the 15-inch refractor of the Harvard College Observatory, obtained photographs of some of the brighter stars, and also some very fair pictures of the moon, that were exhibited at the 1851 Exhibition in London, and also at a meeting of the R.A.S. in May of that year; and some solar and spectroscopic work was also done in Europe.

The important fact of the possibility of thus getting pictures of the heavenly bodies was established, so that with the introduction of the collodion process in 1851, with its great advantages over the difficult and costly daguerreotype, astronomical photography was taken up and soon became firmly established. From this time its history became a record of continual advance, delayed, it is true, from time to time by the want of improvement in instrument or method, when further extensions of the art were attempted, in every case with ultimate success.

Of the early workers with the collodion process, and the more recent workers with the modern gelatine or dry plate process, and the persevering and skilful way they have dealt with the difficulties that always surround a new art, I do not propose to speak, except incidentally. Time would not allow me to do so here, nor is it part of my purpose; my object being rather to deal with the results obtained, to speak of this new art or method of astronomical observation and record, than to give an account of the labours of those who have made it what it now is. I propose to exhibit by the electric lantern such specimens of early and recent work as I have been able to obtain for this purpose,* and from which I think you may form an idea as to the present value of photography as an aid to astronomy, and the probable greater aid it may in future become.

Before doing so, however, although the art of photography is now known to almost every one, I should like to say a few words about the three processes I have mentioned, the daguerreotype, the collodion, and the gelatine or dry plate process, and also to give very shortly a general idea of the instruments and methods in use by the astronomical photographer.

For an account of the daguerreotype one has to refer to the text works, as it is not now in general use. Shortly it may be described

---

* For the loan of some of the photographs exhibited I am indebted to the kindness of Mr. Crookes, Dr. de la Rue, Mr. Lockyer, and Captain Abney, and also to the Brothers Henry of Paris for specimens of their recent work.

as the use of a polished surface of silver (usually supported by a thin backing of copper) that has been exposed to the action of iodine so as to form a film of sensitive iodide of silver, that may be acted upon by light, and afterwards developed by the vapour of mercury. Although not now used for astronomical work, it may be that owing to the intimate connection between the image formed by the silver compound and the backing that supports it, it will be found to have advantages in cases where delicate measurements are required, that the other processes may not have.

The collodion and gelatine (or dry plate) processes are so called from the nature of the medium that is used to carry the sensitive salts of silver during exposure to the action of light. The chief difference between them, beyond the greater sensitiveness of the latter, is that with the collodion plate the process of preparation, exposure, and development must be part of a continuous operation taken in due order and time, with all the necessary apparatus, and the chemicals in a proper state of efficiency, ready to hand, conditions not easy to attend to in astronomical photography, and when long exposures are to be given, hardly to be fulfilled.

With the gelatine process the plates can be prepared beforehand under the best conditions, and almost any time may elapse before use. The exposure of the plate may take place for any length of time, and the development made at any suitable time afterwards; all advantages for astronomical work that are obvious; and in addition, there is the important advantage of the greatest sensitiveness, on which, more than anything else, success so much depends when it becomes a question of dealing with a small amount of light. Each process has its particular advantages of which the specialist may avail himself; but the gelatine dry plate is far beyond the others for nearly every class of astronomical work.

With regard to the instruments and methods of work, while the main principles are adhered to, modifications have to be made to suit different classes of work. These will be briefly mentioned in speaking of the photographs. In all cases there must be an image of the object, and this image must fall exactly on the sensitive plate, and be kept there during exposure.

The image produced by a telescope reflecting or refracting is generally used direct; indeed, with the exception of the sun, where the large amount of light rendered enlargement advantageous, until quite recently the primary image was always used, it being thought that with the small amount of light generally at disposal it was better to get a picture thus and then enlarge it afterwards than to enlarge the first image and so increase the time of exposure and all the trouble that comes from atmospheric and instrumental tremors, and other causes.

We shall see how much has been gained by departing from this old plan and using an enlarged image when we come to examine the photographs of the planets.

In the method of working there is one important difference between that followed by the terrestrial and celestial photographer. Without exception, everything that the latter has to photograph is in continual apparent motion, owing to the rotation of the earth, and in some cases to the proper motion of the object itself. This necessitates the use of an equatorial mounting to carry the photographic apparatus, with clockwork to give it a regular motion in a direction contrary to that of the earth. Even then the difficulties of keeping the telescope moving for one or more hours without allowing deviation of the image on the sensitive plate of a $\frac{1}{1000}$ of an inch during this time, taxes very severely the powers of the observer; for, every such long exposure must be watched not only to correct irregularities of the clock, but other slight though important movements, due to change of refraction and other causes, which, if not immediately corrected, would spoil the picture. These mechanical difficulties, however, are not insurmountable, as will be seen from some of the photographs I shall show; and as instruments improve and workers gain experience they will become less.

There are many technical details of extreme interest to the worker, which it is hardly necessary to name to-night.

The light from the different celestial bodies varies greatly in intensity. Between that from our sun and that from the faintest nebula or star that can be seen, there is such an immense difference, that their relative amounts can hardly be expressed by figures.

Dr. Huggins estimates the light of the faintest nebula that can be seen with a moderately large instrument as equal to $\frac{1}{20000}$ of the light of a single standard candle viewed at a distance of a quarter of a mile, that is, that such a candle a quarter of a mile off is 20,000 times more brilliant than the nebula.

The astronomer, who deals with both, has therefore need of all his art to reduce the light in the one case to that suitable for his purpose; and to utilise every ray he can get in the other, regulating the exposure for the one to a minute fraction of a second, and extending for the other to hours.

Between these two extremes of light-giving power are comprised all other celestial objects.

For the purpose of convenience I will take the photographs, which I propose to show you, in the following order:—(1) those of the sun; (2) the moon; (3) the stars; (4) planets; (5) nebulæ; and (6) comets; giving in nearly every case an early photograph and a recent one for comparison; and, where I can, a specimen of the work of eye and hand that may be directly compared with a photograph of the same object.

With the sun there are two distinct phenomena to observe: (1) the physical aspect of his surface, with the remarkable spots and markings that are frequently visible; and (2) the wonderful prominences and corona that surrounds the sun and becomes visible when he is eclipsed.

The first important photograph of the surface of the sun seems to have been obtained by Dr. de la Rue, July 24th, 1861.

The photograph I show is one copied from a picture, itself a reproduction, without retouch, of the original negative.

(Photographs of the surface of the sun, by Janssen, were also shown.)

Berkowski, in the eclipse of the 28th July, 1851, took by the daguerreotype the first photograph showing the corona and prominences, and Dr. de la Rue in 1860 took the first good photographs of the prominences, and obtained traces of the corona, using the collodion process.

In 1869, Professor Stephen Alexander obtained at Ollumwa the first good photographs of the corona.

(Photographs of the corona and prominences, by General Tennant, Dr. de la Rue, and Captain Abney were shown.)

Since this time, photography has been used at every total eclipse that has been observed, with increasing success.

In speaking of these corona photographs, I must not forget to mention the important work of Dr. Huggins in photographing the uneclipsed corona. He himself has lately given an account of the methods he employs, and I have no doubt that under his skilful direction we shall see the same successful advances as those we can mark in every branch of astronomical photography; although it is a work which many of those who know the great difficulties to be encountered would hardly have dared to attempt.

Photographs of the moon are so easy to obtain that she has received more attention than any other celestial object; yet, strange to say, with less improvement, the pictures taken by Rutherford more than twenty years ago, not being yet surpassed.

Now, however, that we can safely enlarge the primary image without unduly prolonging the exposure we may soon expect photographs of portions of the moon that will be far beyond anything hitherto done, or possible, where the whole image is attempted.

[Photographs of the moon, by Mr. Crookes, Dr. de la Rue, and others, were shown.]

With the stars photography has recently been most successful.

Rutherford, in 1864, completed a photographic objective of $11\frac{1}{4}$ inches aperture and 14 feet focal length, with which he obtained some very fine photographs. Some of his remarks, written in 1864,* in connection with the future of astronomical photography, are so interesting at the present time that I will repeat them. "Since the completion of the photographic objective, but one night has occurred (the 6th of March) with a fine atmosphere, and on that occasion the instrument was occupied with the moon; so that as yet I have not tested its powers upon the close double stars, 2″ being

* 'American Journal of Science and Art,' 2nd series, vol. xxxix. p. 308.

the nearest pair it has been tried upon. This distance is quite manageable provided the stars are of nearly equal magnitude.

"The power to obtain images of the ninth magnitude stars with so moderate an aperture promises to develop and increase the application of photography to the mapping of the sidereal heavens and in some measure to realise the hopes which have so long been deferred and disappointed.

"It would not be difficult to arrange a camera-box capable of exposing a surface sufficient to obtain a map of two degrees square, and with instruments of large aperture we may hope to reach much smaller stars than I have yet taken. There is also every probability that the chemistry of photography will be very much improved and more sensitive methods devised."

In the light of recent work these words are almost prophetic. The sensitive methods have been devised, and the result is that the anticipations formed by Rutherford in 1864 are in 1886 not only fulfilled but exceeded. It is in stellar photography that the astronomer will be most benefited by the immense saving of labour in making charts; a single plate taken in one hour showing in their proper relative place and in their relative photographic magnitudes, all the stars down to the fifteenth magnitude over an area of about six degrees—a result that it would be hardly possible to obtain by the usual method of eye observation and measure.

[Photographs of the stars round Altair, taken in 1883, of part of Orion, taken in 1884 by the speaker, and some of the recent plates of stars and double stars, by the Brothers Henry of Paris, were shown, with a companion plate of similar parts of the sky as shown on Argelander's maps, showing the enormous increase in the number of stars shown by one hour's exposure, in one case ten times as many stars being on the photographic plate as on the same area of the map.]

The planets Jupiter and Saturn were photographed by Dr. de la Rue and others in the early days of photography. I have not been able to obtain copies, or even a sight of any of these earlier ones, but I have some of my own work that will enable you to see the improvement that has been made since 1876.

[Photographs of Saturn from 1877 and of Jupiter from 1878 were shown, including some of the recent work of the Brothers Henry of Paris, showing the great advance obtained by the enlargement of the primary image.]

With nebulæ, although the work has been all done within the last few years, the results have been very satisfactory.

Dr. Draper, in 1880, obtained a very promising picture of the Orion nebula, and I was able in 1883 (after trials commencing in 1879) to get, with a three-foot reflector, some very fair photographs. A comparison with the last drawing will show the chief points of difference.

Other nebulæ have been photographed by myself and others, and the power of the photograph to portray these mysterious shapes has been thoroughly demonstrated.

In a recent photograph of the Pleiades, which the MM. Henry have taken, they have obtained a picture of a nebula near Maia that has not been seen before, and which has since been seen with the great telescope at Pulkowa in Russia—a telescope very much larger than that with which the photograph was taken.

On this plate part of the nebula near Merope is also shown.

[Photographs of Orion with exposures of from 1 to 80 minutes were shown, and also photographs of the drawings made by the Bonds, Lord Rosse, Trouvelot, and others, for comparison.]

Of comets, I have here copies of two of the photographs taken at the Cape of Good Hope, in 1882. These pictures may be called photographs of stars from the immense number that have impressed themselves on the plate.

To others as well as to myself, these photographs came as a revelation of the power of photography in this direction, and it is probably to them that the increased attention lately given to stellar photography is due.

There are other applications of photography to the work of the astronomer besides those I have mentioned.

By the analysis of the light that comes to us from the heavenly bodies the spectroscope tells us what elementary substances exist in those bodies and in this most delicate research photography has played a most important part, especially in dealing with that part of the spectrum that the eye is not able to grasp.

The fleeting image that requires all the care that the mind can give to interpret is recorded by means of photography, and can then be studied under the most favourable conditions.

Such photographs cannot be shown in the same way as those I have shown you to-night, nor can they be rendered intelligible except to the very few who have made the consideration their study.

The recording of the passage of a star past the wires of a transit instrument has always hitherto been done by the eye, but it is quite possible that here photography may come in for this purpose, and render such observations free from personal equation, as the allowance that has to be made for different powers of different brains to record an event is called by astronomers.

There is also a possibility that photography will be available to record the paths of meteors and thus aid in a research that is engaging more attention every day.

The discovery of minor planets by means of photography cannot be helped when such photographs of the heavens as those taken by the MM. Henry are produced consecutively, a comparison from time to time being sufficient to detect the displacement in position that they must undergo, and it is not too much to say that if Uranus had not been discovered by Herschel, and in consequence of the disagreement between the position this planet should have occupied, and those it did occupy owing to the attraction of Neptune, and the subsequent discovery of this planet by the entirely theoretical investigations of

Professors Adams and Leverrier, both would have been discovered eventually by photography. And if there is now, as some suppose there is—and there is nothing against such a supposition—another major planet beyond Neptune, it is most probable that it will be thus discovered.

In thus bringing before you all these wonderful things that can be done by photography I do not wish to imply that it is quite a new thing. Astronomers have known and valued it, but the immense step it has lately taken through the great sensitiveness of the dry plate process has not, I believe, been fully realised.

For many years the art of photography as applied to astronomy has remained very nearly in the state it was when Rutherford wrote those remarks I have quoted. At a bound it has gone far beyond anything that was expected from it, and bids fair to overturn a good deal of the practice that has hitherto existed among astronomers. I hope soon to see it recognised as the most potent agent of research and record that has ever been within the reach of the astronomer; so that the records that the future astronomer will use, will not be the written impressions of dead men's views, but veritable images of the different objects of the heavens recorded by themselves as they existed.

[A. A. C.]

Friday, March 19, 1886.

SIR WILLIAM BOWMAN, Bart. LL.D. F.R.S. Manager and Vice-President, in the Chair.

W. H. M. CHRISTIE, Esq. M.A. F.R.S. Astronomer Royal.

Universal Time.

CONSIDERING the natural conservatism of mankind in the matter of time-reckoning it may seem rather a bold thing to propose such a radical change as is involved in the title of my discourse. But in the course of the hour allotted to me this evening, I hope to bring forward some arguments which may serve to show that the proposal is not by any means so revolutionary as might be imagined at the first blush.

A great change in the habits of the civilised world has taken place since the old days when the most rapid means of conveyance from place to place was the stage-coach, and minutes were of little importance. Each town or village then naturally kept its own time, which was regulated by the position of the sun in the sky. Sufficient accuracy for the ordinary purposes of village life could be obtained by means of the rather rude sun-dials which are still to be seen on country churches, and which served to keep the village clock in tolerable agreement with the sun. So long as the members of a community can be considered as stationary, the sun would naturally regulate, though in a rather imperfect way, the hours of labour and of sleep and the times for meals, which constitute the most important epochs in village life. But the sun does not really hold a very despotic sway over ordinary life, and his own movements are characterised by sundry irregularities to which a well-ordered clock refuses to conform.

Without entering into detailed explanation of the so-called "Equation of Time," it will be sufficient here to state that, through the varying velocity of the earth in her orbit, and the inclination of that orbit to the ecliptic, the time of apparent noon as indicated by the sun is at certain times of the year fast and at other times slow, as compared with 12 o'clock, or noon by the clock. [The clock is supposed to be an ideally perfect clock going uniformly throughout the year, the uniformity of its rate being tested by reference to the fixed stars.] In other words, the solar day, or the interval from one noon to the next by the sun, is at certain seasons of the year shorter than the average, and at others longer, and thus it comes about that by the accumulation of this error of going, the sun is at the beginning of November more than 16 minutes fast, and by the middle of February 14½ minutes slow, having lost 31 minutes, or more than half-an-hour, in the interval. In passing it may be mentioned as a result of this that the afternoons in November are about half-an-hour shorter than the mornings, whilst in February the mornings are half-an-hour shorter than the afternoons. In view of the importance attached by some astronomers to the use of exact local time in civil life, it would be interesting to know how many villagers have remarked this circumstance.

It is essential to bear these facts in mind when we have to consider the extent to which local time regulates the affairs of life, and the degree of sensitiveness of a community to a deviation of half-an-hour or more in the standard reckoning of time. My own experience is that in districts which are not within the influence of railways the clocks of neighbouring villages commonly differ by half-an-hour or more. The degree of exactitude in the measurement of local time in such cases may be inferred from the circumstance that a minute-hand is usually considered unnecessary. I have also found that in rural districts on the Continent arbitrary alterations of half-an-hour fast or slow are accepted not only without protest but with absolute indifference.

Even in this country where more importance is attached to accurate time, I have found it a common practice in outlying parts of Wales (where Greenwich time is about twenty minutes *fast* by local time) to keep the clock half-an-hour fast by railway (i.e. Greenwich) time, or about fifty minutes fast by local time. And the farmers appeared to find no difficulty in adapting their hours of labour and times of meals to a clock which at certain times of the year differed more than an hour from the sun.

There is a further irregularity about the sun's movements which makes him a very unsafe guide in any but tropical countries. He is given to indulging in a much larger amount of sleep in winter than is desirable for human beings who have to work for their living and cannot hibernate as some of the lower animals do. To make up for this he rises at an inconveniently early hour in summer and does not retire to rest till very late at night. Thus it would seem that a clock of steady habits would be better suited to the genius of mankind.

Persons whose employment requires daylight must necessarily modify their hours of labour according to the season of the year, whilst those who can work by artificial light are practically independent of the vagaries of the sun. Those who work in collieries, factories, or mines, would doubtless be unconscious of a difference of half-an-hour or more between the clock and the sun, whilst agriculturists would practically be unaffected by it, as they cannot have fixed hours of labour in any case.

Having thus considered the regulating influence of the sun on ordinary life within the limits of a small community, we must now take account of the effect of business intercourse between different communities separated by distances which may range from a few miles to half the circumference of our globe. So long as the means of communication were slow, the motion of the traveller was insignificant compared with that due to the rotation of the earth, which gives us our measure of time. But it is otherwise now, as I will proceed to explain.

Owing to the rotation of the earth about its axis, the room in which we now are is moving eastward at the rate of about 600 miles an hour. If we were in an express train going eastward at a speed of sixty miles an hour (relatively to places on the earth's surface), the velocity of the traveller due to the combined motions would be 660 miles an hour, whilst if the train were going westward it would be only 540 miles. In other words, if local time be kept at the stations, the apparent time occupied in travelling sixty miles eastward would be 54 minutes, whilst in going sixty miles westward it would be 66 minutes. Thus the journey from Paris to Berlin would apparently take an hour and a half longer than the return journey, supposing the speed of the train to be the same in both cases.

In Germany, under the influence of certain astronomers, the system of local time has been developed to the extent of placing posts along the railways to mark out each minute of difference of time from Berlin. Thus there is an alteration of one minute in time reckoning for every ten miles eastward or westward, and even with the low rate of speed of German trains, this can hardly be an unimportant quantity for the engine-drivers and guards, who would find that their watches appeared to lose or gain (by the station clocks) one minute for every ten miles they have travelled east or west. This would seem to be the *reductio ad absurdum* of local time.

In this country the difficulty as to the time-reckoning to be used on railways was readily overcome by the adoption of Greenwich time throughout Great Britain. The railways carried London (i. e. Greenwich) time all over the country, and thus local time was gradually displaced. The public soon found that it was important to have correct railway time, and that even in the west of England, where local time is about 20 minutes behind Greenwich time, the discordance between the sun and the railway clock was of no practical consequence. It is true that for some years both the local and the railway times were shown on village clocks by means of two minute-hands, but the complication of a dual system of reckoning time naturally produced inconvenience, and local time was gradually dropped. Similarly in France, Austria, Hungary, Italy, Sweden, &c., uniform time has been carried by the railways throughout each country. It is noteworthy that in Sweden the time of the meridian one hour east of Greenwich has been adopted as the standard, and that local time at the extreme east of Sweden differs from the standard by about 36½ minutes.

But in countries of great extent in longitude, such as the United

States and Russia, the time-question was not so easily settled. It was in the United States and Canada that the complication of the numerous time standards then in use on the various railways forced attention to the matter. To Mr. Sandford Fleming, the constructor of the Inter-Colonial Railway of Canada and engineer-in-chief of the Pacific Railway, belongs the credit of having originated the idea of a universal time to be used all over the world. In 1879 Mr. Fleming set forth his views on time-reckoning in a remarkable paper read before the Canadian Institute. In this he proposed the adoption of a universal day, commencing at Greenwich mean noon or at midnight of a place on the anti-meridian of Greenwich, i. e. in longitude 180° from Greenwich. The universal day thus proposed would coincide with the Greenwich astronomical day, instead of with the Greenwich civil day which is adopted for general use in this country.

The American Metrological Society in the following year issued a report recommending that, as a provisional measure, the railways in the United States and Canada should use only five standard times, 4, 5, 6, 7, and 8 hours respectively later than Greenwich, a suggestion originally made in 1875 by Prof. Benjamin Pierce. This was proposed as an improvement on the then existing state of affairs, when no fewer than seventy-five different local times were in use on the railroads, many of them not differing more than 1 or 2 minutes. But the committee regarded this merely as a step towards unification, and they urged that eventually one common standard should be used as railroad and telegraph time throughout the North American continent, this national standard being the time of the meridian 6 hours west of Greenwich, so that North American time would be exactly 6 hours later than Greenwich time.

Thanks to the exertions of Mr. W. F. Allen, Secretary of the General Railway Time Convention, the first great practical step towards the unification of time was taken by the managers of the American railways on November 18, 1883, when the five time standards above mentioned were adopted. Mr. Allen stated in October 1884, that these times were already used on 97½ per cent. of all the miles of railway lines, and that nearly 85 per cent. of the total number of towns in the United States of over 10,000 inhabitants had adopted them.

I wish to call particular attention to the breadth of view thus evinced by the managers of the American railways. By adopting a national meridian as the basis of their time-system, they might have rendered impracticable the idea of a universal time to be used by Europe as well as America. But they rose above national jealousies, and decided to have their time-reckoning based on the meridian which was likely to suit the convenience of the greatest number, thus doing their utmost to promote uniformity of time throughout the world by setting an example of the sacrifice of human susceptibilities to general expediency.

Meanwhile Mr. Sandford Fleming's proposal had been discussed at the Geographical Congress at Venice in 1881, and at a meeting of the Geodetic Association at Rome in 1883. Following on this a special Conference was held at Washington in October 1884, to fix on a meridian proper to be employed as a common zero of longitude and standard of time-reckoning throughout the globe. As the result of the deliberation it was decided to recommend the adoption of the meridian of Greenwich as the zero of longitude, and the Greenwich civil day (commencing at Greenwich midnight and reckoned from 0 to 24 hours) as the standard for time-reckoning. In making this selection the delegates were influenced by the consideration that the meridian of Greenwich was already used by an overwhelming majority of sailors of all nations, being adopted for purposes of navigation by the United States, Germany, Austria, Italy, &c. Further, the United States had recently adopted Greenwich as the basis of their time-reckoning, and this circumstance in itself indicated that this was the only meridian on which the Eastern and Western Hemispheres were likely to agree.

The difficulties in the way of an agreement between the two hemispheres may be appreciated by the remarks of the Superintendent of the American Ephemeris on Mr. Sandford Fleming's scheme for universal time (which was subsequently adopted in its essentials at the Washington Conference):—"A capital plan for use during the millennium. Too perfect for the present state of humanity. See no more reason for considering Europe in the matter than for considering the inhabitants of the planet Mars. No; we don't care for other nations, can't help them, and they can't help us." *

As a means of introducing universal time, it has been proposed by Mr. Sandford Fleming, Mr. W. F. Allen, and others, that standard times based on meridians differing by an exact number of hours from Greenwich should be used all over the world. In some cases it may be that a meridian differing by an exact number of half-hours from Greenwich would be more suitable for a country like Ireland, Switzerland, Greece, or New Zealand, through the middle of which such a meridian would pass, whilst one of the hourly meridians would lie altogether outside of it.

The scheme of hourly meridians, though valuable as a step towards uniform time, can only be considered a provisional arrangement, and though it may work well in countries like England, France, Italy, Austria, Hungary, Sweden, &c., which do not extend over more than one hour of longitude, in the case of such an extensive territory as the United States difficulties arise in the transition from one hour-section to the next which are only less annoying than those formerly experienced, because the number of transitions has been reduced from seventy-five to five, and the change of time has been made so large that there is less risk of its being overlooked. The natural inference from this is that one time-reckoning should be used throughout the

---

* 'Proceedings' of the Canadian Institute, Toronto, No. 143, July 1885.

whole country, and thus we are led to look forward to the adoption in the near future of a national standard time, six hours slow by Greenwich, for railways and telegraphs throughout North America.

We may then naturally expect that by the same process which we have witnessed in England, France, Italy, Sweden, and other countries, railway time will eventually regulate all the affairs of ordinary life. There may of course be legal difficulties arising from the change of time-reckoning, and probably in the first instance local time would be held to be the legal time unless otherwise specified.

It seems certain that when a single standard of time has been adopted by the railways throughout such a large tract of country as North America, where we have a difference of local times exceeding five hours, the transition to universal time will be but a small step.

But it is when we come to consider the influence of telegraphs on business life, an influence which is constantly exercised, and which is year by year increasing, that the necessity for a universal or world time becomes even more apparent. As far as railways are concerned, each country has its own system, which is to a certain extent complete in itself, though even in the case of railways the rapidly increasing inter-communication between different countries makes the transition in time-reckoning on crossing the frontier more and more inconvenient. Telegraphs, however, take no account of the time kept in the countries through which they pass, and the question, as far as they are concerned, resolves itself into the selection of that system of time-reckoning which will give least trouble to those who use them.

For the time which is thus proposed for eventual adoption throughout the world, various names have been suggested. But whether we call it Universal, Cosmic, Terrestrial, or what seems to me best of all, World Time, I think we may look forward to its adoption for many purposes of life in the near future.

The question, however, arises as to the starting-point for the universal or world day. Assuming that, as decided by the great majority of the delegates at Washington, it is to be based on the meridian of Greenwich, it has still to be settled whether the world day is to begin at midnight or noon of that meridian. The astronomers at Rome decided by a majority of twenty-two to eight in favour of the day commencing at Greenwich noon, that is, of making the day throughout Europe begin about midday. However natural it might be for a body of astronomers to propose that their own peculiar and rather inconvenient time-reckoning should be imposed on the general public, it seems safe to predict that a World Day which *commenced* in the middle of their busiest hours would not be accepted by business men. In fact, the idea on which this proposal was founded was that universal time would be used solely for the internal administration of railways and telegraphs, and that accurate local time must be rigidly adhered to for all other purposes. It was conceded, however, that persons who travelled frequently might with advantage use universal time during railway journeys. This attempt to separate the travelling from the stationary public seems to be one that is not likely to meet with success even temporarily, and it is clear that in the future we may expect the latter class to be completely absorbed in the former. Another argument that influenced the meeting at Rome was the supposed use of the astronomical day by sailors. Now it appears that sailors never did use the astronomical day, which begins at the noon *following* the civil midnight of that date, but the nautical day which begins at the noon *preceding*, i. e. twenty-four hours before the astronomical day of the same date, ending when the latter begins. And the nautical day itself has long been given up by English and American sailors, who now use a sort of mongrel time-reckoning, employing civil time in the log-book and for ordinary purposes, whilst, in working up the observations on which the safe navigation of the ship depends, they are obliged to change civil into astronomical reckoning, altering the date where necessary, and interpreting their a.m. and p.m. by the light of nature. It says something for the common-sense of our sailors that they are able to carry out every day without mistake this operation, which is considered so troublesome by some astronomers.

In this connection I may mention that the Board of Visitors of Greenwich Observatory have almost unanimously recommended that, in accordance with the resolution of the Washington Conference, the day in the English 'Nautical Almanac' should be arranged from the year 1891 (the earliest practicable date) to begin at Greenwich midnight (so as to agree with civil reckoning, and remove this source of confusion for sailors), and that a committee appointed by them have drawn up the details of the changes necessary to give effect to this resolution without causing inconvenience to the mercantile marine.

The advantage of making the world day coincide with the Greenwich civil day is that the change of date at the commencement of a new day falls in the hours of the night throughout Europe, Africa, and Asia, and that it does not occur in the ordinary office hours (10 a.m. to 4 p.m.) in any important country except New Zealand. In the United States and Canada the change of date would occur after four in the evening, and in Australia before ten in the morning. This arrangement would thus reduce the inconvenience to a minimum, as the part of the world in which the change of date would occur about the middle of the local day is almost entirely water, whilst on the opposite side we have the most populous continents.

The question for the future seems to be whether it will be found more troublesome to change the hours for labour, sleep, and meals once for all in any particular place, or to be continually changing them in communications from place to place, whether by railway, telegraph, or telephone. When universal or world time is used for railways and telegraphs, it seems not unlikely that the public may find it more convenient to adopt it for all purposes. A business man who daily travels by rail, and constantly receives telegrams from all parts of the world, dated in universal time, would probably find it easier to learn once for all that local noon is represented by 17h. U.T. and midnight by 5h. (as would be the case in the Eastern States of North America), and that his office hours are 15h. to 21h. U.T., than to be continually translating the universal time used for his telegrams into local time.

If this change were to come about, the terms noon and midnight would still preserve their present meaning with reference to local time, and the position of the sun in the sky, but they would cease to be inseparably associated with twelve o'clock.

The introduction of Universal Time would practically involve the adoption of the system of counting the hours in one series from 0 to 24, instead of in the two series 0 to 12 a.m. and p.m., for as applied to Universal Time the terms *ante-meridiem* and *post-meridiem* would be meaningless, except for places on the meridian of Greenwich. The use of the 24 hour system on railways and telegraphs would naturally assist in breaking the spell of habit which associates noon and midnight with 12 o'clock.

It may be mentioned that the Eastern and Eastern Extension Telegraph Companies already use the 24 hour system throughout their extensive lines of telegraph to avoid mistakes of a.m. and p.m., and to save telegraphing these unnecessary letters. In this connection the President of the Western Union Telegraph Company in the United States has stated that the adoption of the 24 hour mode of reckoning would, besides materially reducing the risk of error, save at least 150 million letters annually on the lines of his company. It is also noteworthy that 98 per cent. of the railway managers in the United States, representing 60,000 miles of railway, have expressed themselves in favour of the adoption of the simple notation from 0 to 24 hours.

Considering that the only change which we are called on, in accordance with the Washington Resolutions, to make in our time-reckoning on railways is the adoption of the 24 hour system, it may be hoped that our railway companies will not be behind those of the United States in appreciating the simplification in railway time-tables, which would result from this reform.

[W. H. M. C.]

Friday, April 2, 1886.

WILLIAM HUGGINS, Esq. D.C.L. LL.D. F.R.S. Vice-President, in the Chair.

HOWARD GRUBB, Esq. F.R.S. F.R.A.S.

Telescopic Objectives and Mirrors: their Preparation and Testing.

IT would probably lend an additional interest to a technical subject such as I have to bring before you to-night, could I preface my description of the processes now employed in the construction of telescopic objectives by a short historical account of what has been attempted and achieved in the past, but time will not permit.

A very few words, however, on the history of glass manufacture are necessary.

As I pointed out last Saturday afternoon, Dollond's brilliant discovery, which afforded a means of achromatising objectives, rendered possible their construction of greater size and perfection than formerly, provided suitable material could be obtained. But the chromatic errors being removed, faults in the material hitherto masked by them, were detected, and it was not until after many years that Guinand, a lowly but gifted Swiss peasant, succeeded in producing glass discs of a considerable size and free from these defects.

The secrets of his process have been handed down in his own family to Mons. Feil, of Paris (one of his descendants), and also through M. Bontemps, who for a time was associated with Guinand's son, and afterwards accepted an invitation from Messrs. Chance Bros. & Co., of Birmingham, to assist them in an endeavour to improve that branch of their manufacture. Only these two houses, so far as I am aware, have succeeded in manufacturing optical discs of large size.

TESTING OF OPTICAL GLASS.

Let me here say a few words respecting the testing of optical glass; I mean of the material of the glass, quite apart from the optician's work in forming it into an objective. When received from the glass manufacturer it is sometimes in this state, roughly polished on both sides, and sometimes in this, in which as you see there are small windows only, facets as they are called, polished on the edges. In case of lenses for telescopic objectives, it is always well to have them roughly polished on the sides, to avoid the chance of having to throw away a lens after much trouble and labour has been spent on it.

There are only three distinct points to be looked to in the testing of optical glass; (1) General clearness and freedom from air-bubbles, specks, pieces of "dead metal," &c.; (2) Homogeneity; (3) Annealing.

The first is the least important, and needs no instructions for detection of defects, any one can see these. The second is much more important, and much more difficult to test.

The best test for homogeneity is one somewhat equivalent to Foucault's test for figure of concave mirrors.

The disc of glass should be either ground and polished to form a convex lens, or if that be not convenient, it should be placed in juxtaposition with a convex lens of similar or larger size, and whose excellence has been established by previous experience.

The lens or disc is then placed opposite some small brilliant light—a small gas flame generally suffices—and at such a distance that a conjugate focus is formed at other side and at a convenient distance. When the exact position of this focus is found, the eye is placed as nearly as possible so that the image of flame is formed on the pupil. On looking at it with the eye in this position, the whole lens should appear to be "full of light"; but at the slightest movement to one side the light will disappear and the lens appear quite dark. If the eye be now passed slowly backwards and forwards between the position showing light and darkness, any irregularity of density will be most easily seen.

Of course, like everything else, some experience is necessary.

The rationale of this is very obvious. When the eye is placed exactly at the focus of a perfect lens, the image formed on the pupil is very small, and the slightest movement of the eye will cause the light to appear and disappear. If the eye be not at the focus, the pencil of light will be larger, and consequently it will require a much greater movement of the eye to cause the light to disappear. Now if any portion of the lens be of a different density to the general mass, that portion will have a longer or a shorter focus; consequently while the light flashes off the general area of the lens quickly, it still remains on the defective portions.

By imitating this arrangement and substituting a camera for the eye and forming the focus of a small point of light on the stop of the lens, I have succeeded in photographing veins in glass, and sometimes have found this useful as a record.

The third point—that of proper annealing—is easily tested by the polariscope.

For small discs the usual plan is to hold them between the eye and a polarizing plane, such as a piece of glass blackened at back or a japanned surface, and look at them through the facets, using as an analyser a Nichols prism.

Larger sizes, which are polished on the surfaces, can be more easily examined. It is difficult to describe the appearances, but I will put a few discs into the lantern polariscope and endeavour to point out what amount of polarization may safely be permitted in discs of glass to be used for objectives.

The composition of metallic mirrors of the present day differs very little from that used by Sir Isaac Newton. Many and different alloys have been suggested, some including silver or nickel or arsenic; but there is little doubt that the best alloy, taking all things into account, is made with 4 atoms of copper and 1 of tin, which gives the following proportions by weight: copper 252, tin 117·8.

Calculation of Curves.

Having now obtained the proper material to work upon, the first thing necessary is to calculate the curves to give to the lenses, in order that the objective when finished may be of the required focus, and be properly corrected for the chromatic and spherical aberrations.

As this lecture is intended to deal principally with the technical details of the process, I do not intend to occupy your time for more than a few moments on this head, nor indeed is it at all necessary. In my lecture last Saturday I explained the principles of achromatism, and in many published works full and complete particulars are given as to the calculation of the curves—particulars which are sufficient, and more than sufficient, for the purpose.

Much has been discussed and written concerning the calculation of curves of objectives, and much care and thought has been bestowed by mathematicians on this subject, and so far as the actual constructors are concerned, a certain amount of veil is thrown over this part of the undertaking, as if there were a secret involved, and as if each had discovered some wonderful formulæ by which he was enabled to calculate the curves much more accurately than others.

I am sorry to have to dispel this illusion. Practically the case stands thus. The calculation of the curves which satisfy the conditions of achromatism and desired focus is a most simple one, and can be performed by any one having a very slight algebraical knowledge in a few minutes, provided the refractive indices and dispersive power of the glass be known. Both Messrs. Chance and Feil supply these data quite sufficiently accurately for small-size objectives. Speaking for myself, I am quite content to take the figures as given by these glass manufacturers for any discs up to 10 inches in diameter. If over that size, I grind and polish facets on the disc and measure the refractive and dispersive powers myself.

The calculations of the curves required to satisfy the conditions of spherical aberration are very troublesome, but fortunately these may be generally neglected.

Some years ago the Royal Society commissioned one of its members to draw up tables for the use of opticians, giving the curves required to satisfy the conditions of both corrections for all refractive and dispersive indices.

A considerable amount of labour was expended on this work, but in the end it was abandoned, for it was found that the calculation of these curves was founded on the supposition that all surfaces produced by the opticians were truly spherical; while the fact is, a truly spherical curve is the exception, not the rule. The slightest variation in the form or figure of the curve will produce an enormous variation in the correction for spherical aberration, and it was soon apparent that the final correction for spherical aberration must be left to the optician and not to the mathematician. *Object-glasses cannot be made on paper.* When I tell you that a sensible difference in correction for spherical aberration can be made by half an hour's polishing, corresponding probably to a difference in the first place of decimals in radii of the curves, you will see that it is practically not necessary to enter upon any calculation for spherical aberration. We know about what form gives an approximate correction; we adhere nearly to that, and the rest is done by figuring of the surface.

To illustrate what I mean. I would be quite willing to undertake to alter the curves of the crown or flint lens of any of my objectives by a very large quantity, increasing one and decreasing the other so as to still satisfy the conditions of achromatism, but introducing theoretically a large amount of positive or negative spherical aberration, and yet to make out of the altered lens an object-glass perfectly corrected for spherical aberration.

I am now speaking of ordinary sizes. For very large sizes it is usual to go more closely into the calculations; but I may remark that it is sometimes possible to make a better objective by deviating from the curves which give a true correction for spherical aberration and correcting that aberration by figuring, rather than by strictly adhering to the theoretical curves. So far then as any calculation is required, the ordinary formulæ given in the text-books may be considered amply sufficient.

Having now determined on the curves, we have to consider the various processes which the glass has to undergo from the time it is received in this form from the glass manufacturer to the time when it is turned out a finished objective.

The work divides itself into five distinct operations:—(1) Rough grinding; (2) Fine grinding; (3) Polishing; (4) Centering; (5) Figuring and testing.

1. The rough grinding or approximate shaping of the glass is a very simple process. The glass is cemented on a holder, and is held against a revolving tool supplied with sand and water, and of a shape which will tend to abrade whatever portions are necessary to be removed to produce the required curves. These diagrams will illustrate the various operations.

2. Fine grinding. The tools used for fine grinding are of this

form, and are made of either brass or cast iron. I prefer cast iron, except for very small sizes. They are grooved on the face, in the manner suggested by the late Mr. A. Ross, in order to allow the grinding material to properly distribute itself.

If two spherical surfaces be rubbed together they will, as may be supposed, tend to keep spherical; for the spherical is the only curve which is the same radius every part of its surface. If fine dry abrading powder be used between, the same result will be obtained; but if wet powder be used, the surface will no longer continue spherical, but will abrade away more on the centre and edge than in the zone between. It was to meet this difficulty that the late Mr. A. Ross devised the idea of the distributing grooves. The fine grinding process is the first of the series which calls for any skill on the part of the operator.

That the *modus operandi* of the grinding be the more easily understood, let me explain the principle of the process in a few words.

When two surfaces of unequal hardness are rubbed together with emery powder and water between the two, each little particle of the powder is at any given moment in either of these conditions:—(*a*) Imbedded into the softer surface; (*b*) Rolling between the two surfaces; (*c*) Sliding between the two surfaces.

Those particles which become imbedded in the softer material do no work in abrading it, and but little in abrading the harder. They generally consist of the finer particles, and are kept out of action by the coarser which are rolling or sliding between the surfaces. Further, those that are purely rolling do little or no work. The greater part of the work is performed by those particles which are facetted and which *slide* between the two surfaces.

As the grinder is always of a much softer material than the glass, there is much more friction between the grinder and these particles than between the glass and the same particles, and therefore they partially adhere to the grinder and are carried by it across the face of the glass. This being so, it is now easy to perceive what the best conditions for rapid grinding are. Not too little emery, for then there will not be enough of abrading particles; not too much, for then the particles will roll on each other and tend to crush and disintegrate each other instead of abrading the glass, but just sufficient to form a single layer of particles between the grinder and the glass surface.

In the grinding of the small lenses, I mean up to five or six inches diameter, it is usual to carry out the entire grinding processes by hand; above that size by machinery. Surfaces up to 12 or even 15 inches can be ground by hand; but the labour becomes severe, and for my part I am gradually reducing the size for which the hand grinding is used, as I find the machine work more constant in its effects.

The machinery used is the same as that employed for the polishing operation, and I shall describe it under that head further on.

In the fine grinding operation by hand, the glass is usually cemented on to a holder of this form, having (for smaller sizes) three pieces of cork, to which the lens is attached, this holder being screwed to a spud or nose on top of a post screwed to the floor. The operator having applied the proper quantity of moist emery powder between the grinder and the glass, proceeds to work the former over the latter in a set of peculiar strokes, the amplitude and character of which he varies according to circumstances, at the same time that he changes his position round the post every few seconds.

* * * * * *

Although, as I have shown, the harder material is abraded very much more than the softer, yet the softer (the grinder) suffers considerable abrasion as well as the glass, and the skill of the operator is shown by the facility with which he is able to bring the glass to the curve of the grinder without altering the curve or figure of the latter.

It is even possible for a skilled operator to take a lens of one curve and a grinder of say a deeper curve, and by manipulation to produce a pair of surfaces fitting together and of shallower curves than either.

Measurement of the Curves.

In the early stages of grinding, gauges of the proper radius, cut out of sheet brass or sheet steel, are used for roughly testing the curves of the lenses; but when we get to the finer grinding process, it is necessary to have something much more accurate.

For this purpose a spherometer is used. It is made in various forms, generally with three legs terminating in three hardened steel points, which lie on the glass, and a central screw with fine thread, the point of which can be brought down to bear on the centre of the glass. In this way the versed sine of the curve for a chord equal to diameter of circle formed by these points is measured, and the radius of curve can be easily calculated from this.

I do not find the points satisfactory for regular work. They are apt to get injured or worn, and for ground surfaces are a little uncertain as one or other of the feet may find its way into a deep pit. This particular spherometer has three feet, of about half an inch long, which are hardened steel knife-edges forming three portions of an entire circle. In using this it is laid on the surface to be measured, and the screw with micrometer head is turned till the point is felt to touch the surface of glass. This scale and head can then be read off. The screw in this instrument has fifty threads to the inch, and the head is divided into 100 parts, so that each division is equal to $\frac{1}{5000}$ of an inch. With a little practice it is easy to get determinate measures to $\frac{1}{10}$ of this, or $\frac{1}{50000}$ of an inch, and by adopting special precautions even more delicate measures can be taken, as far probably as $\frac{1}{100000}$ or $\frac{1}{150000}$ of an inch, which I have found to be practically the limit of accuracy of mechanical contact.

To give an idea of the delicacy of the instrument, I bring the screw firstly into contact with the glass. Now the screw is in good contact; but there is so much weight still on the three feet, that if I attempt to turn it round, the friction on the feet oppose me, and it will not stir except I apply such force as will cause the whole instrument to slide bodily on the glass. Now, however, I raise the whole instrument, taking care that my hands touch none of the metal-work, and that the screw be not disturbed. I lay my hands for a moment on part of the glass where centre screw stood, and thus raise its temperature slightly, and on laying the spherometer back in the same place, you now see that it spins on the centre screw, showing how easily it detects what to it is a large lump, caused by expansion of the glass from the momentary contact of my hand.

Flexure.

One of the greatest difficulties to be contended with in the polishing of large lenses is that of flexure during the process.

It may appear strange that in discs of glass of such considerable thickness as are used for objectives, any such difficulty should occur; but a simple experiment will demonstrate the ease with which such pieces of glass can be bent, even under such slight strain as their own weight.

We again take our spherometer and set it upon a polished surface of a disc of glass of about 7½ inches diameter and ¾ inch thick. I set the micrometer head as in the former experiment to bear on the glass, but not sufficiently tight to allow the instrument to spin round. This has now been done while the glass as you see is supported on three blocks near its periphery. I now place one block under the centre of disc and remove the others thus, and you see the instrument now spins round on centre screw.

It is thus evident that not only is this strong plate of glass bending under its own weight, but it is bending a quantity easily measurable by this instrument, which as I shall presently show is quite too coarse to measure such quantities as we have to deal with in figuring objectives.

After this experiment, no surprise will be felt when I say that it is necessary to take very special precautions in the supporting of discs during the process of polishing, to prevent danger of flexure; of course, if the discs are polished while in a state of flexure the resulting surface will not be true when the cause of flexure is removed.

For small-sized lenses no very special precautions are necessary, but for all sizes over 4 inches in diameter I use the equilibrated levers devised by my father, and utilised for the first time on a large scale in supporting the 6-foot mirror of Lord Rosse's telescope. These have been elsewhere frequently described, but I have a small set here as an example.

I have also sometimes polished lenses while floating on mercury.

This gives a very beautiful support, but it is not so convenient, as it is difficult to keep the disc sufficiently steady while the polishing operation is in progress, without introducing other chances of strain.

So far I have spoken of strain or flexure during the process of working the surface; but even if the surface be finished absolutely perfectly, it is evident from the experiment I showed you that very large lenses when placed in their cells must suffer considerable flexure from their own weight alone, as they cannot then be supported anywhere except round the edge.

To meet this, I proposed many years ago to have the means of hermetically sealing the tube, and introducing air at slight pressure to form an elastic support for the objective, the pressure to be regulated by an automatic arrangement according to the altitude. My attention was directed to this matter very pointedly a few years ago from being obliged to use for the Vienna 27-inch objective a crown lens which was according to ordinary rules much too thin.

I had waited some years for this disc, and none thicker could be obtained at the time. This disc was very pure and homogeneous, but so thin that if offered to me in the first instance I would certainly have rejected it. Great care was taken to avoid flexure in the working, but to my great surprise I found no difficulty whatever with it in this respect. This led me to investigate the matter, with the following curious results.

If we call f the flexure for any given thickness t, and f' the flexure for any other thickness t', then $\frac{f}{f'} = \frac{t^2}{t'^2}$ for any given load or weight approximately. But as the weight increases directly as the thickness, the flexure of the discs due to their own weight, which is what we want to know, may be expressed as $\frac{f}{f'} = \frac{t}{t'}$.

Let us now consider the effect of this flexure on the image. In any lens bent by its own weight, whatever part of its surface is made more or less convex or concave by the bending, has a corresponding part bent in the opposite direction on the other surface, which tends to correct the error produced by the first surface. This is one reason why reflectors which have not this second correcting surface are so much more liable to show strain than refractors. If the lens were infinitely thin, moderate flexure would have no effect on the image. The effect increases directly as the thickness. If then the flexure, as I have shown, decreases directly as the thickness, and the effect of that flexure increases directly as the thickness, it is clear that the effect of flexure of any lens due to its own weight will be the same for all thicknesses; in other words, no advantage is gained by additional thickness.

This has reference, of course, only to flexure of the lens in its cell after it has been duly perfected, and has nothing to do with the extra difficulty of supporting a thin lens during the grinding and polishing processes.

Polishing.

The polishing process can be, and is often, conducted precisely in the same manner as the grinding, except that the abrading powders used (oxide of iron, rouge, an oxide of tin putty-powder) is of a finer and softer description, and the surface of the polishing tool is made of a softer material than the metallic grinder.

Very nearly all my polishing is done on the machine I shall presently describe; but before doing so, I will, with your permission, say a few words on the general principles of the polishing process. Various substances are used for the face of the polisher—fine cloth, satin or paper, and pitch. Pitch possesses two important qualities which render it peculiarly suitable for this work, and it is a curious fact that we owe its application for this purpose to the extraordinary perspicuity of Sir Isaac Newton, who we may fairly say was the first to produce an optically perfect surface, and that that material is not only still used for the purpose, but is as far as I know the only substance which possesses the peculiar qualifications necessary to fulfil the required conditions. With skill and care, moderately good surfaces can be obtained from cloth polishers; but it is easy to see why they can never be perfect. There is a certain amount of elasticity in cloth and in its "nap," and there is consequently a tendency to round off the surfaces of the pits left by the grinding powder, and to polish the bottom or floor of these pits at the same time as the upper surface. It is easy to show mathematically that any process which abrades the floors of the pits at the same time as general surfaces even in a very much less degree, can never produce more than an approximation to a perfect surface, and practice agrees with the theory. Paper is said to be much used by the French opticians. I can say nothing about it. I have tried it and failed to produce a perfect surface with it, nor indeed should I expect it. It is of course open to the same objection as cloth. Pitch possesses, as I said, two most important qualities which render it suitable for the work; the first, in its almost perfect inelasticity; the second, a curious quality of subsidence, which we utilise in the process.

If we watch with a microscope, or even a magnifier, the character of two surfaces during the process of polishing, the one with cloth, and the other with pitch, the difference is very striking. With the cloth polisher, the polish appears much quicker, and it would at first sight appear as if the same polishing powder abraded more quickly on the cloth than on the pitch polisher, but I do not believe that such is the case, for if we look at the surface with a magnifier we shall find that while all the surface has assumed a polished appearance, the surface itself has retained some of the form of the original pitted character with the edges rounded off; while in the pitch half-polished surfaces, the floors of the pits are as grey as ever, and the edges are sharp and decisive. In pitch polishing, too, a decided amount of polish appears very quickly, and then for many hours there appears to be little or no further effect. Suddenly, however, the remaining greyness disappears, and the surface is polished. The reason of this is very obvious. The polisher being very inelastic, polishes first only the tops of the hills, and has to abrade away all the material of which these hills are composed before it reaches the valleys or floors of the pits. When it does reach them, the proper polish quickly appears. The second quality of pitch, that of subsidence, is also most valuable.

Pitch can be rendered very hard by continued boiling. By pitch I mean the natural bituminous deposit which comes to us from Archangel, not gas-tar pitch. It can be made so hard that it is impossible to make any impression on it with the finger-nail without splitting it into pieces; and yet even in this hard condition if laid on an uneven surface it will in a few days, weeks, or months subside and take the form of whatever it is resting upon. The cohesion of its particles is not sufficient to enable it to retain its form under the action of gravity; and as this condition is that which science tells us marks the difference between solids and liquids, we must, paradoxical though it may appear, class even the hardest pitch among liquid instead of solid substances.

Now how do we utilise this peculiar quality.

The polishing tool is made by overlaying a metal or wooden disc formed to nearly the required curves by a set of squares of pitch, and while these are still warm pressing them against the glass, the form of which they immediately take.

In the grinding process, I showed you that the regulation of the abrasion was managed partly by the character of the stroke given, and partly by the local touches given to the tool by the stoning process.

In polishing we still retain the same facilities for modifying the stroke, and the same rules I gave apply generally to the polishing process as well as the grinding; but we have not got any process equivalent to that of the local stoning, and even if we had it would be useless, for this very quality of subsidence of the pitch would in a few minutes cause any part of its surface which had been reduced to come into good contact again; we must therefore look for some other means for producing more or less abrasion whenever we require it. This we effect by modifying the size of the squares of pitch in the various zones. Practically it is done in this way by a knife and mallet. Whenever the squares are reduced, the abrasion will be less.

This is a well-known method of regulation; but the rationale is, I think, not generally understood. It is generally explained that there is less abrasion because there is less abrading surface. I do not think this is the true, or at least the entire, explanation. In order to understand the action, you must conceive the pitch to be constantly in a state of subsidence, the amount of that subsidence depending of course on the pressure placed upon it. Now if we reduce the size of the squares in any zone while retaining the same distance from centre to centre of squares, we increase at first the pressure per unit of area on the pitch squares in that zone, and consequently the subsidence will be greater, and the action will not be so tight or severe on that zone.

I know of no substance but pitch and a few of the resins which possess this peculiar quality except perhaps ice, and it is curious to think that the same quality which in ice allows of that gradual creeping and subsidence and consequent formation of glaciers with their characteristic moraines, &c., will in pitch help us to produce accurate optical surfaces.

Polishing Machines.

The two best-known polishing machines are those of the late Earl of Rosse and the late Mr. Lassell, the general forms of which are shown in these diagrams. Time will not permit me to enter into a minute description, of their working, nor is it necessary, as both have been often described.

A few words, however, as to the different character of the strokes given by these machines may be interesting. The stroke of Lord Rosse's machine may be imitated in hand-work by the operator traversing the polisher or mirror in a series of nearly straight strokes, of about one-third the diameter of the glass, to and from himself, at the same time that he keeps walking slowly round the post, and instead of allowing the centre of polisher to pass directly over the centre of mirror, each stroke that he gives he slides a little (about one-tenth diameter) to one side and then a little to the other.

Mr. Lassell's stroke may be imitated by causing the polisher to describe a series of nearly circular strokes a little out of the centre, walking round the post at the same time; thus the centre of polisher will describe a series of epicycloidal or hypocycloidal curves on the speculum.

Many years ago my father devised a machine, figured and described in Nichols' 'Physical Science,' by which either of these motions could be obtained. He appeared to have got better results with Mr. Lassell's strokes, for he afterwards devised a machine which gave the same character of stroke, but over which the operator had greater control, and this machine has been used for many years with great success. Like all machines, however, which give a series of strokes constantly recurring of the same amplitude, it is apt to polish in rings. It is impossible to obtain absolute homogeneity in the pitch patches, and if any one square be a shade harder than the general number, and that square ends its journey at each stroke at the same distance from the centre of speculum or glass, there will almost surely be a change of curvature in that zone. To avoid this I have made a slight modification in the machine, which has increased its efficiency to a great extent. I will now place in the lantern a model of this machine, and first draw you a few curves with the machine in its old state, and afterwards in its improved state.

In order to convey some idea of the relative quantities of material removed by the various processes, I have placed upon the walls a diagram which will illustrate this point in two distinct ways.

The diagram itself represents a section of a lens of about 8 inches aperture and 1 inch thick, magnified 100 times, and shows the relative thickness of material abraded by the four processes.

The quantity removed by the rough grinding process is represented on this diagram by a band 25 inches wide, the fine grinding by one $\frac{8}{10}$ inch wide, the polishing by a line $\frac{1}{50}$ inch wide, while the quantity removed by the figuring process cannot be shown even on this scale as it would be represented by a line only $\frac{1}{10000}$ inch thick.

I have also marked on this diagram the approximate cost of abrasion of a gramme of material by each of the four processes, viz:—

	£	s.	d.	
Rough grinding, about	0	0	1	per gramme.
Fine grinding, „	0	0	7½	„
Polishing, „	0	10	0	„
Figuring, „	48	0	0	„

Figuring and Testing.

By the figuring process, I mean the process of correcting local errors in the surfaces, and the bringing of the surfaces to that form, whatever it may be, which will cause the rays falling on any part to be refracted in the right direction. When an objective has undergone all the processes I have described, and many more which are not so important, and with which I have not had time to deal, and when the objective is centered and placed in its cell, it is, to look at, as perfect as it will ever be, but to look through and use as an objective it may be useless. The fact is that when an objective has gone through all the processes described, and is in appearance a finished instrument, I look upon it as about one-fourth finished. Three-fourths of the work has probably to be done yet. True, sometimes this is by no means the case, and I have had instances of objectives which were perfect on the first trial; but this is, I am sorry to say, the exception and not the rule.

This part of the process naturally divides itself into two distinct heads:—

1. The detection and localisation of faults—what may, in fact, be termed the diagnosis of the objective.

2. The altering of the figures of the different surfaces to cure these faults. This may be called the remedial part.

It may be well here to try to convey some idea of the quantities we have to deal with, otherwise I may be misunderstood in talking of great and small errors.

I have before mentioned that it is possible to measure with the spherometer quantities not exceeding $\frac{1}{50000}$ of an inch, or with special precaution much less even than that; but useful as this instrument is for giving us information as to the general curves of the surface, it is utterly useless in the figuring process, that is, an error which would be beyond the power of the spherometer to detect, would make all the difference between a good and a bad objective.

Take actual numbers and this will be evident. Take the case of a 27-inch objective of 34 feet focus; say there is an error in centre of one surface of about 6 inches diameter, which causes the focus of that part to be $\frac{1}{10}$ of an inch shorter than the rest.

For simplicity sake, say that its surface is generally flat; the centre 6 inches of the surface therefore, instead of being flat, must be convex and of over 1,000,000 inches radius, about 17 miles. The versed sine of this curve, as measured by spherometer, would be only about $\frac{1}{250000}$, 4 millionths of an inch, a quantity mechanically unmeasurable, in my opinion.

If that error was spread over 3 inches only instead of 6 inches, the versed sine would only be about $\frac{1}{1000000}$. Probably the effect on the image of this 3-inch portion of $\frac{1}{10}$ inch shorter focus would not be appreciable on account of the slight vergency of the rays, but a similar error near edge of objective certainly would be appreciable. Until therefore some means be devised of measuring with certainty quantities of 1 millionth of an inch and less, it is useless to hope for any help from mechanical measurement in this part of the process.

If then no known mechanical arrangement be delicate enough to measure these quantities, how, it may be asked, are these errors detected?

The answer to this is, that certain optical arrangements enable us to carry our investigations far beyond the limits of mechanical accuracy. Trials of the objective or mirror as a telescope are really the crucial test, but there are various devices by which defects are detected and localised.

The best object to employ is generally a star of the third or fourth magnitude, when such is available, but as it frequently occurs that no such object is visible, recourse is had to artificial objects. The minute image of the sun reflected from little polished balls of speculum metal, or even a thermometer bulb is a very good object; polished balls of black glass are also used with good effect; but as the sun also is of somewhat fickle disposition in this country, we have frequently to have recourse to artificial light. Small electric lamps, such as this, with their light condensed, and thrown on a polished ball are very useful. In fact, I am never without one of them in working order.

For the detection and localisation of errors it is very useful to be provided with sets of diaphrams which leave exposed various zones of the surface, the foci of which can then be separately measured, but a really experienced eye does not need them.

For concave surfaces, Foucault's test is useful. I shall not trespass on your time to explain this in detail, as it is described very fully in many works, in none better than in Dr. Draper's account of the working of his own reflecting telescope. This diagram will give an idea of the principle of the system, which is really the same as what I have described as useful for detecting want of homogeneity in the substance of the glass.

This system is extremely useful for concave spherical surfaces, but is not available for convex surfaces, and only partially available for concave parabolic surfaces.

The really crucial test is, as I said before, the performance of the objective when used as a telescope; and the appearance of the image not only at the focus, but on each side of it, conveys to the practised eye all the information required for the detection of the errors.

If an objective have but one single fault, its detection is easy; but it generally happens that there are many faults superposed, so to speak. There may be faults of achromatism, and faults of figure in one or all the surfaces; faults of adjustment, and perhaps want of symmetry from some strain or flexure; and the skill of the artist is often severely taxed to distinguish one fault from the rest and localise it properly, particularly if, as is generally the case, there be also disturbances in the atmosphere itself, which mask the faults in the objective, and permit of their detection only by long and weary watching for favourable moments of observation.

It would be impossible in one or a dozen of such lectures as this to enumerate all the various devices that are practised for the localisation of errors, but a few may be mentioned, some of which have never before been made public.

For detection of faults of symmetry, it is usual to revolve one lens on another, and watch the image. In this way it can generally be ascertained whether it is in the flint or crown lens.

With some kinds of glass the curves necessary for satisfying the conditions of achromatism and spherical aberration are such that the crown becomes an equi-convex and the flint a nearly plano-concave of same radius on inside curve as either side of the crown. This form is a most convenient one for the localisation of surface errors in this manner.

The lenses are first placed in juxtaposition and tested. Certain faults of figure are detected. Now calling the surfaces A B C D in the order in which the rays pass through them, place them again together with Canada balsam or castor-oil between the surfaces B and C, forming what is called a cemented objective. If the fault be in either A or D surface, no improvement is seen; if in B or C the fault will be much reduced or modified. Now reverse the crown lens cementing surfaces A and C together. If same fault still shows, it must be in either B or D. If it does not show, it will be in either A or C. From these two experiments the fault can be localised.

It often happens that a slight error is suspected, but its amount is so slight that it appears problematical whether an alteration would really improve matters or not. Or, the observer may not be able to make up his mind as to the exact position of the zone he suspects to be too high or too low, and he fears to go to work and perhaps do harm to an objective on which he has spent months of labour, and which is almost perfect. In many such cases I have wished for some means by which I could temporarily alter the surface and see it so altered before actually proceeding to abrade and perhaps spoil it.

During my trials with the great objective of Vienna, I thought of a very simple expedient, which effects this without any chance at all of injuring the surface. If I suspect a certain zone of an objective is too low, and that the surface might be improved by lowering the rest of it, I simply pass my hand, which is always warmer than the glass, some six or eight times round that particular zone. The effect of this in raising the surface is immediately apparent, and is generally too much at first, but the observer at eye end can then quietly watch the image as the effect goes off, and very often most useful information is thus obtained. The reverse operation, that of lowering any required part of the surface, is equally simple. I take a bottle of sulphuric ether and a camel's-hair brush, and pass the brush two or three times round the part to be lowered, blowing on it slightly at the same time; the effect is immediately perceived, and can always be overdone if required.

So far then for the diagnosis. Now for the remedy. When the fault has been localised, the lens is again put upon the machine and the polisher applied as before, the stroke of the machine and the size of the pitch patches being so arranged as to produce, or tend to produce, a slightly greater action on those parts that have been found to be too high (as before described while treating of the polishing processes).

The regulation of the stroke, excentricity, speed, and general action of the machine, as well as the size and proportion of the pitch squares, and the duration of the period during which the action is to be continued, are all matters the correct determination of which depends upon the skill and experience of the operator, and concerning which it would be impossible to formulate any very definite rules. All thanks are due to the late Lord Rosse and Mr. Lassell, and also to Dr. de la Rue, for having published all particulars of the process which they found capable of communication; but it is a notable fact that, as far as it is possible to ascertain, every one who has succeeded in this line has done so, not by following written or communicated instructions, but by striking out a new line for himself; and I think I am correct in saying that there is hardly to be found any case of a person attaining notable success in the art of figuring optical surfaces by rigidly following directions or instructions given or bequeathed by others.

There is one process of figuring which is said to be used with success among Continental workers. I refer to the method called the process of local touch. In this process those parts, and those parts only, which are found to be high are acted upon by a small polisher.

This action is of course much more severe; and if only it were possible to know exactly what was required, it ought to be much quicker; but I have found it a very dangerous process. I have sometimes succeeded in removing a large lump or ring in this way (by large I mean 3 or 4 millionths of an inch), but I have also and much oftener succeeded in spoiling a surface by its use. I look upon the method of local touch as useful in removing gross quantities, but for the final perfecting of the surface I would not think of employing it.

In small-sized objectives the remedial process is the most troublesome, but in large-sized objectives the diagnosis becomes much the more difficult, partly on account of the rare occurrence of a sufficiently steady atmosphere. In working at the Vienna objective it often happened when the figure was nearly perfect that it was dangerous to carry on the polishing process for more than ten minutes between each trial, and we had then sometimes a week to wait before the atmosphere was steady enough to allow of an observation sufficiently critical to determine whether that ten minutes' working had done harm or good. It must not be supposed either that the process is one in which improvement follows improvement step by step till all is finished. On the contrary, sometimes everything goes well for two or three weeks, and then from some unknown cause, a hard patch of pitch perhaps, or sudden change of temperature, everything goes wrong. At each step, instead of improvement there is disimprovement, and in a few days the work of weeks or months perhaps is all undone. Truly any one who attempts to figure an objective requires to have the gift of patience highly developed.

In view of the extraordinary difficulty in the diagnostic part of the process with large objectives, it is my intention to make provision which I hope may reduce the trouble in the working of the new 28-inch objective for the Royal Observatory, Greenwich.

Two of the greatest difficulties we have to contend with are: 1stly, the want of homogeneity in the atmosphere, through which we have to look in trials of the objective, due to varying hygrometric and thermometric states of various portions; and 2ndly, sudden changes of temperature in the polishing room. The polisher must always be made of a hardness corresponding to the existing temperature. It takes about a day to form a polisher of large size, and if before the next day the temperature changes 10° or 15°, as it often does, that polisher is useless, and a new one has to be made, and perhaps before it is completed another change of temperature occurs. To grapple with these two difficulties I propose to have the polishing chamber underground, and leading from it a long tunnel formed of highly glazed sewer-pipes about 350 feet long, at the end of which is placed an artificial star illuminated by electric light; on the other side of the polishing chamber is a shorter tunnel, forming the tube of the telescope, terminating in a small chamber for eye-pieces and observer. About half-way in the long tunnel there will be a branch pipe connected to the air shaft of the fan, which is used regularly for blowing the blacksmith's fire, and through this when desired a current of air can be sent to "wash it out" and mix up all currents of varying temperature and density. It may be found necessary even to keep this going during observations.

By this arrangement I hope to be able to have trials whenever required, instead of having to wait hours and days for a favourable moment.

Figuring of Plane Mirrors.

There is a general idea that the working of a plane mirror or one of very long radius is a more difficult operation than those of more ordinary radii. This is not exactly the case. There is no greater difficulty in figuring a low curve than a deep one, but the difficulty in the case of absolutely plane mirrors consists simply in the fact that in their figuring there is one additional condition to be fulfilled, viz. that the general radius of curvature must be made accurate within very narrow limits. In figuring a plane mirror to use, for instance, in front of even a small objective, say 4-inch aperture, an error in radius which would cause a difference of focus of $\frac{1}{100}$ of an inch would seriously injure the performance. This would be about equivalent to saying that the radius of curvature of the mirror was about 8 miles, the versed sine of which with the 6-inch spherometer would be about $\frac{1}{50000}$ of an inch. Now what I mean to convey is this; that it would be just as difficult to figure a convex or concave lens of moderate curvature as a flat lens of the same size if it were necessary to keep the radius accurate to that same limit, i. e. one-tenth of a division of this spherometer.

Lick Observatory.

For the final testing of large objectives or mirrors, it is necessary to have them properly mounted, and in such a manner that they can be directed conveniently on any celestial object, and kept so directed by clockwork to enable the observer to devote his whole attention to the testing. I had not intended touching at all on the subject of the mounting of telescopes, but I have been asked to call your attention to this model of a proposed observatory for Mount Hamilton, California, as it embraces some novel features, but I shall do so in only a very few words.

Most here are probably aware that a monster observatory is in course of erection in California, a large sum of money having been left for the purpose by a Mr. Lick. The observatory is already partly complete, and contains some excellent instruments of moderate size, the work already done with which warrants us to hope that the great 36-inch refractive about to be erected will be placed under more favourable conditions for work than any other large telescope in the world.

The 36-inch objective is at present in process of construction by the Messrs. Clark of America, but the mounting has not yet been contracted for.

Some years since, in a paper published in the Transactions of the Royal Dublin Society, I shadowed forth a principle which I considered should be adopted in great telescopes of the future. The trustees of the Lick observatory having invited me to design an instrument for the 36-inch objective, I have put into practical form what I had before given but general principles of, and the design which this model illustrates is the result.

Whether this design will ever be carried out or not I cannot tell, but even as a proposal I trust it may be interesting enough to excuse my introducing it (somewhat irrelevantly perhaps) to your notice to-night.

The design includes the equatorial itself, with its observatory, dome, and provision for enabling the observer to reach the eye end conveniently.

The conditions I laid down for myself in designing this observatory were that it would be possible for the observer single-handed to enter the equatorial room at any time, and that, without using more physical exertion than is necessary for working the smallest-sized telescope, or even a table microscope, he should be able to open the 70-foot dome, turn it round backwards and forwards, point the equatorial to any part of the heavens, revolving it in right ascension and declination to any extent, and finally (the most difficult of all) to bring his own person into a convenient position for observing. I say this last is the most difficult of all, for I think any who have worked with larger instruments will allow that there is generally far more trouble in moving the observatory chair (so called) and placing it in proper position than in pointing the instrument itself. In this instrument the "chair" would require to be 25 feet high, and with its movable platform, ladder, balance-weight, &c., would weigh probably some tons. Even if very perfect arrangements were made for the working of this chair, the mere fact that the observer while attempting to make the most delicate observations is perched upon a small and very unprotected platform 25 feet above the floor and in perfect darkness, tends to reduce his value as an observer to an extent only to be appreciated by those who have tried it.

No matter how enthusiastic a man may be at his work, I would not put a high value on his determinations if made while in a position which calls for constant anxiety for his own personal safety. I would go even further still, and say that even personal comforts or discomforts have much to do with the value of observations.

I propose, therefore, that all the various motions should be effected by water power. There are water engines of various forms now made, some of which have no dead point, and having little vis inertia, are easily stopped and started, and are consequently well adapted for this work.

I propose to use four of them: one for the right ascension motion of the instrument, and one for the declination; one for revolving the dome, and one for raising and lowering the observer himself; but instead of having anything of the nature of a 25-foot chair or scaffold, I propose to make the 70-foot floor of the observatory movable. It is balanced by counterpoise weights, and raised and lowered at will by the observer. Then the observer can without any effort raise and lower the whole floor, carrying himself and twenty people if desired, to whatever height is most convenient for observation; and wherever he is observing, he is conscious that he has a 70-foot floor to walk about on, which even in perfect darkness he can do in safety.

The valves and reversing gear of the water engines are actuated by a piece of mechanism, the motive power of which may be a heavy weight raised into position some time during the previous day by man or water power. By means of a simple electrical contrivance, this piece of machinery itself is under the complete control of the observer, in whatever part of the room he may be, and he carries with him a commutator of a compact and convenient form, with eight keys in four pairs, each pair giving forward and backward movements respectively to

A. Telescope movement in right ascension;
B. Telescope movement in declination;
C. Revolution of dome;
D. Raising of floor.

The remaining operation, opening of shutter, is easily effected without any additional complication.

It is only necessary to anchor the shutter (which moves back horizontally) to a hook in the wall and move the dome in the opposite direction by motion C; the shutter must of course be opened by this motion.

It is very possible that there may be some here who have found what I have had to say on the subject of the figuring of objectives very unsatisfactory. They may have expected, naturally enough, that instead of treating of generalities to such a large extent as I have done, I should have given precise directions, by the following of which rigidly any person could make a telescopic objective.

To those, however, who have followed me in my remarks, the answer to this will probably have already suggested itself. It is the same answer which I give to those who visit my works and ask what the secrets of the process are, or if I am not afraid that visitors will pick up my secrets. All the various processes which I have described up to that of the figuring are I consider purely mechanical processes, the various details of which can be communicated or described as any mechanical process can be; but in the last final and most important process of all there is something more than this. A person might spend a year or two in optical works where large objectives are made, and might watch narrowly every action that was taken, see every part of the process, take notes, and so forth, and yet he could no more expect to figure an objective himself, than a person could expect to be able to paint a picture because he had been sitting in an artist's studio for the same time watching him at his work. Experience and experience only can teach any one the art, and even then it is only some persons who seem to possess the power of acquiring it.

A well-known and experienced amateur in this work declared his conviction that no one could learn the process under nine years' hard work, and I am inclined to think his estimate was not an exaggerated one.

True, it may be said that large objectives can be and are generally turned out by machinery, but what kind of an objective would any machine turn out if left to guide itself, or left to inexperienced hands?

At the risk of being accused of working by what is generally called the rule of thumb, I confess that conditions often arise, to meet which I seem to know intuitively what ought to be done, what crank to lengthen, what tempering is required of the pitch squares; and yet if I were asked I should find it very hard to give a reason for my so doing which would even satisfy myself.

I may safely say that I have never finished any objective over 10 inches diameter, in the working of which I did not meet with some new experience, some new set of conditions which I had not met with before, and which had then to be met by special and newly devised arrangements.

A well-known English astronomer once told me that he considered a large objective, when finished, as much a work of art as a fine painting.

I have myself always looked upon it less as a mechanical operation than a work of art. It is, moreover, an art most difficult to communicate. It is only to be acquired by some persons, and that after years of toilsome effort, and even the most experienced find it impossible to reduce their method to any fixed rules or formulæ.

[H. G.]

Friday, June 11, 1886.

HENRY POLLOCK, Esq. Treasurer and Vice-President, in the Chair.

PROFESSOR DEWAR, M.A. F.R.S. *M.R.I.*

Recent Researches on Meteorites.

PROFESSOR DEWAR explained that Mr. Gerrard Ansdell and himself had been engaged in an examination into the gaseous constituents of meteorites, and he proposed this evening to abstract the results of the investigation, which had been laid before the Royal Society.

The nature of the occluded gases which are present to a greater or less extent in all meteorites, whether belonging to the iron, stony, or carbonaceous classes, has engaged the attention of but few chemists. It is, nevertheless, an especially interesting and important subject, owing to the uncertainty which still exists as to the origin of these celestial bodies.

Graham (Proc. Roy. Soc. vol. xv. (1867) p. 502) was the first who made any experiments in this direction, when he determined the gases occluded in the Lenarto meteoric iron, which yielded 2·85 times its volume of gas, 86 per cent. of which was hydrogen, and 4·5 per cent. carbonic oxide. He was followed in 1872 by Wöhler (Pogg. Ann. vol. cxlvi. p. 297) and Berthelot (Compt. Rend. vol. lxxiv. pp. 48, 119), who estimated approximately the gases contained in the Greenland Ovifak iron. These gases consisted of about equal parts of carbonic acid and carbonic oxide; the celestial origin of this iron is, however, very doubtful.

In the same year (1872) the American chemist, Mallet (Proc. Roy. Soc. vol. xx. p. 365), made a very complete determination of the gases occluded by the Augusta Co., Virginia, meteoric iron, which, however, differed very considerably from Graham's results. He obtained an amount of gas equal to 3·17 times the volume of the iron, made up of 35·83 per cent. of hydrogen, 38·33 per cent. carbonic oxide, 9·75 per cent. of carbonic acid, and 16·09 per cent. of nitrogen.

Wright and Lawrence Smith followed Mallet, and our present knowledge of this interesting subject is principally due to these American chemists. They have taken advantage of the numerous meteoric masses which have fallen from time to time throughout America, and which can easily be obtained in sufficient quantity for complete and accurate observations on their gaseous constituents.

Wright contributed several papers to the 'American Journal' in 1875 and 1876, and, according to his analyses, the total volume of gas occluded and the composition of the same differs considerably in the two principal classes of meteorites. He found the total volume of gas extracted was much greater in the case of the stony meteorites than in the iron ones, the principal characteristics of these gases being, that in the former the carbonic acid greatly predominated, accompanied by a comparatively small amount of carbonic oxide and hydrogen, whereas in the latter the carbonic acid never exceeded 20 per cent., the carbonic oxide being, as a rule, considerably more than this, and the hydrogen sometimes reaching as high as 80 per cent.

It is impossible, however, to arrive at anything more than general conclusions as to the total amount of gas given off by any special meteorite, or its composition, for, as shown by Wright and confirmed by Mr. Ansdell and the lecturer, both the total quantity and composition of the gases vary very considerably according to the temperature at which they are drawn off.

Wright found a notable quantity of marsh-gas in all the stony meteorites which he examined, though not a trace in any of the iron ones; this seemed to be a distinctive difference between the two classes of meteorites, but subsequently Dr. Flight, of the British Museum (Phil. Trans. vol. clxxiii. (1882) p. 885), found marsh-gas in a specimen of the Cranbourne siderite, so that it is evident certain of the iron meteorites also contain this gas.

Lawrence Smith (Amer. Journ. 1876) confined himself principally to an examination of the graphite nodules which are frequently found imbedded in the iron meteorites, and to the nature of the carbon in the so-called carbonaceous meteorites. He extracted the organic or hydrocarbon-like bodies by means of ether, but did not determine the gases given off on heating. Previous to this, Roscoe (Pro. Phil. Soc. Man. 1862) had obtained the same hydrocarbon-like body by exhausting the Alais meteorite with ether, but the quantity he had to work upon was so small that he could not make a very complete examination.

These are some of the principal points that have been made out with regard to the gases occluded by meteorites. The results, however, are so comparatively few, that it was thought worth while to take the opportunity which presented itself, of having several good specimens of meteorites to confirm these results, and, if possible, add something to our present knowledge of the subject.

The investigation may be divided into five parts, having the following objects in view: firstly, the confirmation of previous results by the examination of some well-known meteorite; secondly, the analysis of several whole meteoric stones, whose interior had never been exposed to the effects of the atmosphere, by reason of the characteristic coating of glaze; thirdly, the examination of a celestial graphite nodule, taken from the interior of an iron meteorite; fourthly, the comparison of some meteorite of the carbonaceous class with the above; and fifthly, the examination of different terrestrial graphites.

The method employed for the abstraction of the gases was exactly the same in every case, so that a short description will suffice for all. The temperature was kept as nearly as possible the same in every experiment, but no doubt differences of many degrees occurred in some of the experiments, which was unavoidable in using an ordinary combustion furnace.

The meteorite or graphite, as the case might be, was broken up into a coarse powder, introduced into a convenient length of combustion tubing, and connected up with a Sprengel pump, a small bulb-tube immersed in a freezing mixture intervening, so as to retain any moisture or condensable volatile products that might come off. The tube was first thoroughly exhausted and then heated in an ordinary gas combustion furnace to a low red heat. The gases, during the heating, were gradually drawn off by the Sprengel pump, and when the tube had remained for several minutes at a low red heat it was completely exhausted. The total quantity of gas collected was in every case used for the analysis.

The " Dhurmsala " specimen was an ordinary fragment of a much larger original mass, but in the case of the Pultusk and Mocs meteorites, complete stones were fortunately obtained, weighing respectively 57 and 103 grams, having the characteristic black glaze on their surfaces.

Such a large quantity of water was condensed in the bulb tube in heating the Dhurmsala meteorite, it being the first one examined, that it was thought it might be principally due to the great absorptive power of these porous bodies, and that therefore the moisture might have been condensed in the pores of the meteorite from the surrounding air. The Pultusk and Mocs specimens appeared to be especially adapted for ascertaining whether this was the case, as the complete covering of black glaze would probably prevent the moisture from penetrating to the interior of the stones. The fragments of these stones were therefore transferred as quickly as possible to the combustion tube after they had been broken up. Notwithstanding these precautions, fully as much water was condensed from them as from the Dhurmsala specimen, which seems to suggest that the water is really combined in some form in the stone and not obtained directly from the surrounding atmosphere, although it must be admitted that the glaze on both the stones was not of a very glossy character, and did not have the appearance of being absolutely impervious to moisture.

The pumice-stone was examined merely with a view to comparing the gases occluded by a porous body of volcanic origin with those contained in meteorites. The sample taken was a fresh piece of stone, which had not been dried or purified in any way.

It is evident that it differs considerably from the meteoric stones, the total occluded gas being very small, only about half its volume, the carbonic acid at the same time being much less, with a proportionate increase in the carbonic oxide.

The general method of analysis was as follows, and the accuracy of the results was confirmed by varying in some cases the method of separating the gases. The carbonic acid was first removed from the mixture by caustic potash, the carbonic oxide being then absorbed by subchloride of copper, and the remainder of the gases exploded with excess of oxygen. The carbonic acid formed was again removed by caustic potash, and the excess of oxygen by alkaline pyrogallate, the residue being taken as nitrogen. The relative quantities of marsh-gas and hydrogen were calculated from the total diminution after explosion, and the amount of carbonic acid formed:—

	Sp. Gr.	Occluded Gases in vols. of the Meteorite.	Percentage Composition.				
			CO_2.	CO.	H.	CH_4.	N.
Dhurmsala	3·175	2·51	63·15	1·31	28·48	3·9	1·31
Pultusk	3·718	3·54	66·12	5·40	18·14	7·65	2·69
Mocs	3·67	1·94	64·50	3·90	22·94	4·41	3·67
Pumice-stone ..	2·50	0·55	39·50	18·50	25·4	..	16·60

It will be seen that the above numbers are quite confirmatory of Wright's results, the carbonic acid in the three meteorites examined being by far the largest constituent, while marsh-gas in considerable quantity was found in all. The percentage of this latter gas is somewhat higher than that found by Wright in the stony meteorites he examined, but this is probably due to the fact that a rather higher temperature was employed by the lecturer to drive off the gases. This supposition seems to be confirmed on considering the analysis of the Pultusk meteorite; for whereas Wright's abstracted gas only reached 1·75 times the volume of the stone, the total quantity of gas obtained by the lecturer was twice as much or equal to 3·54 times its volume.

It is therefore unquestionable that marsh-gas is given off on heating these meteoric stones, but whether it exists as such occluded in the material, or whether it is formed by some chemical decomposition of some organic constituent of the mass, is by no means clear.

Wright came to the conclusion "that the marsh-gas really existed as such in the stony meteorites, as the temperature at which it was driven off would be too low for its formation," at the same time he thinks it quite possible that "at very much higher temperatures, in the reaction by which the carbonic acid is broken up by the iron, a portion of the carbon might combine with the hydrogen present to form marsh-gas."

Knowing the great absorptive power for gases possessed by porous bodies generally, it was thought advisable to determine directly what this absorptive power was in the case of these stony meteorites, which are of such an eminently porous nature.

For this purpose the powdered Dhurmsala meteorite, from which the gases had been removed, was left in moist air under a bell-glass, for different periods of time as tabulated below, the gases being drawn off at a low red heat after each period:—

	Occluded Gas in vols. of the Meteorite.	CO_2.	CO.	H.	N.
After 24 hours	0·61	54·0	..	42·4	3·6
After 6 days more	2·47	47·0	5·0	47·0	1·0
After 8 days more	0·63	96 1	2·0	1·5	..

The absorption of water and gases evidently went on tolerably rapidly for the first seven days, but after the second heating of the meteorite, its porosity seems to have been affected in some way, for after a further period of eight days, it was found to take up only about a fourth of the quantity of gas which it had absorbed in the previous six days.

The actual amount of water given off after this exposure to a moist atmosphere was considerably less than what was obtained in the original heating of the meteorite, and from this it was inferred that the water is chemically combined in the stone. It would be difficult to explain, otherwise than by chemical combination, the power by which the water is retained by these meteorites, as it is not given off until a very high temperature is reached. In any case it is clear that the hydrogen must come from the action of water on the iron-nickel alloy, or finely disseminated carbon. Greville Williams has pointed out that the large amount of hydrogen obtained from heating finely divided zinc-dust is not due to free hydrogen, but to the action of the zinc on the hydrated oxide of zinc.

To pass on to the consideration of the various graphites examined. The celestial graphite was a perfect oblong nodule weighing 30 grams, which had been taken from the interior of a mass of the Toluca meteoric iron. It had a uniform dull-black colour, except at one end where there was a slight incrustation of sulphide of iron. Its fracture showed a uniform dull-black, compact mass; it was easily pounded up in a porcelain mortar, and formed a fine granular powder without any lustre.

On extracting the gases from this specimen a considerable quantity of marsh-gas was obtained, so that it appeared most important to compare it with some samples of terrestrial graphites, more especially as the occluded gases had never, as far as the lecturer was aware, been determined in these bodies.

For this purpose four samples of native graphites were obtained. The Cumberland graphite was a magnificent specimen of the original Borrodale, and had been in a private cabinet for over fifteen years. It had the characteristic dense homogeneous structure and brilliant external lustre of the graphites coming from this district. The Siberian example was from the Alexandref Mine; its structure was columnar and striated, with little external lustre; it was rather more easily broken up than the Borrodale, but formed the same dull black powder. The specimen from Ceylon was of the type usual to that island: highly lustrous and flaky, breaking up very easily, and forming small shining plates when ground up. The last sample, which was from the same cabinet as the others, but whose origin was unfortunately unknown, had a dull external surface, was exceedingly porous, and much more brittle than any of the previous ones, grinding up very easily into a dull black powder. It had more the appearance of the celestial graphites, which was heightened by having slight incrustations of sulphide of iron on its surface. Its low specific gravity also shows it to be some exceptional variety.

It seemed most important in connection with this subject to examine some matrix with which the graphites are usually found associated. These rocks are very variable, but consist principally of a kind of decomposed trap or gneiss. A good specimen was obtained of semi-decompósed gneiss from Canada with a considerable quantity of graphite disseminated throughout the mass, and also several samples of Ceylon graphite imbedded in its matrix, which in this case consisted of felspar and quartz.

The results, as tabulated below, confirm Wright's analyses of several trap rocks, in which he found principally carbonic acid and hydrogen. The small quantity of marsh-gas no doubt comes from the disseminated graphite, but the presence of the hydrogen is more difficult to explain and requires further investigation.

	Sp. Gr.	Occluded Gases in vols. of the Graphite.	CO_2.	CO.	H.	CH_4.	N.
Celestial graphite ..	2·26	7·25	91·81	..	2·50	5·40	0·1
Borrodale " ..	2·86	2·60	36·40	7·77	22·2	26·11	6·66
Siberian " ..	2·05	2·55	57·41	6·16	10·25	20·83	4·16
Ceylon " ..	2·25	0·22	66·60	14·80	7·40	3·70	4·50
Unknown " ..	1·64	7·26	50·79	3·16	2·50	39·53	3·49
Gneiss " ..	2·45	5·32	82·38	2·38	13·61	0·47	1·20
Felspar	2·59	1·27	94·72	0·81	2·21	0·61	1·40

On comparing these samples of graphite, it will be seen that the Borrodale and the Siberian give off about the same total volume of gas, that the celestial and the unknown graphites closely approximate each other in this respect, yielding more than double the volume of the others, and that the Ceylon sample stands alone in yielding a very minute quantity. All the terrestrial samples, except that from Ceylon, are alike in giving off a very considerable quantity of marsh-gas, though they differ somewhat in the actual quantity, and it is evident that although the celestial graphite contains a considerable amount, it is very much less than that yielded by the terrestrial samples.

A few tentative experiments were made to ascertain the absorbing power for gases of this celestial graphite. For this purpose dry carbonic acid, marsh-gas, and hydrogen were respectively drawn through the tube containing the previously exhausted graphite for twelve hours in the cold, the gases being pumped out at a low red heat after each treatment with the dry gas. After the carbonic acid treatment the volume of gas collected was only 1·1 times that of the graphite, containing 98·4 per cent. of carbonic acid; after the marsh-gas the volume of the gas was only 0·9 that of the graphite, containing 94·1 per cent. carbonic acid; and after the hydrogen the volume of the gas collected was only 0·17 times that of the graphite, containing 95·0 per cent. of carbonic acid. It is therefore evident that the large quantity of gas occluded in celestial graphites cannot be explained by any special absorptive power of this variety of carbon. In view of the large and varying percentages of marsh-gas in the gaseous products of all these graphites, it appeared of especial interest to ascertain whether the quantity of marsh-gas extracted coincided in any way with the hydrogen obtained by their combustion. All the samples were therefore submitted to ultimate analysis, with the following results:—

	Percentage Composition.		
	Hydrogen.	Carbon.	Ash.
Celestial graphite	0·11	76·10	23·50
Borrodale ,, ...	0·11	94·76	4·85
Siberian ,, ...	0·17	79·07	20·00
Ceylon ,, ...	0·017	90·90	9·08
Unknown ,, ...	0·246	78·51	21·26

These analyses do not seem to point to any very definite conclusion as to the origin of the marsh-gas. The unknown graphite, which contains the largest percentage of marsh-gas, certainly comes out far the highest in hydrogen, and the hydrogen in the Ceylon graphite also bears a certain relation to the small quantity of marsh-gas it contains, but the first three samples are very similar to each other in the amount of hydrogen they contain.

In order to get some further insight into the origin of this marsh-gas in the celestial graphite, about 2 grams of the original nodule were very finely ground up and digested for several hours with strong nitric acid. After being thoroughly washed from every trace of nitric acid and dried at 110° C., it was again submitted to analysis, with the result that the amount of hydrogen remained exactly the same as before, proving that it existed in the form of some very stable compound in the graphite.

To clear up this matter still further, about 10 grams of the original nodule were digested with pure ether in the way described by Lawrence Smith for extracting the hydrocarbon-like bodies. It was allowed to stand for twenty-four hours with excess of ether, and then filtered, and washed with more ether. The graphite thus treated was dried at 110° C., and the gases extracted from it.

For the purpose of comparing one of the terrestrial graphites with the above in regard to its behaviour with ether, the specimen of unknown origin was selected, as yielding the largest quantity of marsh-gas. The residue, after digestion with ether, was dried, and the gases pumped out as before.

It will be seen that by this treatment with ether the volume of gas given off by the celestial graphite, and also the marsh-gas, have been reduced to rather more than one-half, while with regard to the unknown graphite, although the total volume of gas remains about the same (probably due to a rather higher temperature being employed), the marsh-gas has also been reduced to rather less than one-third the original amount, and the hydrogen has correspondingly increased.

	Occluded Gases in vols. of the Graphite.	CO_2.	CO.	H.	CH_4.	N.
Celestial graphite before treatment with ether ..	7·25	91·81	..	2·50	5·40	0·1
Celestial graphite after treatment with ether	3·50	81·50	10·63	1·41	2·12	0·74
Unknown graphite before treatment with ether ..	7·26	50·79	3·16	2·50	39·53	3·49
Unknown graphite after treatment with ether ..	7·15	64·86	5·67	14·37	12·96	2·00

These experiments prove that either the ether did not dissolve out all the actual carbonaceous compounds present, or that the marsh-gas was subsequently formed during the heating of the graphite.

As Dr. de la Rue had kindly placed at Professor Dewar's disposal a splendid specimen of the Orgueil meteorite, the opportunity was taken of comparing the gases occluded by this typical specimen of the carbonaceous class with those obtained from the stony meteorites and the graphites. This meteorite has been so thoroughly examined by Clöez and Pisani (Compt. Rend. vol. lix. (1864) pp. 37, 132) with regard to its chemical inorganic constituents, that nothing need be said as to its general composition. The investigation was therefore confined to the gases given off on heating which had not previously been determined.

During the heating of the meteorite a large quantity of water, on which floated numerous small pieces of sulphur, collected in the bulb tube immersed in the freezing mixture. This water was strongly acid, and indeed smelt strongly of sulphurous acid. On evaporating it to dryness with a drop of hydrochloric acid, abundance of ammoniacal salts were found in the residue. In the cool anterior part of the combustion-tube a considerable sublimate had collected, which proved to be principally sulphate of ammonium with traces of sulphides and sulphites, and a large quantity of free sulphur. A very large quantity of gas was given off, having the following composition:—

	Sp. gr.	Occluded Gases in vols. of the Meteorite.	Percentage Composition.				
			CO_2.	CO.	CH_4.	N.	SO_2.
Orgueil meteorite ..	2·567	57·87	12·77	1·96	1·50	0·56	83·00

Sulphurous acid is evidently the main constituent of the gases given off; but if this gas, which has been formed from the decomposition of the sulphate of iron, be eliminated, the meteorite yields 9·8 times its volume of gas, having very much the same composition as that from some of the stony meteorites, viz.:—

CO_2, 76·05; CO, 11·67; CH_4, 8·93; N, 3·33.

Clöez found the organic matter in this meteorite to be composed of carbon 63·45, hydrogen 5·98, oxygen 30·57, which is nearly in the proportions of a terrestrial humus substance. It is known that such substances break up by the action of heat into gases of the nature found above, at the same time, however, a quantity of the carbonic acid undoubtedly comes from the presence of the carbonates of magnesium and iron. The operation by which terrestrial carbon has been changed into graphite is by no means clear. As a rule the transition of one kind of carbon into another necessitates the action of a very high temperature. If, therefore, a really high temperature is in all cases necessary, it is difficult to explain how compounds of carbon came to resist decomposition, and should come to be found associated with all natural graphites.

It may be assumed that the graphite resulted from the action of water, gases and other agents, on the carbides of the metals, and that during the chemical interactions which took place, a portion of the carbon became transformed into organic compounds.

In either case it points to the conclusion that the method of formation of the meteoric and terrestrial graphites was similar, and it is perfectly possible they may after all have come from a common source.

It is proposed to continue this investigation, and in order to acquire further information, to examine the gases given off from meteorites at definite temperatures, and especially the gases from such as can be found coated with an impervious glaze, and to examine more particularly into the presence of water in such bodies, and the source of the nitrogen found in the same.

Since the above analyses of different graphites were made, a sample of the artificial graphite which results from the action of oxidising agents on the cyanogen compounds present in crude caustic soda has been examined. The following analysis shows that this artificial variety of graphite is characterised by giving a very large yield of marsh-gas.

CO_2	45·42
CO	39·88
CH_4	4·43
H	8·31
N	2·00

Occluded gases in volumes of the graphite = 53·13.

Meteorites, no doubt, have an exceedingly low temperature before they enter the earth's atmosphere, and the question had been raised as to what chemical reactions could take place under such conditions. It resulted from Professor Dewar's investigations that at a temperature of about − 130° C. liquid oxygen had no chemical action upon hydrogen, potassium, sodium, phosphorus, hydriodic acid, or sulphydric acid. It would appear, therefore, that as the absolute zero is approached even the strongest chemical affinities are inactive.

The lecturer exhibited at work the apparatus by which he had recently succeeded in solidifying oxygen. The apparatus is illustrated in the accompanying diagram,* where a copper tube is seen passing through a vessel kept constantly full of ether and solid carbonic acid; ethylene is sent through this tube, and is liquefied by the intense cold; it is then conveyed by the tube, through an indiarubber stopper, into the interior lower vessel; the outer one is filled with ether and solid carbonic acid. A continuous copper tube, about 45 feet long, conveying oxygen, passes first through the outer vessel, and then through that containing the liquid ethylene; the latter evaporates through the space between the two vessels, and thus intense cold is produced, whereby oxygen is liquefied in the tube to the extent occasionally of 22 cubic centimetres at one time. The temperature at which this is effected is about − 130° C., at a pressure of 75 atmospheres, but less pressure will suffice. When the oxygen is known to be liquid, by means of a gauge near the oxygen inlet, the valve A is opened, and the liquid oxygen rushes into a vacuum in the central glass tube below; some liquid ethylene at the bottom of the next tube outwards is also caused to evaporate into a vacuum at the same moment, and instantly some of the liquid oxygen in the central tube becomes solid, owing to the intense cold of the double evaporation.

* This illustration appeared in 'Industries' of July 16, 1886, and is kindly lent by the proprietors.

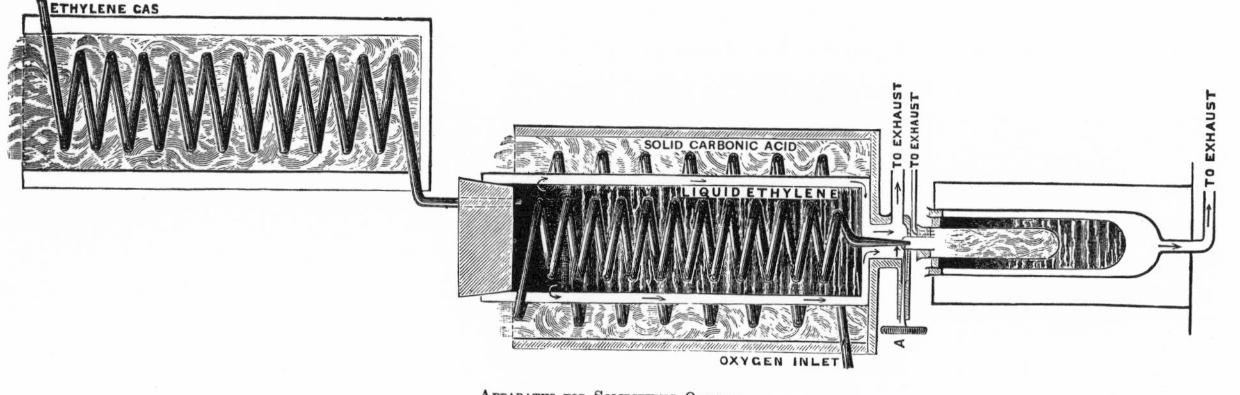

Apparatus for Solidifying Oxygen.

The outer glass vessel serves to keep moisture from settling on the sides of the ethylene tube. By means of the electric lantern and a lens, an image of this part of the apparatus was projected upon the screen, this being the first time that the experiment had been shown on a large scale in public.

Performing the experiment, the temperature reached was a little below 200° C., that is only 50° to 70° above the absolute zero of temperature, and in the experiment about 5 lbs. of liquid ethylene were employed.

With reference to the main subject, the lecturer said that meteorites came from regions of intense cold into our atmosphere; most of them weigh but a few ounces or pounds, but exceptional meteorites weigh several hundredweight. A spherical body 3 feet in diameter, moving at the rate of 18 miles a second at the height of 23 miles, where the barometric pressure is only two-tenths of an inch, produces locally a compression pressure 5600 times greater than that of the surrounding air. Descending vertically, it would pass through the whole atmosphere in 15 seconds. The velocity stated in these data is relatively low as compared with that of planetary bodies. Meteorites travel at the rate of about 36 miles in a second. The velocity of a shot from a 100-ton gun is about half a mile per second.

Meteorites reach the earth covered with a thin and very remarkable glaze, due to the fusion of their external surface during their brief passage through the atmosphere. A velocity of 145 feet per second in air gives an increase of 10° temperature, and the rate continues as the square of the velocity. The surface temperature of a body moving at the rate of 39 miles per second would reach 2,000,000°.

The lecturer placed a piece of iron against a rotating emery wheel, the friction of which caused showers of sparks to be thrown out. These were so hot that some of the little globules of iron composing them were fused into a plate of glass placed to catch them. Great similarities exist between the flight of these globules and the flight of meteorites, the heat and light in both cases being partly due to friction and partly to chemical action. That chemical action has an influence, he proved by applying oxygen gas to the sparks, thereby causing them to burn more brilliantly, and by applying carbonic acid to them, thus reducing their brilliancy. When a piece of meteorite was applied to the emery wheel in place of the piece of iron, the sparks were far less abundant, and of a dull red colour. The glaze of meteorites can be imitated to some extent by cooling a piece of meteorite to 200° C., and then dropping it for a moment into the electric furnace; the temperature explains the glazing of a meteorite, and that it has a motion of rotation must also be considered in estimating the amount of friction, and therefore of heat, to which it is subjected in its passage through the atmosphere of the earth. An enormous amount of its energy, however, is expended in heating the air, and aërial vibrations thus set up explain the noises made by the passage of meteorites.

What is the origin of the gases in meteorites? Their presence agrees with the discovery of Dr. Huggins, that comets give a hydrocarbon spectrum. The origin of terrestrial graphite is far from being agreed upon by geologists; in some places it is evidently transformed coal, in other cases they cannot say whether it comes from vegetable or primitive sources. Whatever the origin of the graphite in meteorites, it contains similar impurities to those in terrestrial graphite; the nodules in celestial graphite are similar to those of terrestrial graphite, and might as well have come from some body like the earth as from any other source. Another conclusion is that the marsh-gas is not occluded in meteorites, but is a product of distillation by heat, just as the gas might be distilled from shales. The graphite of meteorites has no power of occluding marsh-gas, therefore the inference is that the marsh-gas and hydrogen in them are the result of the decomposition of organic bodies. In the spectrum of one of the comets, Dr. Huggins once photographed a peculiar band in the ultra-violet, which band indicated the presence of cyanogen. One meteorite on the table contained chloride of ammonium, therefore it contained a compound of nitrogen, and such would account for the production of cyanogen.

[J. D.]

Friday, January 21, 1887.

William Huggins, Esq. D.C.L. LL.D. F.R.S. Manager and Vice-President, in the Chair.

Sir William Thomson, LL.D. F.R.S. *M.R.I.*

PROFESSOR OF NATURAL PHILOSOPHY IN THE UNIVERSITY OF GLASGOW.

The Sun's Heat.

From human history we know that for several thousand years the sun has been giving heat and light to the earth as at present, possibly with some considerable fluctuations, and possibly with some not very small progressive variation. The records of agriculture, and the natural history of plants and animals within the time of human history, abound with evidence that there has been no exceedingly great change in the intensity of the sun's heat and light within the last three thousand years; but for all that, there may have been variations of quite as much as 5 or 10 per cent., as we may judge by considering that the intensity of the solar radiation to the earth is $6\frac{1}{2}$ per cent. greater in January than in July; and neither at the equator nor in the northern or southern hemispheres has this difference been discovered by experience or general observation of any kind. But as for the mere age of the sun, irrespective of the question of uniformity, we have proof of something vastly more than three thousand years in geological history, with its irrefragable evidence of continuity of life on the earth in time past for tens of thousands, and probably for millions of years.

Here, then, we have a splendid subject for contemplation and research in Natural Philosophy or Physics—the science of dead matter. The sun, a mere piece of matter of the moderate dimensions which we know it to have, bounded all round by cold ether,* has been doing work at the rate of four hundred and seventy-six thousand million million million horse-power for three thousand years; and at

* The sun warms and lights the earth by wave motion, excited in virtue of his white-hot temperature, and transmitted through a material commonly called the luminiferous ether, which fills all space as far as the remotest star, and has the property of transmitting radiant heat (or light) without itself becoming heated. I feel that I have a right to drop the adjective luminiferous, because the medium, far above the earth's surface, through which we receive sun-heat (or light), and through which the planets move, was called ether 2000 years before chemists usurped the name for "sulphuric ether," "muriatic ether," and other compounds, fancifully supposed to be peculiarly ethereal; and I trust that chemists of the present day will not be angry with me if I use the word ether, pure and simple, to denote the medium whose undulatory motions constitute radiant heat (or light).

possibly a higher, certainly not much lower, rate for a few million years. How is this to be explained? Natural philosophy cannot evade the question, and no physicist who is not engaged in trying to answer it can have any other justification than that his whole working time is occupied with work on some other subject or subjects of his province by which he has more hope of being able to advance science.

It may be taken as an established result of scientific inquiry that the sun is *not* a burning fire, and *is* merely a white hot fluid mass cooling, with some little accession of fresh energy by meteors occasionally falling in, but of very small account in comparison with the whole energy of heat which he gives out from year to year. Helmholtz's form of the meteoric theory of the origin of the sun's heat, may be accepted as having the highest degree of scientific probability that can be assigned to any assumption regarding actions of prehistoric times. The essential principle of the explanation is this; at some period of time, long past, the sun's initial heat was generated by the collision of pieces of matter gravitationally attracted together from distant space to build up his present mass; and shrinkage due to cooling gives, through the work done by the mutual gravitation of all parts of the shrinking mass, the vast heat-storage capacity in virtue of which the cooling has been, and continues to be, so slow.

In some otherwise excellent books it is "paradoxically" stated that the sun is becoming hotter because of the condensation.* Paradoxes have no place in science. Their removal is the substitution of true for false statements and thoughts, not always so easily effected as in the present case. The truth is, that it is because the sun is becoming less hot *in places of equal density* that his mass is allowed to yield gradually under the condensing tendency of gravity and thus from age to age cooling and condensation go on together.

An essential detail of Helmholtz's theory of solar heat is that the sun must be fluid, because even though given at any moment hot enough from the surface to any depth, however great, inwards, to be brilliantly incandescent, the conduction of heat from within through solid matter of even the highest conducting quality known to us, would not suffice to maintain the incandescence of the surface for more than a few hours, after which all would be darkness. Observation confirms this conclusion so far as the outward appearance of the

---

* [Note of February 21, 1887.—The "paradox" referred to here, is, as I now find, merely a mis-statement (faulty and manifestly paradoxical through the omission of an essential condition) of an astonishing and most important conclusion of a paper by J. Homer Lane, which appeared in the 'American Journal of Science, for July 1870 (referred to more particularly on p. 11 below). In Newcomb's 'Popular Astronomy,' first edition, p. 508, the omission is supplied in a footnote, giving a clear popular explanation of the dynamics of Lane's conclusion; and the subject is similarly explained in Ball's 'Story of the Heavens,' pp. 501, 502, and 503, with complete avoidance of the "paradox." And now I take this opportunity of correcting my hasty correction of the "paradox" by the insertion of the five words in italics added to line 6 of the paragraph.—W. T.]

sun is concerned, but does not suffice to disprove the idea which was so eloquently set forth by Sir John Herschel, and which prevailed till thirty or forty years ago, that the sun is a solid nucleus inclosed in a sheet of violently agitated flame. In reality, the matter of the outer shell of the sun, from which the heat is radiated outwards, must in cooling become denser, and so becoming unstable in its high position must fall down, and hotter fluid from within must rush up to take its place. The tremendous currents thus continually produced in this great mass of flaming fluid constitute the province of the newly-developed science of solar physics, which, with its marvellous instrument of research—the spectroscope—is yearly and daily giving us more and more knowledge of the actual motions of the different ingredients, and of the splendid and all-important resulting phenomena.

To form some idea of the amount of the heat which is being continually carried up to the sun's surface and radiated out into space, and of the dynamical relations between it and the solar gravitation, let us first divide that prodigious number (476×10^{21}) of horse-power by the number ($6 \cdot 1 \times 10^{18}$) of square metres * in the sun's surface, and we find 78,000 horse-power as the mechanical value of the radiation per square metre. Imagine, then, the engines of eight ironclads applied, by ideal mechanism of countless shafts, pulleys, and belts, to do all their available work of, say 10,000 horse-power each, in perpetuity driving one small paddle in a fluid contained in a square metre vat. The same heat would be given out from the square metre * surface of the fluid as is given out from every square metre of the sun's surface.

But now to pass from a practically impossible combination of engines, and a physically impossible paddle and fluid and containing vessel, towards a more practical combination of matter for producing

---

* A square metre is about 10¾ (more nearly 10·764) square feet, or a square yard and a fifth (more nearly 1·196 square yards). The metre is a little less than 40 inches (39·37 inches = 3·281 feet = 1·094 yards). The kilometre, which we shall have to use presently, being a thousand metres, is a short mile as it were (·6214 of the British statute mile). Thus in round numbers 62 statute miles is equal to 100 kilometres, and 161 kilometres is equal to 100 statute miles. The awful and unnecessary toil and waste of brain power involved in the use of the British system of inches, feet, yards, perches, or rods, or poles, "chains," furlongs, British statute miles, nautical miles, square rod (30¼ square yards)! rood (1210 square yards)! acre (4 roods), may be my apology, but it is only a part of my reason, for not reckoning the sun's area in acres, his activity in horse-power per square inch or per square foot, and his radius, and the earth's distance from him in British statute miles, and for using exclusively the one-denominational system introduced by the French ninety years ago, and now in common use in every civilised country of the world, except England and the United States of North America. The British ton is 1·016 times the French ton, or weight of a cubic metre of cold water (1016 kilogrammes). The French ton, of 1000 kilogrammes, is ·9842 of the British ton. Thus for many practical reckonings, such as those of the present paper, the difference between the British and the French ton may be neglected.

the same effect: still keep the ideal vat and paddle and fluid, but place the vat on the surface of a cool, solid, homogeneous globe of the same size (697,000 kilometres radius) as the sun, and of density (1·4) equal to the sun's mean density. Instead of using steam-power, let the paddle be driven by a weight descending in a pit excavated below the vat. As the simplest possible mechanism, take a long vertical shaft, with the paddle mounted on the top of it so as to turn horizontally. Let the weight be a nut working on a screw-thread on the vertical shaft, with guides to prevent the nut from turning—the screw and the guides being all absolutely frictionless. Let the pit be a metre square at its upper end, and let it be excavated quite down to the sun's centre, everywhere of square horizontal section, and tapering uniformly to a point in the centre. Let the weight be simply the excavated matter of the sun's mass, with merely a little clearance space between it and the four sides of the pit, and with a kilometre or so cut off the lower pointed end to allow space for its descent. The mass of this weight is 326 million tons. Its heaviness, three-quarters of the heaviness of an equal mass at the sun's surface, is 244 million tons solar surface-heaviness. Now a horse-power is, per hour, 270 metre-tons, terrestrial surface-heaviness; or 10 metre-tons, solar surface-heaviness, because a ton of matter is twenty-seven times as heavy at the sun's surface as at the earth's. To do 78,000 horse-power, or 780,000 metre-tons solar surface-heaviness per hour, our weight must therefore descend at the rate of one metre in 313 hours, or about 28 metres per year.

To advance another step, still through impracticable mechanism, towards the practical method by which the sun's heat is produced, let the thread of the screw be of uniformly decreasing steepness from the surface downwards, so that the velocity of the weight, as it is allowed to descend by the turning of the screw, shall be in simple proportion to distance from the sun's centre. This will involve a uniform condensation of the material of the weight; but a condensation so exceedingly small in the course even of tens of thousands of years, that, whatever be the supposed material, metal or stone, of the weight, the elastic resistance against the condensation will be utterly imperceptible in comparison with the gravitational forces with which we are concerned. The work done per metre of descent of the top end of the weight will be just four-fifths of what it was when the thread of the screw was uniform. Thus, to do the 78,000 horse-power of work, the top end of the weight must descend at the rate of 35 metres per year: or 70 kilometres per 2000 years.

Now let the whole surface of our cool solid sun be divided into squares, for example as nearly as may be of one square metre area each, and let the whole mass of the sun be divided into long inverted pyramids or pointed rods, each 697,000 kilometres long, with their points meeting at the centre. Let each be mounted on a screw, as already described for the long tapering weight which we first considered; and let the paddle at the top end of each screw-shaft revolve in a fluid, not now confined to a vat, but covering the whole surface of the sun to a depth of a few metres or kilometres. Arrange the viscosity of the fluid and the size of each paddle so as to let the paddle turn just so fast as to allow the top end of each pointed rod to descend at the rate of 35 metres per year. The whole fluid will, by the work which the paddles do in it, be made incandescent, and it will give out heat and light to just about the same amount as is actually done by the sun. If the fluid be a few thousand kilometres deep over the paddles, it would be impossible, by any of the appliances of solar physics, to see the difference between our model mechanical sun and the true sun.

To do away with the last vestige of impracticable mechanism in which the heavinesses of all parts of each long rod are supported on the thread of an ideal screw cut on a vertical shaft of ideal matter, absolutely hard and absolutely frictionless: first, go back a step to our supposition of just one such rod and screw working in a single pit excavated down to the centre of the sun, and let us suppose all the rest of the sun's mass to be rigid and absolutely impervious to heat. Warm up the matter of the pyramidal rod to such a temperature that its material melts and experiences as much of Sir Humphry Davy's "repulsive motion" as suffices to keep it balanced as a fluid, without either sinking or rising from the position in which it was held by the thread of the screw. When the matter is thus held up without the screw, take away the screw or let it melt in its place. We should thus have a pit from the sun's surface to his céntre, of a square metre area at the surface, full of incandescent fluid, which we may suppose to be of the actual ingredients of the solar substance. This fluid, having at the first instant the temperature with which the paddle left it, would at the first instant continue radiating heat just as it did when the paddle was kept moving; but it would quickly become much cooler at its surface, and to a distance of a few metres down. Currents of less hot fluid tumbling down, and hotter fluid coming up from below, in irregular whirls, would carry the cooled fluid down from the surface, and bring up hotter fluid from below, but this mixing could not go on through a depth of very many metres to a sufficient degree to keep up anything approaching to the high temperature maintained by the paddle; and after a few hours or days, solidification would commence at the surface. If the solidified matter floats on the fluid, at the same temperature, below it, the crust would simply thicken as ice on a lake thickens in frosty weather; but if, as is more probable, solid matter, of such ingredients as the sun is composed of, sinks in the liquid when both are at the melting temperature of the substance, thin films of the upper crust would fall in, and continue falling in, until, for several metres downwards, the whole mass of mixed solid and fluid becomes stiff enough (like the stiffness of paste or of mortar) to prevent the frozen film from falling down from the surface. The surface film would then quickly thicken, and in the course of a few hours or days become less than red-hot on its upper surface, the whole pit full of fluid would go on cooling with extreme slowness until, after possibly about a million million million years or so, it would be all at the same temperature as the space to which its upper end radiates.

Let precisely what we have been considering be done for every one of our pyramidal rods, with, however, in the first place, thin partitions of matter impervious to heat separating every pit from its four surrounding neighbours. Precisely the same series of events as we have been considering will take place in every one of the pits.

Suppose the whole complex mass to be rotating at the rate of once round in twenty-five days, which is, about as exactly as we know it, the time of the sun's rotation about his axis.

Now at the instant when the paddle stops let all the partitions be annulled, so that there shall be perfect freedom for currents to flow unresisted in any direction, except so far as resisted by the viscosity of the fluid, and leave the piece of matter, which we may now call the Sun, to himself. He will immediately begin showing all the phenomena known in solar physics. Of course the observer might have to wait a few years for sunspots, and a few quarter-centuries to discover periods of sunspots, but they would, I think I may say probably, all be there just as they are, because I think we may feel that it is most probable that all these actions are due to the sun's own substance, and not to external influences of any kind. It is, however, quite possible, and indeed many who know most of the subject think it probable, that some of the chief phenomena due to sunspots arise from influxes of meteoric matter circling round the sun.

The energy of chemical combination is as nothing compared with the gravitational energy of shrinkage, to which the sun's activity is almost wholly due. A body falling forty-six kilometres to the sun's surface *or through the sun's atmosphere*, has as much work done on it by gravity, as corresponds to a high estimate of chemical energy in the burning of combustible materials. But chemical combinations and dissociations may, as urged by Lockyer, in his book on the 'Chemistry of the Sun,' just now published, be thoroughly potent determining influences on some of the features of non-uniformity of the brightness in the grand phenomena of sun-spots, hydrogen flames, and corona, which make the province of solar physics. But these are questions belonging to a very splendid branch of solar science to which only allusion can be made at the present time.

What concerns us as to the explanation of sun-light and sun-heat may be summarised in two propositions:—

(1) Gigantic currents throughout the sun's liquid mass are continually maintained by fluid, slightly cooled by radiation falling down from the surface, and hotter fluid rushing up to take its place.

(2) The work done in any time by the mutual gravitation of all the parts of the fluid, as it shrinks in virtue of the lowering of its temperature, is but little less than (so little less than, that we may regard it as practically equal to) the dynamical equivalent of the heat that is radiated from the sun in the same time.

The rate of shrinkage corresponding to the present rate of solar radiation has been proved to us, by the consideration of our dynamical model, to be 35 metres on the radius per year, or one ten-thousandth of its own length on the radius per two thousand years. Hence, if the solar radiation has been about the same as at present for two hundred thousand years, his radius must have been greater by one per cent. two hundred thousand years ago than at present. If we wish to carry our calculations much farther back or forward than two hundred thousand years, we must reckon by differences of the reciprocal of the sun's radius, and not by differences simply of the radius, to take into account the change of density (which, for example, would be three per cent. for one per cent. change of the radius). Thus the rule, easily worked out according to the principles illustrated by our mechanical model, is this:—

Equal differences of the reciprocal of the radius correspond to equal quantities of heat radiated away from million of years to million of years.

Take two examples—

(1) If in past time there has been as much as fifteen million times the heat radiated from the sun as is at present radiated out in one year, the solar radius must have been four times as great as at present.

(2) If the sun's effective thermal capacity can be maintained by shrinkage till twenty million times the present year's amount of heat is radiated away, the sun's radius must be half what it is now. But it is to be remarked that the density which this would imply, being 11·2 times the density of water, or just about the density of lead, is probably too great to allow the free shrinkage as of a cooling gas to be still continued without obstruction through overcrowding of the molecules. It seems, therefore, most probable that we cannot for the future reckon on more of solar radiation than, if so much as, twenty million times the amount at present radiated out in a year. It is also to be remarked that the greatly diminished radiating surface, at a much lower temperature, would give out annually much less heat than the sun in his present condition gives. The same considerations led Newcomb to the conclusion "that it is hardly likely that the sun can continue to give sufficient heat to support life on the earth (such life as we now are acquainted with, at least) for ten million years from the present time."

In all our calculations hitherto we have for simplicity taken the density as uniform throughout, and equal to the true mean density of the sun, being about 1·4 times the density of water, or about a quarter of the earth's mean density. In reality the density in the upper parts of the sun's mass must be something less than this, and something considerably more than this in the central parts, because of the pressure in the interior increasing to something enormously great at the centre. If we knew the distribution of interior density we could easily modify our calculations accordingly; but it does not seem probable that the correction could, with any probable assumption as to the greatness of the density throughout a considerable proportion of the sun's interior, add more than a few million years to the past of solar heat, and what could be added to the past must be taken from the future.

In our calculations we have taken Pouillet's number for the total activity of solar radiation, which practically agrees with Herschel's. Forbes * showed the necessity for correcting the mode of allowing for atmospheric absorption used by his two predecessors in estimating the total amount of solar radiation, and he was thus led to a number 1·6 times theirs. Forty years later Langley,† in an excellently worked out consideration of the whole question of absorption by our atmosphere, of radiant heat of all wave-lengths, accepts and confirms Forbes's reasoning, and by fresh observations in very favourable circumstances on Mount Whitney, 15,000 feet above the sea-level, finds a number a little greater still than Forbes (1·7, instead of Forbes' 1·6, times Pouillet's number). Thus Langley's measurement of solar radiation corresponds to 133,000 horse-power per square metre, instead of the 78,000 horse-power which we have taken, and diminishes each of our times in the ratio of 1 to 1·7. Thus, instead of Helmholtz's twenty million years, which was founded on Pouillet's estimate, we have only twelve millions, and similarly with all our other time reckonings based on Pouillet's results. In the circumstances, and taking fully into account all possibilities of greater density in the sun's interior, and of greater or less activity of radiation in past ages, it would, I think, be exceedingly rash to assume as probable anything more than twenty million years of the sun's light in the past history of the earth, or to reckon on more than five or six million years of sunlight for time to come.

We have seen that the sun draws on no external source for the heat he radiates out from year to year, and that the whole energy of this heat is due to the mutual attraction between his parts acting in conformity with the Newtonian law of gravitation. We have seen how an ideal mechanism, easily imagined and understood, though infinitely far from possibility of realisation, could direct the work done by mutual gravitation between all the parts of the shrinking mass, to actually generate its heat-equivalent in an ocean of white-hot liquid covering the sun's surface, and so keep it white-hot while

* 'Edin. New Phil. Journal,' vol. xxxvi. 1844.

† 'American Journal of Science,' vol. xxvi. March, 1883.

constantly radiating out heat at the actual rate of the sun's heat-giving activity. Let us now consider a little more in detail the real forces and movements actually concerned in the process of cooling by radiation from the uttermost region of the sun, the falling inwards of the fluid thus cooled, the consequent mixing up of the whole mass of the sun, the resulting diminished elastic resistance to pressure in equi-dense parts, and the consequent shrinkage of the whole mass under the influence of mutual gravitation. I must first explain that this "elastic resistance to pressure" is due to heat, and is, in fact, what I have, in the present lecture, called "Sir Humphry Davy's repulsive motion" (p. 5). I called it so because Davy first used the expression "repulsive motion" to describe the fine intermolecular motions to which he and other founders of the Kinetic Theory of Heat attributed the elastic resistance to compression presented by gases and fluids.

Imagine, instead of the atoms and molecules of the various substances which constitute the sun's mass, a vast number of elastic globes, like schoolboys' marbles or billiard balls. Consider first, anywhere on our earth a few million such balls put into a room, large enough to hold a thousand times their number, with perfectly hard walls and ceiling, but with a real wooden floor; or, what would be still more convenient for our purpose, a floor of thin elastic sheet steel, supported by joists close enough together to prevent it from drooping inconveniently in any part. Suppose in the beginning the marbles to be lying motionless on the floor. In this condition they represent the atoms of a gas, as for instance, oxygen, nitrogen, or hydrogen, absolutely deprived of heat, and therefore lying frozen, or as molecular dust strewn on the floor of the containing vessel.

If now a lamp be applied below the oxygen, nitrogen, or hydrogen, the substance becoming warmed by heat conducted through the floor, will rise from its condition of absolutely cold solid, or of incoherent molecular dust, and will spread as a gas through the whole enclosed space. If more and more heat be applied by the lamp the pressure of the gas outwards in all directions against the inside of the enclosing vessel will become greater and greater.

As a rude mechanical analogue to this warming of a gas by heat conducted through the floor of its containing vessel, from a lamp held below it, return to our room with floor strewn with marbles, and employ workmen to go below the floor and strike its underside in a great many places vehemently with mallets. The marbles in immediate contact with the floor will begin to jump from it and fall sharply back again (like water in a pot on a fire simmering before it boils). If the workmen work energetically enough there will be more and more of commotion in the heap, till every one of the balls gets into a state of irregular vibration, up and down, or obliquely, or horizontally, but in no fixed direction; and by mutual jostling the heap swells up till the ceiling of the room prevents it from swelling any further. Suppose now the floor to become, like the walls and ceiling, absolutely rigid. The workmen may cease their work of hammering, which would now be no more availing to augment the motions of the marbles within, than would be a lamp applied outside to warm the contents of a vessel, if the vessel were made of ideal matter impermeable to heat. The marbles being perfectly elastic will continue for ever * flying about in their room striking the walls and floor and ceiling and one another, and remaining in a constant average condition of denser crowd just over the floor and less and less dense up to the ceiling.

In this constant average condition the average velocity of the marbles will be the same all through the crowd, from ceiling to floor, and will be the same in all directions, horizontal, or vertical, or inclined. The continually repeated blows upon any part of the walls or ceiling will in the aggregate be equivalent to a continuous pressure which will be in simple proportion to the average density of the crowd at the place. The diminution of pressure and density from the floor upwards will be precisely the same as that of the density and pressure of our atmosphere calculated on the supposition of equal temperature at all heights, according to the well-known formula and tables for finding heights by the barometer.

In reality the temperature of the atmosphere is not uniform from the ground upwards, but diminishes at the rate of about 1° C. for every 162 metres of vertical ascent in free air, undisturbed by mountains, according to observations made in balloons by the late Mr. Welsh, of Kew, through a large range of heights. This diminution of temperature upwards in our terrestrial atmosphere is most important and suggestive in respect to the constitution of the solar atmosphere, and not merely of the atmosphere or outer shell of the sun, but of the whole interior fluid mass with which it is continuous. The two cases have so much in common that there is in each case loss of heat from the outer parts of the atmosphere by radiation into space, and that in consequence circulating currents are produced through the continuous fluid, by which a thorough mixing up and down is constantly performed. In the case of the terrestrial atmosphere the lowest parts receive by contact heat from the solid earth, warmed daily by the sun's radiation. On the average of night and day, as the air does not become warmer on the whole, it must radiate out into

---

* To justify this statement I must warn the reader that the ideal perfectly elastic balls which we are imagining must be supposed somehow to have such a structure that each takes only a definite average proportion of its share of the kinetic energy of the whole multitude, so that on the average there is a constant proportion of energy in the translatory motions of the balls ; the other part being the vibratory or rotational motions of the parts of each ball. For simplicity also we suppose the balls to be perfectly smooth and frictionless, so that we shall not be troubled by need to consider them as having any rotatory motions, such as real balls with real frictional collisions would acquire. The ratio of the two kinds of energy for ordinary gases, according to Clausius, to whom is due this essential contribution to the kinetic theory, is—of the whole energy, three-fifths translational to two-fifths vibrational.

space as much heat as all that it gets, both from the earth by contact, and by radiation of heat from the earth, and by intercepted radiation from the sun on its way to the earth. In the case of the sun the heat radiated from the outer parts of the atmosphere is wholly derived from the interior. In both cases the whole fluid mass is kept thoroughly mixed by currents of cooled fluid coming down and warmer fluid rising to take its place, and to be cooled and descend in its turn.

Now it is a well-known property of gases and of fluids generally (except some special cases, as that of water within a few degrees of its freezing temperature, in which the fluid under constant pressure contracts with rise of temperature) that condensation and rarefactions, effected by augmentations and diminutions of pressure from without, produce elevations and lowerings of temperature in circumstances in which the gas is prevented from either taking heat from or giving heat to any material external to it. Thus a quantity of air or other gas taken at ordinary temperature (say 15° C. or 59° F.) and expanded to double its bulk becomes 71° C. cooler; and if the expansion is continued to thirty-two times its original bulk it becomes cooled 148° farther, or down to about 200° C. below the temperature of freezing water, or to within 73° of absolute cold. Such changes as these actually take place in masses of air rising in the atmosphere to heights of eight or nine kilometers, or of twenty or twenty-five kilometers. Corresponding differences of temperature there certainly are throughout the fluid mass of the sun, but of very different magnitudes because of the twenty-seven fold greater gravity at the sun's surface, the vastness of the space through which there is free circulation of fluid, and last, though not least, the enormously higher temperature of the solar fluid than of the terrestrial atmosphere at points of equal density in the two. This view of the solar constitution has been treated mathematically with great power by Mr. J. Homer Lane, of Washington, U.S., in a very important paper read before the National Academy of Sciences, of the United States in April 1869, and published with further developments in the 'American Journal of Science,' for July 1870. Mr. Lane, by strict mathematical treatment finds the law of distribution of density and temperature all through a globe of homogeneous gas left to itself in space, and losing heat by radiation outwards so slowly that the heat-carrying currents produce but little disturbance from the globular form.

One very remarkable and important result which he finds is, that the density at the centre is about twenty * times the mean density; and this, whether the mass be large or small, and whether of oxygen, nitrogen, or hydrogen, or other substance; provided only it be of one kind of gas throughout, and that the density in the central parts is not too great to allow the condensation to take place, according to

* Working out Lane's problem independently, I find 22½ as very nearly the exact number.

the ordinary gaseous law of density, in simple proportion to pressure for the same temperatures. We know this law to hold with somewhat close accuracy for common air, and for each of its two chief constituents, oxygen and nitrogen, separately, and for hydrogen, to densities of about two hundred times their densities at our ordinary atmospheric pressure. But when the compressing force is sufficiently increased, they all show greater resistance to condensation than according to the law of simple proportion, and it seems most probable that there is for every gas a limit beyond which the density cannot be increased by any pressure however great. Lane remarks that the density at the centre of the sun would be "nearly one-third greater than that of the metal platinum," if the gaseous law held up to so great a degree of condensation for the ingredients of the sun's mass; but he does not suggest this supposition as probable, and he no doubt agrees with the general opinion that in all probability the ingredients of the sun's mass, at the actual temperatures corresponding to their positions in his interior, obey the simple gaseous law through but a comparatively small space inwards from the surface; and that in the central regions they are much less condensed than according to that law. According to the simple gaseous law, the sun's central density would be thirty-one times that of water; we may assume that it is in all probability much less than this, though considerably greater than the mean density, 1·4. This is a wide range of uncertainty, but it would be unwise at present to narrow it, ignorant as we are of the main ingredients of the sun's whole mass, and of the laws of pressure, density, and temperature, even for known kinds of matter at very great pressures and very high temperatures.

The question, Is the sun becoming colder or hotter? is an exceedingly complicated one, and, in fact, either to put it or to answer it is a paradox, unless we define exactly where the temperature is to be reckoned. If we ask, How does the temperature of equi-dense portions of the sun vary from age to age? the answer certainly is that the matter of the sun of which the density has any stated value, for example, the ordinary density of our atmosphere, becomes always less and less hot, whatever be its place in the fluid, and whatever be the law of compression of the fluid, whether the simple gaseous law or anything from that to absolute incompressibility. But the distance inwards from the surface at which a constant density is to be found diminishes with shrinkage, and thus it may be that at constant depths inwards from the bounding surface the temperature is becoming higher and higher. This would certainly be the case if the gaseous law of condensation held throughout, but even then the effective radiational temperature, in virtue of which the sun sheds his heat outwards, might be becoming lower, because the temperatures of equi-dense portions are clearly becoming lower under all circumstances.

Leaving now these complicated and difficult questions to the scientific investigators who are devoting themselves to advancing the science of solar physics, consider the easily understood question, What is the temperature of the centre of the sun at any time, and does it rise or fall as time advances? If we go back a few million years to a time when we may believe the sun to have been wholly gaseous to the centre, then certainly the central temperature must have been augmenting; again, if, as is possible though not probable at the present time, but may probably be the case at some future time, there be a solid nucleus, then certainly the central temperature would be augmenting, because the conduction of heat outwards through the solid would be too slow to compensate the augmentation of pressure due to augmentation of gravity in the shrinking fluid around the solid. But at a certain time in the history of a wholly fluid globe, primitively rare enough throughout to be gaseous, shrinking under the influence of its own gravitation and its radiation of heat outwards into cold surrounding space, when the central parts have become so much condensed as to resist further condensation greatly more than according to the gaseous law of simple proportions, it seems to me certain that the early process of becoming warmer, which has been demonstrated by Lane, and Newcomb, and Ball, must cease, and that the central temperature must begin to diminish on account of the cooling by radiation from the surface, and the mixing of the cooled fluid throughout the interior.

Now we come to the most interesting part of our subject—the early history of the Sun. Five or ten million years ago he may have been about double his present diameter and an eighth of his present mean density, or ·175 of the density of water; but we cannot, with any probability of argument or speculation, go on continuously much beyond that. We cannot, however, help asking the question, What was the condition of the sun's matter before it came together and became hot? It may have been two cool solid masses, which collided with the velocity due to their mutual gravitation; or, but with enormously less of probability, it may have been two masses colliding with velocities considerably greater than the velocities due to mutual gravitation. This last supposition implies that, calling the two bodies A and B for brevity, the motion of the centre of inertia of B relatively to A, must, when the distances between them was great, have been directed with great exactness to pass through the centre of inertia of A; such great exactness that the rotational momentum, or "moment of momentum," * after collision was no more than to let

* This is a technical expression in dynamics which means the importance of motion relatively to revolution or rotation round an axis. Momentum is an expression given about a hundred and fifty years ago (when mathematicians and other learned men spoke and wrote Latin) to signify translational importance of motion. Moment of a couple, moment of a magnet, moment of inertia, moment of force round an axis, moment of momentum round an axis, and corresponding verbal combinations in French and German, are expressions which have been

the sun have his present slow rotation when shrunk to his present dimensions. This exceedingly exact aiming of the one body at the other, so to speak, is, on the dry theory of probability, exceedingly improbable. On the other hand, there is certainty that the two bodies A and B at rest in space if left to themselves undisturbed by other bodies and only influenced by their mutual gravitation, shall collide with direct impact, and therefore with no notion of their centre of inertia, and no rotational momentum of the compound body after the collision. Thus we see that the dry probability of collision between two neighbours of a vast number of mutually attracting bodies widely scattered through space is much greater if the bodies be all given a rest, than if they be given moving in any random directions and with any velocities considerable in comparison with the velocities which they would acquire in falling from rest into collision. In this connection it is most interesting to know from stellar astronomy, aided so splendidly as it has recently been by the spectroscope, that the relative motions of the visible stars and our sun are generally very small in comparison with the velocity (612 kilometers per second) which a body would acquire in falling into the sun, and are comparable with the moderate little velocity (29·5 kilometres per second) of the earth in her orbit round the sun.

To fix the ideas, think of two cool solid globes, each of the same mean density as the earth, and of half the sun's diameter; given at rest, or nearly at rest, at a distance asunder equal to twice the earth's distance from the sun. They will fall together and collide in exactly half a year. The collision will last for about half an hour, in the course of which they will be transformed into a violently agitated incandescent fluid mass flying outward from the line of the motion before the collision, and swelling to a bulk several times greater than the sum of the original bulks of the two globes.* How far the fluid mass will fly out all round from the line of collision it is impossible to say. The motion is too complicated to be fully investigated by any known mathematical method; but with sufficient patience a mathematician might be able to calculate it with some fair approximation to the truth. The distance reached by the extreme circular

---

introduced within the last sixty years (by scientists speaking as now, each his own vernacular) to signify the importance of the special subject referred to in each case. The expression moment of momentum is highly valuable and convenient in dynamical science, and it constitutes a curious philological monument of scientific history.

* Such incidents seem to happen occasionally in the universe. Laplace says some stars "have suddenly appeared, and then disappeared, after having shone for several months with the most brilliant splendour. Such was the star observed by Tycho Brahe in the year 1572, in the constellation Cassiopeia. In a short time it surpassed the most brilliant stars, and even Jupiter itself. Its light then waned away, and finally disappeared sixteen months after its discovery. Its colour underwent several changes; it was at first of a brilliant white, then of a reddish yellow, and finally of a lead-coloured white, like to Saturn." (Harte's translation of Laplace's 'System of the World.' Dublin, 1830.)

fringe of the fluid mass would probably be much less than the distance fallen by each globe before the collision, because the translational motion of the molecules constituting the heat into which the whole energy of the original fall of the globes become transformed in the first collision, is probably about three-fifths of the whole amount of that energy. The time of flying out would probably be less than half a year, when the fluid mass must begin to fall in again towards the axis. In something less than a year after the first collision the fluid will again be in a state of maximum crowding round the centre, and this time probably even more violently agitated than it was immediately after the first collision; and it will again fly outward, but this time axially towards the places whence the two globes fell. It will again fall inwards, and after a rapidly subsiding series of quicker and quicker oscillations it will subside, probably in the course of two or three years, into a globular star of about the same dimensions, heat, and brightness as our present sun, but differing from him in this, that it will have no rotation.

We supposed the two globes to have been at rest when they were let fall from a mutual distance equal to the diameter of the earth's orbit. Suppose, now, that instead of having been at rest they had been moving in opposite directions with a velocity of two (more exactly 1·89) metres per second. The moment of momentum of these motions round an axis through the centre of gravity of the two globes perpendicular to their lines of motion is just equal to the moment of momentum of the sun's rotation round his axis. It is an elementary and easily proved law of dynamics that no mutual action between parts of a group of bodies, or of a single body, rigid, flexible, or fluid, can alter the moment of momentum of the whole. The transverse velocity in the case we are now supposing is so small that none of the main features of the collision and of the wild oscillations following it, which we have been considering, or of the magnitude, heat, and brightness of the resulting star, will be sensibly altered; but now, instead of being rotationless, it will be revolving once round in twenty-five days and so in all respects like to our sun.

If instead of being at rest initially, or moving with the small transverse velocities we have been considering, each globe had a transverse velocity of three-quarters (or anything more than ·71) of a kilometre per second, they would just escape collision, and would revolve in ellipses round the centre of inertia in a period of one year, just grazing each other's surface every time they came to the nearest points of their orbits.

If the initial transverse velocity of each globe be less than, but not much less than, ·71 of a kilometre per second, there will be a violent grazing collision, and two bright suns, solid globes bathed in flaming fluid, will come into existence in the course of a few hours, and will commence revolving round their common centre of inertia in long elliptic orbits in a period of a little less than a year. Tidal interaction between them will diminish the eccentricities of their orbits, and if continued long enough will cause the two to revolve in circular orbits round their centre of inertia with a distance between their surfaces equal to 6·44 diameters of each.

Suppose now, still choosing a particular case to fix the ideas, that twenty-nine million cold solid globes, each of about the same mass as the moon, and amounting in all to a total mass equal to the sun's, are scattered as uniformly as possible on a spherical surface of radius equal to one hundred times the radius of the earth's orbit, and that they are left absolutely at rest in that position. They will all commence falling towards the centre of the sphere, and will meet there in two hundred and fifty years, and every one of the twenty-nine million globes will then, in the course of half an hour, be melted, and raised to a temperature of a few hundred thousand or a million degrees Centigrade. The fluid mass thus formed will, by this prodigious heat, be exploded outwards in vapour or gas all round. Its boundary will reach to a distance considerably less than one hundred times the radius of the earth's orbit on first flying out to its extreme limit. A diminishing series of out and in oscillations will follow, and the incandescent globe thus contracting and expanding alternately, in the course it may be of three or four hundred years, will settle to a radius of forty* times the radius of the earth's orbit. The average density of the gaseous nebula thus formed would be $(215 \times 40)^{-3}$, or one six hundred and thirty-six thousand millionth of the sun's mean density; or one four hundred and fifty-four thousand millionth of the density of water; or one five hundred and seventy millionth of that of common air at an ordinary temperature of 10° C. The density in its central regions, sensibly uniform through several million kilometres, is (see note on p. 11) one twenty thousand millionth of that of water; or one twenty-five millionth of that of air. This exceedingly small density is nearly six times the density of the oxygen and nitrogen left in some of the receivers exhausted by Bottomley in his experimental measurements of the amount of heat emitted by pure radiation from highly heated bodies. If the substance were oxygen, or nitrogen, or other gas or mixture of gases simple or compound, of specific density equal to the specific density of our air, the central temperature would be 51,200° Cent., and the average translational velocity of the molecules 6·66 kilometres per second, being $\sqrt{\frac{3}{7}}$ of 10·2, the velocity acquired by a heavy body falling unresisted from the outer boundary (of 40 times the radius of the earth's orbit) to the centre of the nebulous mass.

The gaseous nebula thus constituted would in the course of a few million years, by constantly radiating out heat, shrink to the size of our present sun, when it would have exactly the same heating and lighting efficiency. But no motion of rotation.

* The radius of a steady globular gaseous nebula of any homogeneous gas is 40 per cent. of the radius of the spherical surface from which its ingredients must fall to their actual positions in the nebula to have the same kinetic energy as the nebula has.

The moment of momentum of the whole solar system is about eighteen times that of the sun's rotation; seventeen-eighteenths being Jupiter's and one-eighteenth the Sun's, the other bodies being not worth taking into account in the reckoning of moment of momentum.

Now instead of being absolutely at rest in the beginning, let the twenty-nine million moons be given each with some small motion, making up in all an amount of moment of momentum about a certain axis, equal to the moment of momentum of the solar system which we have just been considering; or considerably greater than this, to allow for effect of resisting medium. They will fall together for two hundred and fifty years, and though not meeting precisely in the centre as in the first supposed case of no primitive motion, they will, two hundred and fifty years from the beginning, be so crowded together that there will be myriads of collisions, and almost every one of the twenty-nine million globes will be melted and driven into vapour by the heat of these collisions. The vapour or gas thus generated will fly outwards, and after several hundreds or thousands of years of outward and inward oscillatory motion, may settle into an oblate rotating nebula extending its equatorial radius far beyond the orbit of Neptune, and with moment of momentum equal to or exceeding the moment of momentum of the solar system. This is just the beginning postulated by Laplace for his nebular theory of the evolution of the solar system; which, founded on the natural history of the stellar universe, as observed by the elder Herschell, and completed in details by the profound dynamical judgment and imaginative genius of Laplace, seems converted by thermodynamics into a necessary truth, if we make no other uncertain assumption than that the materials at present constituting the dead matter of the solar system have existed under the laws of dead matter for a hundred million years. Thus there may in reality be nothing more of mystery or of difficulty in the automatic progress of the solar system from cold matter diffused through space, to its present manifest order and beauty, lighted and warmed by its brilliant sun, than there is in the winding up of a clock* and letting it go till it stops. I need scarcely say that the beginning and the maintenance of life on the earth is absolutely and infinitely beyond the range of all sound speculation in dynamical science. The only contribution of dynamics to theoretical biology is absolute negation of automatic commencement or automatic maintenance of life.

I shall only say in conclusion:—Assuming the sun's mass to be composed of materials which were far asunder before it was hot, the immediate antecedent to its incandescence must have been either two bodies with details differing only in proportions and densities from

---

* Even in this, and all the properties of matter which it involves, there is enough, and more than enough, of mystery to our limited understanding. A watch-spring is much farther beyond our understanding than is a gaseous nebula.

the cases we have been now considering as examples; or it must have been some number more than two—some finite number—at the most the number of atoms in the sun's present mass, a finite number (which may probably enough be something between 4×10^{57} and 140×10^{57}) as easily understood and imagined as number 4 or 140. The immediate antecedent to incandescence may have been the whole constituents in the extreme condition of subdivision—that is to say, in the condition of separate atoms; or it may have been any smaller number of groups of atoms making minute crystals or groups of crystals—snowflakes of matter, as it were; or it may have been lumps of matter like a macadamising stone; or like this stone * (Fig. 1), which you might mistake for a macadamising

FIG. 1.

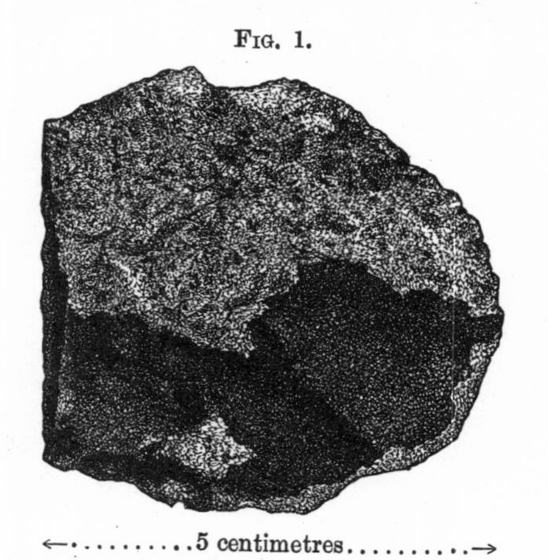

←.........5 centimetres.........→

stone, and which was actually travelling through space till it fell on the earth at Possil, in the neighbourhood of Glasgow, on April 5, 1804; or like that * (Fig. 2) which was found in the Desert of Atacama, in South America, and is believed to have fallen there from the sky—a fragment made up of iron and stone, which

* These three meteorites are in the possession of the Hunterian Museum of the University of Glasgow, and the woodcuts, Figs. 1, 2, and 3, have been executed from the actual specimens kindly lent for that purpose by the keeper of the museum, Professor Young. The specimen represented by Fig. 1 is contained in the Hunterian collection, that by Fig. 2 in the Eck collection, and that by Fig. 3 in the Lanfine collection—the scale of dimensions is shown for each. It may be remarked that Fig. 2 represents a section of the meteorite taken in the plane of the longest rectangular axes; the bright markings being large and well-formed crystals of olivine, embedded in a matrix of iron. In Fig. 3 is depicted the beautiful Widmanstätten marking characteristic of all meteoric iron, and so well shown in the well-known Lenarto meteorite.

looks as if it has solidified from a mixture of gravel and melted iron in a place where there was very little of heaviness; or this splendidly crystallised piece of iron (Fig. 3), a slab cut out

FIG. 2.

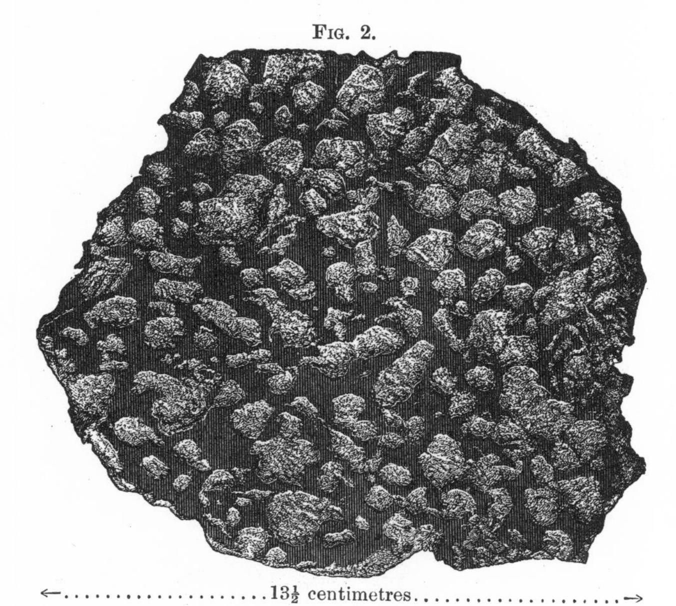

←....................13½ centimetres....................→

FIG. 3.

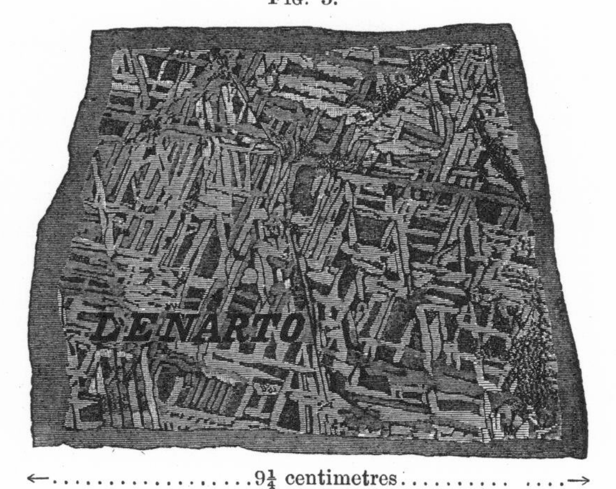

←..................9¼ centimetres.......... ....→

of the celebrated aërolite which fell at Lenarto, in Hungary; or this wonderfully-shaped specimen (of which two views are given

FIG. 4.

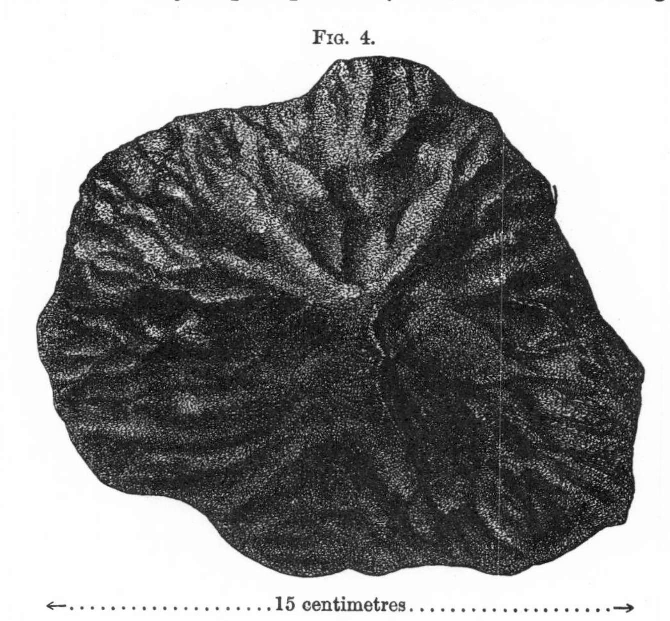

←.....................15 centimetres.....................→

FIG. 5.

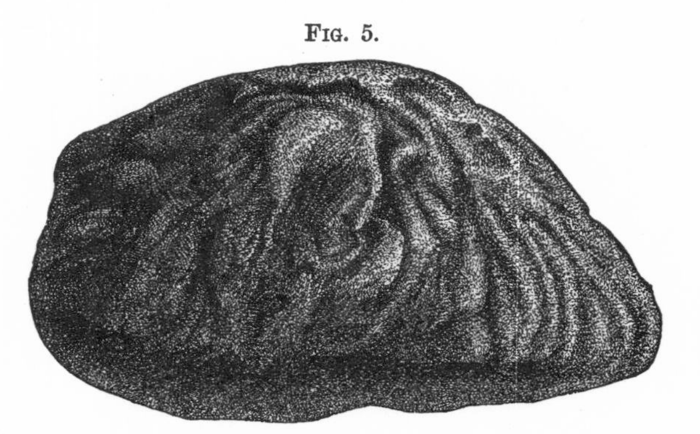

in Figs. 4 and 5), a model of the Middlesburgh meteorite (kindly given me by Professor A. S. Herschel), having corrugations showing how its melted matter has been scoured off from the front part of its surface in its final rush through the earth's atmosphere when it was seen to fall on March 14, 1881, at 3.35 P.M.

For the theory of the sun it is indifferent which of these varieties of configurations of matter may have been the immediate antecedent of his incandescence, but I can never think of these material antecedents without remembering a question put to me thirty years ago by the late Bishop Ewing, Bishop of Argyll and the Isles: "Do you imagine that piece of matter to have been as it is from the beginning; to have been created as it is, or to have been as it is through all time till it fell on the earth?" I had told him that I believed the sun to be built up of meteoric stones, but he would not be satisfied till he knew or could imagine, what kind of stones. I could not but agree with him in feeling it impossible to imagine that any one of such meteorites as those now before you has been as it is through all time, or that the materials of the sun were like this for all time before they came together and became hot. Surely this stone has an eventful history, but I shall not tax your patience by trying just now to trace it conjecturally. I shall only say that we cannot but agree with the common opinion which regards meteorites as fragments broken from larger masses, and we cannot be satisfied without trying to imagine what were the antecedents of those masses.

[W. T.]

Friday, June 3, 1887.

EDWARD WOODS, Esq. Pres. Inst. C.E. Manager and Vice-President, in the Chair.

DAVID GILL, LL.D. F.R.S.

HER MAJESTY'S ASTRONOMER AT THE CAPE OF GOOD HOPE.

The Applications of Photography in Astronomy.

LITTLE more than a year ago Mr. Ainslie Common delivered a lecture in this place on the subject of "Photography as an aid to Astronomy." Given by one who is a consummate master of the art of celestial photography, that lecture (complete as to history and full of suggestion as it was) would, under ordinary circumstances, have precluded further reference to the subject in these Friday evening lectures for some years to come.

But the past year has witnessed such developments of the subject, and the importance of photography in astronomy has been so much advanced by the conclusions of the recent Astro-photographic Congress, as to afford a reasonable apology for the present lecture.

On the 16th of April last there was held at Paris a Congress attended by upwards of fifty astronomers and physicists, representing nearly every civilised nation in the world. It was convened for the purpose of considering a scheme of international co-operation in the work of charting the sky on a large scale. Or, rather, its object was to obtain a series of pictures, which, taken within a comparatively limited period of time, and with the necessary precautions, would enable astronomers of the present day to hand down to future generations a complete record of the positions and magnitudes of all the stars in the heavens to a given order of magnitude. The labours of that Conference are now concluded, certain important resolutions have been adopted, and the way has been so far cleared for giving these resolutions practical effect.

It seems of importance therefore to lay before the members of the Royal Institution some account of the history of this remarkable Congress, to illustrate and explain the grounds of the conclusions which it has arrived at, and otherwise to bring the history of photographic astronomy up to the present date. I pass over the already well told early history of celestial photography, except in so far as it relates to star charting. It was Warren de la Rue who first called attention to the means furnished by photography for charting groups of stars. In his Report to the British Association at Manchester in 1861 on the progress of celestial photography, he indicates a photographic object-glass of short focus as the instrument best suited for the purpose, and he states that, by mounting such a lens and camera on an equatorial stand provided with clockwork, he has photographed such groups of stars as the Pleiades, the chief difficulty being not to fix the images of the stars, but to distinguish them from the specks which are found on the plates or rather in the collodion.

In 1864 Rutherford of New York completed a telescope of 11½ inches aperture and 14 feet focal length, specially constructed for celestial photography, and obtained fine photographs of stars to the 9th order of magnitude. His remarks, although quoted in Mr. Common's lecture last year, have such importance on the present subject that I venture to repeat them.

" The power to obain imagest of the 9th magnitude stars with so moderate an aperture promises to develop and increase the application of photography to the mapping of the sidereal heavens and in some manner to realise the hopes which have so long been deferred and disappointed.

" It would not be difficult to arrange a camera-box, capable of exposing a surface sufficient to obtain a map of two degrees square, and with instruments of large aperture we may hope to reach much smaller stars than I have yet taken. There is also every probability that the chemistry of photography will be very much improved and more sensitive methods devised."

Mr. Common well remarks that in the light of recent work these words are almost prophetic.

But Rutherford did not stop here. In the eyes of an astronomer a picture of stars is of comparatively little importance unless it is capable of accurate measurement. Recognising this important feature of the case, Rutherford devised a suitable apparatus, which he applied to the measurement of two of his photographs of the Pleiades. These measures having been put into the hands of Dr. Gould, that astronomer compared them with those of the same group of stars made by Bessel with his celebrated heliometer, and found a satisfactory accordance.*

Encouraged by these results, Dr. Gould, when he went to the Argentine Republic in 1870 to found the Cordoba Observatory, which has since been rendered so famous by his labours, took Rutherford's telescope with him. Unfortunately one of the lenses was broken in transport, and such delay was incurred in replacing it, that the proposed work could not be begun till 1875. But thanks to the clear skies of Cordoba, and the marvellous activity of the observatory under Dr. Gould's direction, 1350 photographs were obtained in course of a few years, containing representations of all the principal star-clusters of the southern hemisphere, besides a special series of plates taken for the purpose of determining the parallax (or distance) of several of the more remarkable stars in the southern hemisphere.

---

* Astron. Nach. No. 162, vol. xlviii. Dec. 1866.

This fine series of pictures is now being submitted to measurement by Dr. Gould, and the results are awaited with the greatest interest by all astronomers.

The first of Dr. Gould's plates were taken with the old wet collodion process, but the work was afterwards greatly facilitated by employment of the more sensitive modern dry plates.

It was, in fact, the introduction of the gelatine dry plate process in 1876, which really paved the way for the rapid development of celestial photography. The convenience of the manipulation and the great increase of sensitiveness of the plates at once placed a new power in the hands of astronomers. Draper photographed the nebula of Orion in 1880; and after trials, commencing in 1879, Common succeeded in obtaining the exquisite photographs of that object which have been exhibited more than once in this theatre.

In 1882 appeared the splendid comet of that year. At the Royal Observatory, Cape of Good Hope, we were not at the time engaged in photographic operations. Several photographers in the Cape Colony found it possible to obtain impressions of the comet, but they were unable to secure pictures of scientific value, because they were unprovided with means to follow the diurnal motion. I had no available camera belonging to the observatory, and no experience in the development of modern dry plates. In these circumstances, I applied to Mr. Allis, a skilful photographer in my neighbourhood, who eagerly consented to co-operate with me in the work. I arranged means to attach his camera to the stand of an equatorial telescope, and the telescope itself was employed to follow the nucleus of the comet accurately during the whole time of exposure by the aid of the driving clock and with small corrections given by hand. The lens employed had an aperture of only 2 inches, and a focal length of 11 inches; but the result was a series of pictures, one of which, obtained after an exposure of two hours, is now on the screen.

The photograph shows a very satisfactory delineation of the tail and envelope of the comet.

Important and useful as these results were, there was another feature of the pictures which seemed to me still more so. In forwarding copies of these photographs to the Royal Astronomical Society of London and to the Paris Academy of Sciences, I drew particular attention to the large number of stars shown upon the plate, and insisted upon the importance of the means thus offered to photograph comparatively large areas of the sky and thus rapidly make charts of the entire heavens.

The one step wanting was now provided, and the new and more sensitive dry plate rendered the former suggestions of de la Rue and Rutherford now valuable and practicable.

Formerly the old collodion wet plates required large instruments (with small field) and long exposure to depict stars even to the 9th magnitude, and astronomers trusted entirely to the accuracy of their driving clocks, which could not follow a star with perfect accuracy during a long exposure. Now the modern rapid dry plates in conjunction with the large fields of the photographic objective overcame the first of these difficulties, and the plan of employing a guiding telescope overcame the second.

The use of a guiding telescope was not even a new device, for it had been employed long before by Hartnup and others, who, in their early attempts to photograph the moon, kept the image of a lunar spot by hand upon the cross wires of the finder of the telescope during the long exposures then necessary.

There was thus nothing really new either in my suggestion or in the *modus operandi*, only the result was a fortunate one, for Mr. Common says that "these photographs came to him as a revelation of the power of photography for the purpose of star-charting," * and Admiral Mouchez tells me that these Cape photographs and my suggestions first directed his attention and that of the brothers Henry to the application of photography to the work of star-charting, which had for many years been carried on at Paris by the older methods of astronomy.

Common was amongst the first to take up the work in England, and here on the screen is one of his photographs with a 4-inch lens, executed in December 1883. But being engaged in other researches, Common made no attempt to commence a systematic survey of the heavens.

Isaac Roberts, of Liverpool, was also early at work in the same field, and after preliminary experiments he acquired a powerful telescope, with which he began a systematic survey of the northern heavens.

It required some time to find the necessary means and apparatus to begin the realisation of my ideas at the Cape, but at last the work was started in the beginning of 1885 on the following definite plan, viz. to complete the cartography of the heavens from 20° south of the Equator to the South Pole, and so as certainly to include all stars to the 9th magnitude.

The reasons for the adoption of this plan were the following:—

The celebrated astronomer Argelander charted the heavens on this scale from the North Pole to the Equator, and the work has recently been extended to 20° south of the Equator by Schönfeld, the pupil and successor of Argelander.

Argelander's *Durchmusterung*, as it is called, has furnished, ever since the date of its publication, the nomenclature of all the fainter stars employed in the daily operations of astronomy; it has furnished the working catalogues which are essential for the more exact determination of the places of all these stars; it has given us the first accurate data for determining the distribution of the stars according to magnitude and apparent position in the heavens, and is the first

* 'Proc. Royal Institution,' vol. xii. part iii. p. 734.

solid existing basis for founding any theory as to the constitution of the stellar universe. To complete the Durchmusterung for the remaining portion of the heavens was therefore the most pressing need of modern astronomy. I commenced the work in 1885 by the aid of photography. I hope in two or three years, if I have the honour of lecturing again in this theatre, I shall then be able to tell you that the work in question is finished.

I should here explain that mere pictures of the stars are of comparatively little value, or rather of about the same value to an astronomer as a series of charts of parts of the world would be to a sailor if there were no lines of latitude or longitude marked upon them.

The every-day useful part of the Durchmusterung is the catalogue giving the positions and magnitude of all the stars. That work is rapidly advancing in the hands of my able and enthusiastic friend, Professor Kapteyn of Groningen, who, with the aid of three assistants, has undertaken to devote five or six years of his life to the measurement of the Cape photographs and the computation of the results.

When this has been done, as I venture to think it will be within five years, astronomers will be in possession of that preliminary survey of the whole heavens which is necessary for the more refined and elaborate researches which must follow as results of the Paris Congress.

But to return to the work that was meanwhile being done in Paris by the brothers Paul and Prosper Henry.

These astronomers had been engaged since 1871 in the construction of charts of the Ecliptic by the older processes of observation, but when they reached that portion of the heavens where the Milky Way crosses the Ecliptic, the number of stars became so overwhelming that the task of charting seemed almost too great for human patience and skill. But fortunately the time had come when dry plate photography could be called in to aid, and this aid was in the hands of men singularly competent to develop such an opportunity to the fullest extent. The brothers Henry had long aspired to be not only distinguished practical astronomers, but, following the traditions of Huyghens and the Herschels, they desired also to be the artists of their own optical means. Bound together by strong brotherly affection and common tastes, gifted alike with practical talents of a high order, and with an energy and determination of character that permit no obstacle to success, these men thus happily united have devoted the spare hours of their busy astronomical duties at the Paris Observatory, first to the study of optics, and afterwards to the grinding and polishing of lenses and specula, which have won for them a now world-wide reputation as opticians of the highest rank.

I had the pleasure, a few weeks ago, of visiting the modest workshop attached to their house at Montrouge, and I shall not soon forget that visit, nor the many lessons moral as well as practical which I learned.

Every detail of their process of working has been evolved by themselves; they employ no assistant, and their every appliance is simple and practical in a degree which I can only compare with the simple and practical character of the men who designed it.

Such were the men above all others to develop the application of photography to the charting of the heavens. They had high appreciation of the value of the work which they were about to undertake, they had the fullest knowledge of the requirements of the case, and they had the practical skill which enabled them to perfect the necessary apparatus. Their first attempts were made with a telescope of six inches aperture (the object-glass being specially ground for photographic work), and the tube was temporarily adapted to an existing equatorial stand.

With an exposure of forty-five minutes, pictures of stars were obtained to the 12th magnitude, in which the star discs were quite round and sharply defined.

Fully appreciating the beauty of this result, and seeing its importance, Admiral Mouchez boldly faced many administrative difficulties, and accepted without delay the proposals of the brothers Henry to construct an object-glass of thirteen inches aperture and about eleven feet focal length, as well as the offer of M. Gautier to mount the same on a suitable stand. The new instrument was mounted in May 1885. A photograph of the complete instrument is now on the screen.

Both from an optical as well as a mechanical point of view, the new instrument was admirably adapted for its intended work, and the results obtained by the brothers Henry, and rapidly published and circulated by Admiral Mouchez, at once astonished and delighted the astronomical world.

I now show a few of the more remarkable of these star pictures on the screen.

After such results as these there was no longer room for doubt or delay. The exquisite precision of these pictures, the sharpness and roundness of the images of the stars, and the results of actual measurement on the plates, proved that all necessary accuracy had been attained.

The means of rapidly obtaining the data for an accurate survey of the heavens on a very large scale were now within the reach of astronomers, and the time for decisive action had arrived.

The work, however, was too extensive to be undertaken at a single observatory, or even by a single country, and it was agreed on all hands that international co-operation was essential for its execution in a sufficiently short space of time.

I need not enter into the details of preliminary consultation or correspondence, but at last a time was fixed, and invitations were issued by Admiral Mouchez, Director of the Paris Observatory, under the auspices of the Paris Academy of Sciences, for an International Congress of Astronomers to be held at Paris.

A preliminary committee having arranged the general order of business, the Congress was opened on the 16th April, and its thoroughly representative character will be understood from the following statement of the nationalities of the members present.

France	20	Austria	2	Spain	1
England and Colonies	8	Sweden	2	Switzerland	1
Germany	6	Denmark	2	Portugal	1
Russia	3	Belgium..	1	Brazil	1
Holland	3	Italy	1	Argentine Republic	1
U.S. America ..	3				

Before the Conference, a great many people, I will not say astronomers, held that the chief object was to photograph as many stars as possible, and simply preserve these plates or issue photographic copies of them, so that astronomers of the future, by merely comparing one of these originals or copies with a similar photograph of the same part of the sky taken 50 or 100 years hence, would find out what stars had changed in position or magnitude, or whether any new star had appeared.

There is no doubt this was the view of the popular writers—it is very easily understood, and it appeals very directly to the imagination. Such a project alone would no doubt have had great importance and would probably in the future have brought to light a great many very interesting isolated facts.

But for the broader and more refined purposes of astronomy, for the discussion of such great questions as the motion of the solar system in space, the common movement of large groups of stars, the accurate determination of precession, and the general refinement of astronomy of precision, these mere pictures would have no value.

It was essential for these larger and more permanently important ends that all data should be provided for the most refined determination of the *absolute* position of any star upon any plate. This view was endorsed by the Congress.

The objects of the survey of the heavens to be carried out were defined ultimately thus:—"To make a photographic chart of the sky for the present epoch, and to obtain the data for determining the positions and magnitudes of all the stars to the 14th magnitude," as that magnitude is at present defined in France.

At present there are no exact determinations of stellar magnitude to that order of faintness, and the considerations which really guided the Conference were, that stars which are called 14th magnitude are photographed by the Henrys with an exposure of about 15 minutes of time. With such an exposure the time required for the work contemplated by the Congress would not be too great, but to demand long exposures would lead to the loss of many plates by interruptions from clouds, &c., and would unduly prolong the time required for completion of the whole work. As it is, the number of stars photographed to 14th magnitude will number about 20 millions.

It was seriously urged that stars to the 15th or even 16th

magnitude and higher should be photographed, but it was felt that there was real danger of failure in an attempt to do too much.

It no doubt produces a strong effect on the imagination to be told that astronomers are to be engaged on making charts of the sky which will contain 60 or 100 millions of stars, or photographing stars on their plates which cannot be seen at all in the most powerful telescopes. There is thus a strong temptation to yield to this demand for sensation, to produce a few astonishing plates with the loss of much precious time, and to sacrifice the real progress of astronomy to the love of the marvellous. Besides, what are you to do with pictures of 100 millions of stars when you have got them? What would be the use of pictures of all these stars, unless at some future time a sufficient number of astronomers were to arise to compare similar photographs, taken, say one hundred years hence, with the photographs taken in our day? I am happy to think that the number of men who devote themselves to the pursuit of astronomy is on the increase, but I have no desire that the number of men in Great Britain who occupy themselves exclusively with astronomy will ever correspond with that in the floating island of Laputa, as described by Dean Swift, where all the men were exclusively occupied with astronomy, and had to be flapped on the head with little bladders containing parched peas to arouse them from their abstract occupations. And yet, unless something of this sort happens, I see no adequate prospect of the utilisation of pictures of 100 millions of stars.

The Congress, therefore, very wisely limited their chart plates to the 14th magnitude. But, as was well said by M. Bouquet de la Grye, it was not necessary to summon fifty or sixty astronomers to a Congress to arrange for taking mere photographs of stars—a number of photographers provided with instruments like the Henrys could have done all that without a congress. It was very strongly felt that the true *raison d'être* of the Conference was to secure astronomical data, precise and exact as the operations of astronomers should be.

Accordingly they resolved that—

"In addition to the duplicate series of plates giving all the stars to the 14th magnitude, there should be a series of plates of shorter exposure to insure a greater accuracy in the micrometric measurement of the standard stars, and to render the construction of a catalogue possible. The plates intended for the formation of the catalogue shall contain all the stars to the 11th magnitude inclusive." That is to say, it was determined to catalogue the absolute places of stars to the 11th magnitude.

But no photographic plate of itself gives us any information about the absolute places of stars, though it gives the means to determine the relative positions of the stars on the limited area of each plate; you must trust to the old-fashioned meridian observations to determine the absolute places of the brighter stars on each plate, and then measure the position of the fainter stars relative to these standard stars.

Now if a plate is exposed long enough to get satisfactory pictures of stars to the 14th magnitude, the images of the standard stars of the 7th, 8th and 9th magnitude will not have the highest perfection, and consequently the places of the fainter stars cannot be measured relative to the ill-defined standard stars with the highest precision.

This will be evident if we examine actual photographs.

One illustrates a short exposure, the other a long exposure. The short exposure gives sharp definition of the brighter stars, the long exposure brings into view a much greater number of stars, but the sharp definition of the brighter stars is completely lost. Therefore, if we wish to have determinations of *absolute* positions, we cannot have long exposures.

The meaning of the series of plates of short exposure, and showing stars only to the 11th magnitude, is thus explained:

Of stars to 11th magnitude there are about 1½ millions in the sky, and a catalogue containing all these stars may be considered complete for the practical purposes of astronomy, because that magnitude is the faintest which can be measured with accuracy in the larger class of equatorials usually employed in working observatories.

I need not enter into detail about the technical means which are to be taken for eliminating the various sources of error, such as contraction of the photographic film in course of development, and so forth. All these points have been considered by the Congress, or put into the hands of specialists when it appeared that any particular point required further special study, and they are too technical to be entered upon here. The chart of stars to the 14th magnitude will be of importance for many purposes, such as the search for minor planets, and the trans-Neptunian planet, for variable stars, and for data as to the law of distribution of stars of the higher order of magnitude. But I do not hesitate to say that the work which astronomers of future generations will be most grateful for, and which will most powerfully conduce to the progress of astronomy, will *not* be the chart but the catalogue.

And now, Ladies and Gentlemen, I have dragged you through what I fear has so far been a weary account, to bring you to an apparently very uninteresting conclusion.

Catalogues and figures are not matters of much popular interest, and yet from such uninviting material has been built up the fair structure of the exact astronomy of the present day; and out of such materials have been evolved the facts which appeal so strongly to the minds of men, and most strongly so because men know that the conclusions rest not on mere imaginings alone, but on solid facts and figures also.

But now as to the practical execution of this useful work. After all the preliminary details of the operations have been fully discussed—when the instruments have been designed and made, and the mode of working and the methods of measurement and reduction have been devised, the practical execution of the work becomes one long round of routine labour, requiring skilled and careful superintendence it is true, but still routine work of a very trying character.

Such work never has been, and never will be, the occupation of the amateur or single-handed astronomer. Essential as such work is to the progress of astronomy, it can only be executed at regular Government establishments, and therefore the conclusions of the Conference will have to be submitted to the various Governments, and the necessary votes of money must be secured. France has already definitely sanctioned the funds for four photographic telescopes of the kind which the Conference has decided to adopt for the work. And we cannot doubt that the modest claims which will be made on England's treasury for her share in this great work will be liberally responded to.

But there are other applications of photography to astronomy which have a daily growing importance. It was desirable that the Conference should recognise this work, and establish relations with those engaged upon it.

Accordingly the following resolution was passed:—

"The Congress expresses the desirability of there being a special committee which shall occupy itself with the applications of photography to astronomy, other than the construction of the chart. It recognises the importance of these applications and the relations which it is desirable to establish between different kinds of work. The Congress request Messrs. Common and Janssen to undertake the realisation of this proposition."

At first sight this may appear a somewhat barren resolution—but indeed it is not so. It must be remembered that the Congress was convened for the purpose of discussing a special object, it had arrived at definite conclusions and recommendations in connection with that object, and it was felt that to go beyond that object might imperil the adoption of its recommendations by the various Governments.

But in the hands of men like Common and Janssen the resolution of the Conference is not likely to be a barren one, indeed it is certain that it will not be so, for they are already taking steps to unite fellow-workers in this field.

Their Committee will associate itself with those who are engaged upon the Charts, and will follow up in detail and with special instruments and methods the subjects of interest which from time to time will be encountered by the routine workers.

So remarkable has been the progress of the miscellaneous application of photography to astronomy within the past year, that some account of it is essential to bring the history of the subject up to date.

For example, we have the recent work of Professor Pritchard, of Oxford.

He has applied photography during the past year to the most refined and difficult problem of practical astronomy, viz. the determination of the annual parallax (i. e. the distance) of the fixed stars.

He has selected for experiment the interesting double star 61 Cygni. One of the original negatives of the series is now on the screen. This star, as is well known, was selected by Bessel, on account of its large proper motion, as the most suitable star for his first experiment. It was probable, because its large apparent motion among the stars was so great, its real distance from us would be less than that of stars of less apparent motion. Bessel's observations with the Königsberg heliometer proved this to be the case, and his discussion of these observations first convinced astronomers that the measurement of interstellar spaces was a problem not entirely beyond their reach.

Prof. Pritchard has now photographed the star during a whole year, and within a few days he promises that we shall have the results of his measurement of the plates. It will be of great interest to compare his results with previous independent determinations of the parallax of the same star made by other astronomers with different means, but it will be still more interesting for the future of astronomy to compare the amount of accuracy which the photographic method affords, as compared with the older existing methods. From preliminary results published by Prof. Pritchard we are led to expect a very high accuracy from the new process.

So far, however, as present experience goes, we shall not be able to apply this new method to the measurement of the parallax of very bright stars, because, when the plates have been exposed long enough to obtain pictures of the faint comparison stars, the discs of the brighter stars become too large and ill-defined for exact measurement. It may be that this obstacle will yet be overcome, but at present it has still to be faced.

On the question of the comparative merits of refractors and reflectors as the proper instruments for photographic use, very elaborate comparison has been instituted, and much discussion has been held.

From the simple facts, that the best work yet done has been done in stellar photography by refractors, and that they are in many ways more convenient and simple in use than reflecting telescopes, the Paris Congress unanimously adopted the refractor as the instrument to be adopted for the international star charts. But here is a very remarkable picture taken with the Oxford reflector, which shows star discs very sharp and very round over a very large field of view, viz. eighty minutes of radius.

In the photography of special objects, such as star clusters and nebulæ, much has been done.

Common's exquisite photograph of the great nebula of Orion you have seen before in this theatre, and for exquisite beauty of detail it has never been excelled. But of this we may be sure that, if Mr. Common is spared in health and strength to complete the great reflector of five feet aperture upon which he is now engaged, that photograph, beautiful as it is, will be far surpassed. Here is another photograph of the same object by Mr. Roberts. So short is the focus of his telescope, so sensitive are the plates he has employed, that the detail of the brighter parts has been completely burnt out, but a great deal of new found detail is brought to light.

Here is another photograph of the same object by Professor Pickering, taken with a four lens objective of eight inches aperture and very short focus, and including a field of 5 degrees square. Exposure 82 min. This shows what can be done with such a combination.

In 1885 the brothers Henry, photographing the Pleiades on November 16th, discovered a new nebula, near the bright star Maia in the group. Here on the screen is a copy of the original negative by which the discovery was made. You observe the nebula like a filmy projection from one of the stars.

After the nebula had been discovered by photography it was found to be visible in the great telescope of 30 inches aperture at Pulkowa. But to *discover* is one thing, to see after discovery is another.

Strangely enough this new nebula was really photographed a fortnight before its discovery at Paris, by Professor Pickering at Cambridge in America. In exhibiting the photograph to the National Academy of Sciences five days before Henry's discovery, Professor Pickering pointed out the "wing" attached to the star, but there was only one plate shown, the impression was that the mark was due to a defect in the gelatine film.

Here, however, is another picture of the Pleiades taken at Cambridge with the same instrument and an exposure of eighty-two minutes, which shows nebulosity about more than one star of the group.

And here is a copy of a negative by Mr. Roberts, of Liverpool, with an exposure of three hours. The star discs are of course large and ill defined; but the quantity of nebula, invisible to the eye in the largest telescopes, is quite surprising.

These photographs appear to prove conclusively that the nebula and the stars in this group are one system; the doctrine of chances renders it almost an impossibility to suppose that such a symmetrical arrangement of nebulous matter with respect to the stars could exist by chance, if the stars were projected in front of a far distant background of nebulous matter.

Here is a photograph of the stars surrounding the celebrated variable star η Argus, taken at the Cape with the telescope of 9 inches aperture, generously presented to me for such work by Mr. James Nasmyth. The nebula surrounding this star is very faint compared with the Orion nebula, and it seems to be deficient in actinic rays, and besides, the telescope is intended for stellar photography by its long focal length, and not for nebulæ, which require a shorter proportional focus—i. e. more intrinsically brilliant image.

Still there is the nebula, and I believe this is the only existing photograph of the object. The exposure was $2\frac{3}{4}$ hours, and yet although the original negative has been enlarged four diameters the star discs remain well defined. The corresponding region of the sky is less than the moon's apparent diameter, and of the many thousands of stars visible on the photograph not a single one is visible to the naked eye. The star η Argus was in 1843 nearly the brightest star in the heavens; in fact, second only to Sirius. It is now between the 7th and 8th magnitude.

Here is a star-cluster in Argus. The star discs are not so sharply defined, but the original negative has been much magnified to bring out the star discs.

Here is a photograph, also taken at the Cape, of the wonderful star cluster ω Centauri. It is the finest globular cluster in the heavens, and I do not know that I have ever seen the separation of the central stars so distinctly with the eye as they are shown in this photograph. Perhaps by photographing we shall learn what motions occur in each cluster. This negative has been enlarged four diameters from the original.

Here is a photograph of the well-known cluster in Hercules, taken by Mr. Roberts, of Liverpool, and a still more wonderful one by the Henrys, of Paris. They must be magnified more highly to give any idea of their quality.

When the objects are bright, such as bright double stars, or planets, or the moon, we can enlarge the image produced by the telescope, by aid of a secondary magnifier.

Because of the greater size of the original pictures thus produced, the granulation of the photographic film interferes to a less extent with the detail of the picture. Of course, this advantage is purchased at the cost of a longer exposure, because the same amount of light is spread over a larger area of the sensitive plate, and consequently the same area of the film receives less intense light. With very bright objects, such as the sun, moon, and planets, this is of little consequence, and may be an advantage, as permitting more accurate regulation of the exposure.

Here is a picture of the sun photographed by M. Janssen at Meudon, near Paris. The exposure is less than 1/1000th of a second of time. And here is an enlarged photograph of the same spot, showing an amount of detail which no artist could convey by hand and eye, nor could he emulate the absolute accuracy of the photograph.

Here are some photographs of the planet Jupiter, taken at Paris, the original image being magnified 18 times.

Here is another showing the remarkable red spot—you even have before your eyes evidence of the rotation of Jupiter on its axis by the change in the position of the spot during the same evening.

This spot appeared in 1878 and measured about 30,000 miles in length by 7000 miles in breadth. It became of a deep red colour in 1879, and for the three following years was a most striking feature in the planet. It almost faded entirely in 1883, but has again become nearly as bright as it was in 1882.

Miss Clerke tells the story most admirably and suggestively in the last edition of her 'History of Astronomy,' to which work I would refer those of my hearers in whom these beautiful photographs may excite a sufficient interest.

To enter fully into the matter would demand a lecture to itself—and the minute hand of that inexorable clock warns me that I must move on.

Here are some photographs of Saturn, which illustrate the remarkable progress of celestial photography.

Here are some photographs of double stars : one of these, a photograph of γ Virginis, taken at Greenwich, is probably the finest photograph of a double star in existence. The star discs measure less than $1''$ in diameter.

Last of all I come to the most recent revelations of the power of photography as an aid to astronomy. Dr. Henry Draper, in 1872, was the first to photograph the lines in the spectrum of a star, but his admirable investigations were interrupted by death in 1882. In 1886, his widow placed in the hands of Professor Pickering, of Harvard College Observatory in America, not only an ample sum of money for the purchase of costly apparatus, but also made a liberal provision for carrying on the work of photographic spectroscopy as a memorial to her husband. So noble a gift, and the execution of so pious a purpose, could not have been placed in abler or more active hands.

Within the past few weeks we have received the first-fruits of the Henry Draper Memorial Fund.

When I began preparation of this lecture, I cabled to Professor Pickering a request for some glass copies of his original negatives. He kindly complied, and they arrived this morning. Time only permits me to show them rapidly, but those who remember Dr. Huggins's lectures on stellar spectra in this theatre, will recognise the enormous importance of such pictures as these.

The ingenuity of the adopted methods, the extraordinary success attained, the promise of rich harvest, exceeding our highest previous expectations, which the results afford, are themes upon which one could dilate for hours.

Here we have the spectra of the distant stars whose actual discs we can never hope to see, registering in these rhythmical lines the story of their constitution and temperature, with an accuracy and precision which not many years ago we should have been glad to obtain in the records of the spectrum of our own sun.

And this is not all ; not only have we such results for a few stars, but we are promised " that the complete work will include a catalogue of the spectra of all the stars of the 6th magnitude and brighter, a more extensive catalogue of spectra of stars brighter than the 8th magnitude, and a detailed study of the spectra of the bright stars." These are Prof. Pickering's own words.

What Prof. Pickering promises, we know from long experience, that he will perform. We may also well say with him, that "a field of work and promise is open, and there seems to be an opportunity to erect to the name of Dr. Henry Draper a memorial such as heretofore no astronomer has received."

There is in England wealth enough and to spare. Many a rich man dies puzzled how to dispose of his money; and there is many a living man who would gladly give for such an object if he knew how to do so. There is field enough in astronomy, and there are men enough in England to do the work. Let us hope they will receive aid such as Prof. Pickering has received; and having done so, they will give an equally good account of their stewardship.

The miscellaneous applications of photography to astronomy offer a field so full of promise, so certain of immediate reward to those who are possessed of the necessary originality and the means to carry out their ideas, that there is more hope of private enterprise in that direction than in the more routine work of star-charting.

But tempting as these fields are, brilliant and interesting as are the discoveries to be found in them, there is in the work instituted by the Paris Congress an element that cannot be overlooked and which compels attention—it is this: the question of the lapse of time. Every year which passes after that work has been carried out, increases its value and importance; every year that we neglect in doing it will be a reproach to the astronomers of the day. Into all the great problems which that work is destined to solve, the element of time enters—and time lost now in such work can never be recalled.

Of the Congress itself I would say a few last words. Its proceedings were characterised by an earnest spirit of work and entire absence of international jealousy. Our reception by the French was cordial and hospitable in the highest degree, the decisions of the Congress were almost unanimous, and were marked by a moderation and judgment which must render them acceptable to the responsible authorities of the various Governments.

Lastly, I would add that the good will which pervaded the meetings, the general success of the Congress as a whole, were in no small degree due to the genial influence of the single-hearted, earnest-minded man who convened it—Admiral Mouchez.

[D. G.]

Friday, January 25, 1889.

Colonel J. A. Grant, C.B. C.S.I. F.R.S. Vice-President, in the Chair.

Professor G. H. Darwin, M.A. LL.D. F.R.S. *M.R.I.*

*Meteorites and the History of Stellar Systems.**

The great advances which have been recently made in the art of celestial photography have now made it possible to study the details of structure of some of the nebulæ, which formerly only appeared as a chaotic luminosity, even when viewed through a powerful telescope. To illustrate this new method of research, a photograph of the great nebula in Andromeda, by Mr. Isaac Roberts, was exhibited; it showed that the nebula consists of a bright central condensation, surrounded by several faintly luminous concentric rings.

A short sketch of the Nebular Hypothesis of Laplace and Kant was then given. This is a mechanical theory of the evolution of a nebula into a star with attendant planets.

It was then pointed out that Mr. Roberts's photograph exhibits exactly the condition which Laplace imagined, and it thus confirms the substantial truth of his hypothesis.

But many points in the evolution of a planetary system are still involved in much obscurity, and there is in particular one difficulty, so fundamental that some astronomers have been led by it virtually to throw over the nebular hypothesis.

That theory attributes the annulation of the nebula to the gradual diminution and ultimate vanishing of gaseous pressure at the equator of a rotating mass of gas. Thus it is the very essence of the hypothesis that the nebula should consist of continuous gas.

Now there is in the solar system at present no trace of the all-pervading gas from which it is supposed to have been evolved, whilst there is much evidence that the space surrounding the sun and planets is peopled by countless loose stones or meteorites flying about in various directions.

This latter view is confirmed by the recent spectroscopic researches of Mr. Lockyer, who has been led to suggest that the luminous gas, which undoubtedly forms the visible portion of nebulæ, is gas volatilised from the solid state, and rendered incandescent by violent impacts between meteoric stones. These gases, he says, cool quickly, cease to be luminous and condense, but the collisions being incessant, the whole nebula shines with a steady light.

---

* See Phil. Trans. R.S. vol. 180A (1889), pp. 1–69.

M*

It appears, then, to be probable that the immediately antecedent state of the sun and planets was not a continuous gas, but was a swarm of loose stones. Here, then, there arises a dilemma; for on the one hand the meteoric theory denies the continuity of the matter forming nebulæ, whilst on the other hand the nebular hypothesis demands such continuity.

The object of this lecture was to show that there is, however, a way in which these two apparently conflicting ideas may be reconciled.

In order to prepare the way for the suggested reconciliation, a sketch was then given of the Kinetic Theory of Gases, according to which a gas consists of a great number of elastic molecules, moving at high speed in all directions at hazard, and continually coming into collision with one another by chance.

According to this theory a pigmy, of a size comparable with the average distance between adjacent molecules, would be conscious of the blows he received from individual molecules, and he would have lost the sense of gaseous pressure, which arises from impacts too numerous and too rapid for discrimination. Thus, what is called gaseous pressure is a question of the magnitude of the observer.

The suggestion, then, of this lecture was that celestial nebulæ are of such large dimensions, that meteorites might be treated as molecules and that their collisions might impart to a nebula, as a whole, the quasi-gaseous mechanical properties demanded by the nebular hypothesis.

But if such a suggestion is to rise above the level of mere conjecture, it demands a careful examination in detail. It is accordingly necessary not only to consider the details of an individual collision, but also to examine whether a meteoric medium is of sufficiently fine texture to fulfil the conditions imputed to it.

The kinetic theory of gases requires that the molecules of gas should be perfectly elastic, and, although meteorites are certainly not perfectly elastic, it was maintained that the sudden volatilisation of gas, at the point of contact of two of them at the moment of collision, would act as a violent explosive between them, and would impart to them a virtual elasticity of considerable perfection.

The investigation of the degree of fineness of grain necessary to admit the applicability of the theory involved numerical calculation, and the requisite data had necessarily to be derived from the solar system.

If the sun's mass were broken up into iron meteorites, each weighing say a pound, the dimensions of each meteorite would be known, and their number would be four followed by thirty zeros.

These iron stones were then supposed to be distributed in a swarm extending beyond the present orbit of the planet Neptune. To give numerical precision, the swarm was taken to extend as far beyond Neptune as Saturn now is from the Sun.

These supposed conditions were adopted merely by way of an example which should represent a nebula of extreme tenuity; for if the meteorites were not too sparsely distributed to impart quasi-gaseous properties to the whole in this supposed case, the nebula would *à fortiori* possess those properties when it had shrunk to smaller dimensions. The swarm was supposed to be arranged in a perfect sphere, and what may be described as the layers of equal density of population were taken to be concentric spheres, but the density of population would necessarily be much greater towards the middle than towards the outside.

The whole crowd of stones would arrange itself automatically into a steady condition, in which the population had no tendency to shift, although, of course, the dance and collisions between the constituents of the crowd would be incessant. When this steady condition was submitted to calculation, it was possible to discover the average velocity of the stones, the average density of population, and the average frequency of collision at each point of the swarm.

It will naturally occur to the reader to inquire as to the source of the great velocity of the stones; it arose from gravitation, the stones having fallen in from a great distance towards a centre of aggregation.

If somewhere in space there were an aggregation of meteorites, and if a stone were released from a state of rest at a very great distance, it would fall towards the swarm under the influence of gravitation. On reaching the swarm it would have acquired a certain velocity, and would penetrate to some uncertain distance, until it happened to strike another stone. Henceforth its path would be zigzag, as it happened to strike, and it became incorporated as a member or molecule of the swarm.

The supposed visitant from outside space imported energy of motion into the swarm, and besides increased the total mass of the swarm. Thus, if it be imagined that the swarm is increased by the addition of stone after stone, each being let fall from a distance, it is clear that, in the course of accretion, the energy of agitation of the meteorites continually increases. When stones have ceased to fall in, the materials of the nebula were collected, and by means of incessant collisions the swarm gradually attained the steady condition above referred to.

By reasoning of this kind it was possible to discover how fast the stones were moving, but it is proper to add that an important correction had to be applied to allow for the fact that at each collision between two stones some speed is lost. In the process of settling down into the steady condition, each stone loses, by imperfect elasticity, three-tenths of the speed it would have if it were a fresh arrival from space.

It makes no material difference in the result by whatever process the stones were collected together, and the account given above of the formation of a swarm was not intended as a contribution to its history, but was only meant to explain the mechanical principles involved.

By this line of argument it may be concluded that when the solar swarm extended half as far again as the planet Neptune, the average velocity of the stones was three miles a second.

It was next necessary to find out how often the stones came into collision, how far they travelled from one collision to the next, and whether the collisions could be frequent enough to impart to the whole nebula the gaseous property demanded by the nebular hypothesis.

Even a microscopic animal in our atmosphere is not aware of the individual impacts of molecules on his body, and his sensation is still that of gaseous pressure. But it must clearly be a giant who would not be aware of the individual blows of meteorites in a meteoric nebula, but would only realise their average effects.

It would not be easy to explain the exact reasoning by which it is possible to determine how large the giant must be in order to act as a judge of the gaseous property of the meteoric swarm, nor of how a comparison of his dimensions with the texture of a meteoric swarm is to be made, and it must suffice to say that the comparison is best clothed in a form which may appear something quite different, but which is really substantially the same.

It may be stated, then, that a meteoric nebula would behave sufficiently like a gas to allow the nebular hypothesis to be true, if the average path of a meteorite between two collisions were only a short portion of that curved orbit which it would describe under the action of gravitation if it could move through the swarm without ever colliding with another stone.

These explanations led on to the numerical values derivable from calculation, on the hypothesis that the solar nebula, consisting of 1 lb. iron stones, was distributed in a swarm extending half as far again as the present distance of the planet Neptune from the Sun.

It appeared, then, that at the middle of the swarm a meteorite would, on the average, come into collision every 13 hours, and would travel 140,000 miles between collisions; at the distance of the small planets called the asteroids, it would collide every 17 hours, and would travel 190,000 miles between; at the distance of Uranus the collisions would be at intervals of 25 days, and the path 6,000,000 miles; and lastly, at the distance of Neptune, the interval would be 190 days, and the path 28,000,000 miles.

It may also be shown that the path described between collisions forms a larger portion of the whole curved orbit of a meteorite the further we go from the middle of the swarm. Even at the distance of the planet Neptune the collisions were, relatively speaking, so frequent that, on the average, gravity only sufficed to draw the meteorite aside from the straight path by 1-66th of the length of path it had traversed, before it was deflected into a fresh orbit by collision with another stone. The fraction 1-66th was then the numerical value of the criterion of the applicability of quasi-gaseous properties to the swarm, and this fraction is so small that it may be concluded that the swarm passes the proposed test.

It followed, therefore, that if meteorites possess a virtual elasticity, a swarm of meteorites provides a gas-like medium of fine enough structure to satisfy the demands of the nebular hypothesis.

The result of this discussion then appeared to justify the opinion that the meteoric theory may be reconciled with Laplace's hypothesis, and that they may both be held to be true.

After this discussion of the proposed modification of Laplace's hypothesis, it was natural to turn to the series of events which may be supposed to have occurred after the nebular stage of evolution.

At the various centres of condensation, which now form the sun, planets, and satellites, the swarm of meteorites must be supposed to have become denser, and the collisions too frequent to allow the gases to condense again, so that by degrees all solid matter in the neighbourhood of such centres would be volatilised. Away from these condensations there were still many free meteorites, but the majority of those which formed the swarm in primitive times would have been absorbed, and the absorption would still go on gradually.

The collisions amongst the free meteorites became rarer and less violent, and finally, when relative motion was nearly annulled, almost ceased to occur. The residue of the meteoric swarm then consisted of sparse flights of meteorites moving in streams. There is evidence of the existence of such streams at the present day in the zodiacal light, in falling stars, and in comets. But these are the dregs and sawdust of the solar system, and merely give a memento of the myriads which must have existed in early times, before the sun and planets were formed.

The subject of this lecture is a large one, and the limits of time rendered it impossible to do more than speak of the more prominent features of the problem. The value of the investigation of which some account has been given, will appear very different to different minds. To some it will stand condemned as altogether too speculative; others may think that it is better to risk error on the chance of winning truth. It was, however, contended by the lecturer that the line of thought flowed in the channel of truth, and that by its aid many other interesting problems might perhaps be solved with sufficient completeness to throw further light on the evolution of nebulæ and of planetary systems.

[G. H. D.]

Friday, May 24, 1889.

SIR JAMES CRICHTON BROWNE, M.D. LL D. F.R.S. Vice-President, in the Chair.

The Rev. S. J. PERRY, D.Sc. F.R.S. F.R.A.S. Director of Stonyhurst College Observatory.

The Solar Surface during the Last Ten Years.

THE solar surface is a subject on which so much has been written of late years, that it would be highly unsatisfactory to attempt in one short hour even a brief enumeration of the results obtained and the theories advanced by the many eminent men who have devoted attention to this branch of solar physics. The end which I propose to myself this evening, and which, I venture to think, is most in accordance with precedent in these discourses, is to lay before you, in as clear a manner as I am able, the results obtained at the observatory to which I am attached, in so far as they enter into our present subject, and to touch upon the work of others in such a manner only as to complete the picture, by showing the bearing of our labours on the general result.

For the last ten years I have been anxiously endeavouring to make Stonyhurst as efficient an observatory for solar physics as the means at my disposal would admit, so I was naturally desirous not to undertake any work that would be a mere repetition of what was being done better elsewhere. From the outset, therefore, I excluded from my programme a daily photograph of the sun, as this had been already undertaken at the Royal Observatory, assisted by other Government observatories in India and at Mauritius, and the most I could have hoped for in this direction would have been to fill up a few gaps in an almost perfect series. My choice, therefore, lay between drawings made at the telescope by aid of a solar eye-piece, and the use of a sketch-board on which the sun's image could be projected. The object in view being to procure the most complete and faithful representation of nature, I had no hesitation in rejecting all forms of solar eye-pieces, as by this method of observation too much is left to the imagination of the draughtsman; although a solar prism is not unfrequently of great advantage in supplementing other methods of attack, especially when delicate details require verification. The sketch-board, on the other hand, may be so used as to leave very little indeed to the ideas or bias of the observer. To effect this the following method has been adopted. A circle ten and a half inches in diameter is first traced upon a piece of drawing paper, which is pinned to a board just rigid enough to hold the paper firmly, and then the whole is clamped on to the eye-end of the telescope. The eye-piece and board are each capable of fine adjustment, so that the sharpest image of the sun may be made just to fill the 10½-inch circle, and a marked diameter of the picture is brought into precise coincidence with the direction of the daily motion. The clock-work of the equatorial then keeps the image fixed in position on the paper, whilst an accurate outline is traced of the umbra and penumbra of every spot visible on the disk. When the sky is clear this outline can be made as correct as the finest point of a hard pencil can delineate it; and even when, as so often happens, the transparency of the atmosphere is changing every moment, a short time at the instrument will generally enable the observer to verify the perfect accuracy of his sketch. The details are then filled in as quickly as the nature of the sky permits, each portion of the drawing being over and over again brought into coincidence with the projected image, in order to detect and remove the slightest difference between them. By this means the final picture gives the advantage of all the best moments of seeing that occur during the progress of the observation, and not merely the result at one single moment, which may be far from the best for definition even on the finest day. Immediately the spots have been completed the faculæ are traced, and their details reproduced as nearly as possible. By the advice of Prof. Stokes, P.R.S., a red pencil is used in drawing the faculæ, and thus the difference between bright and dark markings and their connection with each other, stand out much more boldly than if the same black pencil were used throughout. Finally, when the sky is good for definition, the general surface of the sun is carefully scrutinised for some time, and any peculiarities noted.

The sun's image having thus been sketched and examined, the drawing-board is replaced by an automatic spectroscope of six prisms of 60°, through each of which the light may be made to pass twice, and to which may be added a Hilger-Christie half-prism, raising the total available dispersion to about 36 prisms of 60°. The chromosphere is measured with the slit radial, a dispersion of six prisms being generally used; and when the definition is very good, a sweep is made round the limb, with 12 prisms and a tangential slit, to study the forms of the prominences and the direction of the currents in the chromosphere. Spot-spectra have been occasionally observed with the same instrument; but lately (in a room adjoining the equatorial dome) a large grating has been mounted, with which it is proposed to take daily photographs and eye-measurements of the spectra of sun-spots. A heliostat and a 5½-inch object-glass of Alvan Clark are used in connection with the grating spectroscope. The combination of this solar observatory with an establishment supplied with a complete set of self-recording meteorological and magnetic instruments, affords a ready opportunity of comparing solar results with the daily photographic records of terrestrial phenomena.

As it is most important, before referring to any conclusions that may be drawn from our observations, to test severely the fidelity of which solar drawings are capable, I will throw upon the screen a number of sun-pictures drawn at Stonyhurst, and enable you to contrast them with sketches of the same spots made by experienced astronomers in England, and at Brussels, Palermo, and Kalocsa, and also with photographs taken at Dehra Dun and at Meudon; and I think these few examples will amply suffice to vindicate a high character for solar drawings. But I have not been satisfied with this ready comparison, and lately this solar work has been put to a more rigid test of accuracy by placing side by side the measurements of areas obtained from drawings and photographs; and here again the hand-sketch by projection seems to bear well the scrutiny.

The method of projection not only permits the area covered by spots and faculæ to be accurately determined, but it also furnishes precise data for finding the heliographic co-ordinates of any mark upon the solar surface. These, however, are now given so fully in the annual publications of the Royal Observatory, and each individual spot can be so readily identified, that an independent calculation would be a mere waste of time and energy. The results, therefore, dependent on position alone, are deduced immediately from the Greenwich Tables.

The decade of years which we are now considering, covers almost an entire solar cycle, that period of eleven years the proof of whose existence was the fruit of the unwearied labours of Baron Schwabe, in his daily observations from 1826 to 1868. The present cycle falls much below that which preceded it in the extent of its spotted area, but it is remarkable for the duration of its maximum period. The last minimum occurred about November 1878, and therefore, if we accept the mean values 3·7 and 7·4 as the number of years from minimum to maximum and from maximum to minimum respectively, we obtain $1878 \cdot 9 + 3 \cdot 7 = 1882 \cdot 6$ as the date of the maximum of the cycle, and $1882 \cdot 6 + 7 \cdot 4 = 1890 \cdot 0$ as the epoch of the approaching minimum. Many facts seem to support this conclusion. Thus, the greatest sun-spot area occurred on April 21, 1882, and the largest individual spot was at its maximum on November 18 of the same year, when it covered almost one four-hundredth of the visible hemisphere. The total spot area in April reached, however, the much higher figure of one one-hundred-and-eightieth, or about 6,000,000,000 square miles, which on the following day had diminished to something under 4,000,000,000, showing the marvellous activity of the solar forces at that epoch. In the same year also the mean monthly amount of umbra was greater in April and in November than at any other time of the cycle, making it highly probable that the disturbances then penetrated more deeply below the surface of the photosphere. Again, if we consider the mean latitude of the spotted area, its value at the maximum seems also to favour the claims of 1882; and these are still more strengthened by the number of times in which the spectral lines have been found contorted in the observations of the chromosphere. But, on the other hand, it may be urged, that if we reckon the days without spots, we find very little change from 1880 to the beginning of 1885; and the monthly mean area of sun-spots was slightly in excess in July 1883 of what it had been in the previous year; and even as late as June 1885 we meet with days on which the spotted area was larger than the mean extent for any month in 1882 or 1883. May it not be possible that the great comet of 1882, which passed so deep within the limits of the corona, and which seems to be but one of a numerous family, may have exercised some disturbing influence, and been a partial cause of this prolonged maximum?

The limits are rather wide for the periods separating maximum from maximum, but there is much more steadiness in the lapse of time between successive minima, and if we may judge from the present rapid increase in the number of days without spots, the next minimum should be fast approaching, and may not be far removed from the computed epoch of January 1890. Thus, against one spotless day in 1884, we have 10 for 1885, 61 for 1886, 106 for 1887, 160 for 1888, and more than half the days that have been fit for observation since the beginning of the present year. The important law connecting the mean latitude of sun-spots with their extent of area, first published by Carrington, and afterwards so vividly represented in curves by Prof. Spörer, has received a fresh verification in the present cycle. In close relation to this is the distribution of spots in the two hemispheres, and this can be well illustrated by aid of curves and by mapping each spot in its true heliographic position. Thus we find that the greater activity of the southern hemisphere since 1883 is strongly marked, and there seems to be very little evidence that successive spots tend to form along meridians, although examples of this are not wholly wanting. A general glance at spot distribution shows that an outbreak may be expected on any meridian, but that at certain times spots congregate more thickly in one longitude than in another. Thus in 1881 and 1882 the longitudes most favoured lie between 180° and 270°, whilst in 1887 more spots appeared between 0° and 90°. The preponderance of southern spots in this year is very striking.

The variability of the sun-spot area which we have been so far considering, is a point of the greatest interest in solar physics, and it may even become of great practical importance, if the supposed connection between certain terrestrial phenomena and solar cycles is once clearly established. But apart from this, the study of spots in themselves, their wonderful changes, and their individual history, may be highly instructive and teach us much concerning the nature of the solar surface. The spots seen first in March and April 1884, and which reappeared in May and June and July, furnish excellent examples of the formation of companion groups, the principal spot acquiring a regularity of outline when it stands alone in middle life, and regaining new companions as it hurries on to its final extinction.

The last few years have also afforded some striking instances of the repulsion and proper motion of spots, so clearly brought out in the researches of Carrington. Perhaps the most remarkable change of this nature recorded at Stonyhurst occurred between May 27 and 28 in 1884, when a spot of considerable size, situated between two others, moved through a great part of the distance separating its two companions.

Another class of observations consists in noting the variety of tints connected with the bridges and other bright portions of intricate spots, and which are seen only from time to time. Thus, in 1884, on the 5th of May and the 6th of July, there were outbursts of a crimson hue in the centre of spots. Such appearances vary in duration as much as in colour. On one occasion a reddish brown tint, very noticeable on the bright separations of the umbra of a spot, did not remain constantly visible, but was intermittent. At another date a yellowish green tint remained visible for several days on the bridge crossing a spot. In each case the coloured portion was carefully compared with other parts of the disk, so as to ascertain whether the effect might not be due to the object-glass or eye-piece.

Some useful hints in view of future theory can certainly be gained by a careful study of the peculiar forms assumed occasionally by the penumbra of spots, not when the whole area is in a state of violent commotion, for then it is conceivable that any form may possibly present itself, but when the umbra is regular and the spot quiet. I would especially draw attention to some examples of multiple penumbræ observed in 1882 and 1883, which seem to indicate that such disturbances are due to successive impulses from a common centre. Or, again, I would point to the penumbral matter extending from the nucleus of certain spots in 1884, which might reasonably be adduced as evidence of an apparent outflow of a floating substance. It is doubtless fascinating to argue in support of some exhaustive theory, complete in itself, and able alone to offer a satisfactory explanation of every observed appearance ; but may it not be possible that several true causes concur in the production of solar phenomena ? And if this be admitted, then a line too hard and fast may easily stand in the way of the true explanation of spot formation. Might not the last mentioned observations suggest the question, whether absorbent vapours may not sometimes be cast up from the seething mass beneath, although a down-rush be the prevailing feature of a sun-spot ?

The study of the bright markings on the solar surface is perhaps somewhat less interesting, although almost equally important as that of the spot area, but the state of the sky interferes much more seriously with the former class of observation than with the latter. The clustering of faculæ about nascent spots, their spreading as the spots gets older, their lingering long after the parent spots have disappeared, the apparition of fresh spots in their midst, are facts of constant recurrence. That faculæ are the first evidence of a coming disturbance has never been observed at Stonyhurst, but a region where faculæ abound has always been found to have previously been occupied by a group of spots. Such an observation as that of the birth of the great November spot of 1882, and other similar instances are not wanting, in which a few dots, in a region of perfect calm, suddenly developed into a vast centre of disturbance, is a strong proof that the law which regards faculæ as the constant forerunners of spots, cannot be accepted as universal. Do not such instances as these give some additional weight to the opinion that large spots owe their origin rather to an external cause, as the prime mover, than to the internal forces of heat and pressure and chemical affinity? Might not the latter forces, however great, hold each other almost in equilibrium until their energies are freed by the advent of an external, though perhaps lesser, disturbing force? The periodicity of sun-spots, their total absence for months, nay almost even for years, combined with their enormous development and rapid changes about the epoch of maximum area, seems to preclude the possibility of internal forces, in a gaseous body, being the sole cause of such phenomena; although there appears to be no reason why these forces should not exercise an overwhelming influence when called into play by an external agent.

The distribution of faculæ, especially of small isolated patches, is much more general than that of spots, many being visible even near the sun's poles, and this has been forced upon my attention more and more of late as we have approached nearer the minimum of the cycle.

Another point, which is often accepted as established, is the lagging of faculæ, and it is sometimes adduced as a proof that faculæ are cast up from a lower level, for thus, in possessing a less linear velocity than the surrounding photosphere, they would naturally be left behind. The observations of the present cycle can scarcely be said to add to the stringency of this argument, as out of more than 4000 cases recorded in the Greenwich Tables, 74 per cent. lean neither way, and of the remainder 5·7 per cent. show the faculæ preceding, against 20·3 per cent., which alone are in direct confirmation of the assumed law.

The general surface, with its ever varying aspect, can never be adequately represented by the pencil, but in this case recourse must be had to the instantaneous photograph. And yet it is true to say that even in this branch of the subject much may be done by the method of projection, and light may thus be thrown on many points which could scarcely be settled by photography alone. I refer particularly to any possible variation of tint in definite regions of the solar disk, to rapid and continuous changes, and to such appearances as the smugged areas in the photographs, which suggest at first bad definition, but which are in reality evidence of a new form of disturbance not indicated by other phenomena. Watching the image of the sun as it depicts itself upon the drawing-board, we soon perceive that no part of the surface is long at rest. Everywhere are seen small dark shadowy objects, attracting special attention now in one place and then in another, sometimes forming groups, and at other times almost companionless, affecting no special zone, and combining with no other form of solar marking, except occasionally with a few bright faculæ.

These dim ill-defined objects seem first to have been observed in 1875 by Trouvelot, who gave them the name of veiled spots. They soon forced themselves into notice when our daily sun-work was started at Stonyhurst, and ever since they have never failed to be carefully watched. At first they were divided, for convenience, into three classes, but on further examination there seems to be no need of more than a single distinction. When they first catch the eye all present much the same general appearance, resembling small fragments of ill-defined penumbra, but their position on the disk, and still more their duration, soon enable the observer to distinguish the class to which they belong. Those of the first class appear in all heliographic latitudes, and never remain visible for more than two or three minutes; whilst the others, which have been called sub-permanent spots, are confined exclusively to the spot zones on either side of the equator, and this class may remain on the disk for two or more days. Sub-permanent spots are not always to be found on the surface of the sun, their tint is a shade less dull than that of the other veiled spots, and occasionally there is almost the appearance of an umbra in their midst, though this rarely could be mistaken for a true umbra. For even when their shading is in some parts more intense than in others, the whole remains always ill-defined, and the limits of its several parts are hard to distinguish. Frequently these dim objects show themselves in considerable numbers in the neighbourhood of fully developed spots, but then the latter are generally approaching their time of dissolution. They may, perhaps, aptly be described as imperfectly developed, or penumbral spots, and consequently be included in the ordinary spotted area. But it is quite otherwise with the first class of veiled spots, which except for their diminished brilliancy would have nothing in common with the fully developed sun-spots. Seen in all solar latitudes, they are never absent from the sun, being, with good definition, as frequent and as visible at the epoch of spot maximum as at minimum, but catching the eye more readily when markings more intense are absent from the surface. The most striking characteristic of this class of spots is the rapidity with which they invariably disappear; but although no individual spot ever lasts more than about three minutes, the first seen may be joined in quick succession by a multitude of others similar to itself, and thus transform vast areas, and give to portions of the solar surface that blurred appearance which is so marked a feature in Janssen's magnificent sun-pictures. The general distribution of these faint objects, their evanescent character, and their ill-defined appearance, seem to connect them more immediately than any other feature of the solar surface with the vertical convection currents which form so important a part of the sun's internal economy.

One exceptional observation of the formation of veiled spots may be of interest, as showing the strange phenomena we may occasionally witness by a lengthened examination of the solar image. At 10h. 15m. one morning a group of spots was visible on the sun, and presented no very special feature; less than half an hour later the leader of the group had apparently shot out a number of minute bodies, and then five minutes sufficed for all these to be transformed into veiled spots, which disappeared as rapidly as usual.

Instances have not been wanting of other moving bodies seen upon the solar surface, always rapid in their course, and sometimes disappearing without crossing the limb. I refer, of course, only to observations in which every precaution was taken to test the objective nature of these bodies, as false images might so easily deceive a tired retina, especially after exposure to a strong light. But there would be little use in dwelling upon even well-established cases of this nature, as they all probably find their explanation in the passage of bodies between the sun and the observer, and promise very little additional light for an inquiry into the nature of the solar photosphere.

Before concluding these few words on a series of solar observations of the last ten years, I may be expected by some to add one more to the already long list of solar theories, or at least to pass in rapid review the most plausible theories that up to the present time have been advanced for reducing to harmony all the well-established facts regarding the sun with which we are acquainted. But considering the scope of my discourse, which has dwelt mostly on the work done at a single observatory, and, even so, has only treated in detail one method of examining the solar photosphere, I should not be warranted in advancing any theory of my own, or in judging of the theories of others, without first considering the many important facts connected with our subject which the spectroscope has taught us and also extending my researches to the solar chromosphere and the corona. It is only by diligently collating the facts laboriously accumulated by telescope and spectroscope in every known region of the sun that we may hope at last to build up a complete and satisfactory solar theory. Much has already been achieved in this direction, and we are in possession not only of reliable data, but also of important deductions therefrom, which may serve as a solid foundation for some future superstructure.

I may have added, perhaps, some little towards the completion of the edifice, if I have convinced you this evening, that by the persevering and judicious use of the pencil we may yet hope to throw some light on questions of solar physics, that might not so readily have been secured by any other means. [S. J. P.]

Friday, May 30, 1890.

WILLIAM HUGGINS, Esq. D.C.L. LL.D. F.R.S., Vice-President, in the Chair.

A. A. COMMON, ESQ. F.R.S. Treas. R.A.S. *M.R.I.*

Astronomical Telescopes.

BEFORE speaking of the enormous instruments of the present day, with their various forms and complicated machinery, it will be well to give some little time to a consideration of the principles involved in the construction of the telescope, the manner in which it assists the eye to perceive distant objects, and in a brief and general way to the construction and action of the eye as far it affects the use of the telescope, all as a help to consider in which way we may hope to still further increase our sense of vision.

I will ask you to bear with me when I mention some things that are very well known, but which if brought to mind may render the subject much more easy. Within pretty narrow limits the principles involved in the construction of the telescope are the same whatever form it ultimately assumes. I will take as an illustration the telescope before me, which has served for the finder to a large astronomical telescope, and of which it is really a model. On examination we find that it has, in common with all refracting telescopes, a large lens at one end and several smaller ones at the other; the number of these small lenses varies according to the purpose for which we use the telescope. Taking out this large lens we find that it is made of two pieces of glass; but as this has been done for a purpose to be presently explained which does not affect the principle, we will disregard this, and consider it only as a simple convex lens, to the more important properties of which I wish first of all particularly to draw your attention, leaving the construction of telescopes to be dealt with later on.

Stated shortly, such a lens has the power of refracting or bending the rays of light that fall upon it: while they are passing through the lens the course they take is altered; if we allow the light from a star to fall upon the lens, the whole of the parallel rays coming from the star on to the front surface are brought by this bending action to a point at some constant distance behind, and can be seen as a point of light by placing there a flat screen of any kind that will intercept the light. For all distant objects the distance at which the crossing of the rays takes place is the same. It depends entirely on the substance of the lens and the curvature we give to the surfaces, and not at all upon the aperture or width of the lens. The brightness only of the picture of the star, depends upon the size of the lens, as that determines the amount of light it gathers together. If, instead of one star we have three or four stars together, we will find that this lens will deal with the light from each star just as it did with the light of the first one, and just in proportion to the angular distance they are apart in the sky, so will the pictures we see of them be apart on our screen. So if we let the light from the moon fall on our lens, all the light from the various parts of the moon's surface will act like the separate stars, and produce a picture of the whole moon (in the photographic camera the lens produces in this manner a picture of objects in front of it, and this picture we see on the ground glass). When we attempt to get pictures of near objects that do not send rays of light that are parallel, we find that as the rays of light from them do not fall on the lens at the same angle to the axis, the picture is formed further away from the lens. The nearer the object whose picture we wish to throw upon the screen is to the lens, the further the screen must be moved. If we try this experiment we shall find, when we have the object at the same distance as the screen, the picture is then of the same size as the object, and the distance of the screen from the lens is twice that which we have found as the focal length; on bringing the object still nearer the lens, we find we must move the screen further and further away, until when the object is at the focus the picture is formed at an infinite distance away, or, what is more to our purpose, the rays of light from an object at the focus of a convex lens after passing through the lens are parallel, exactly as we have seen such parallel rays falling on the glass come to a focus, so that our diagram answers equally well whatever the direction of the rays; and this holds good in other cases where we take the effect of reflection as well as refraction.

We can also produce pictures by means of bright concave surfaces acting by reflection on the light falling upon them. Such a mirror or concave reflecting surface as I have here will behave exactly as the lens, excepting, of course, that it will form the picture in front instead of behind. The bending of the rays in the case of the convex lens is convergent, or towards the axis, for all parallel rays; if we use the reverse form of lens—that is, one thicker at the edge than in the middle—we find the reverse effect on the parallel rays; they will now be divergent, or bend away from the axis; and so with reflecting surfaces if we make the concavity of our mirror less and less, till it ceases and we have a plane, we shall get no effect on the parallel rays of light except a change of direction after reflection. If we go beyond this and make the surface convex we shall then have practically the same effect on the reflected rays as that given to the refracted ray by the concave glass lens.

As regards the size of the picture produced by lenses or mirrors of different focal length, the picture is larger just as the focal length is greater, and the angular dimension is converted into a linear one on the screen in due proportion. Now, as we shall assume that the eye sees all things best at the distance of about nine inches, we may say that the picture taken with a lens of this focal length gives at once the proper and most natural representation we can possibly have of anything at which we can look. Such a picture of a landscape, if placed before the eye at the distance of nine inches, would exactly cover the real landscape point for point all over. A picture taken with a lens of shorter focal length, say four inches, will give a picture as true in all the details as the larger one, but if this picture is looked at, at nine inches distance, it is not a true representation of what we see; in order to make it so, we must look at it with a lens or magnifier. With a larger picture one can look at this at the proper distance, which always is the focal distance of the lens with which it was obtained, when we will see everything in the natural angular position that we have in the first case.

But if, instead of looking at this larger picture, which we may consider taken with a lens of say ninety inches focal length, at a distance of ninety inches, we look at it at a distance of nine inches, we have practically destroyed it as a picture by reducing the distance at which we are viewing it, and we have converted it into what is for that particular landscape a telescopic picture; we see it, not from the point at which it was taken, but just as if we were at one-tenth of the distance from the particular part that we examine. A telescope with a magnifying power of ten, would enable us to see the landscape just as we see it in the photograph, when we examine it in the way I have mentioned.

Having thus seen how a lens or mirror acts, we will turn our attention to the eye. Here we find an optical combination of lenses that act together in the same way as the single convex lens of which we have been speaking. We will call this combination the lens of the eye. It produces a picture of distant objects which in the normal eye falls exactly in focus upon the retina. We are conscious that we do see clearly at all distances beyond about nine inches.

At less than this distance objects becomes more and more indistinct as they are brought nearer to the eye. From what we have seen of the action of the lens in producing pictures of near and distant objects, we know that some movement of the screen must be made in order to get such pictures sharply focussed, a state of things necessary to perfect vision. We might therefore suppose that the eye did so operate by increasing when necessary the distance between the lens and retina, but we know that the same effect is produced in another way; in fact, the only other way. The eye by a marvellous provision of nature, secures the distinctness of the picture on the retina of all objects beyond a distance of about 9 inches, by slightly but sufficiently varying the curvature of one of the lenses; by an effort of will, we can make the accommodating power of the eye slightly greater, and so see things clearly a little nearer; but at about the distance of 9 inches, the normal eye is unconscious of any effort in thus accommodating itself to different distances. The picture produced by the lens of the eye, whose focal length we will assume to be six-tenths of an inch, falls on the retina, which we will assume further to be formed of a great number of separate sensible points, which, as it were, pick up the picture where it falls on these points, and through the nervous organisation, produce the sense of vision. Possibly when these points are affected by light, there may be some connective action, either produced by some slight spherical aberration of the lens or otherwise; but I do not wish to go any further in this matter than is necessary to elucidate my subject. What I am concerned with now is the extent to which the sensibility of the retina extends. Experiment tells us that it extends to the perception of two separate points of light whose angular distance apart is one minute of arc, or in other words, at the distance we can see best, two points whose distance apart is about 1/400 of an inch.

This marvellous power can be better appreciated when we remember that the actual linear distance apart of two such points on the retina is just a little more than 1/6000 of an inch.

In dealing with the shape of small objects the difference between a circle, square, and triangle, can be detected when the linear size of their images on the retina is about 1/2000 of an inch. It may be therefore fairly taken that these separate sensible points of the retina are somewhere about 1/12,000 part of an inch apart from each other. Wonderfully minute as must this structure be, we must remember, as we have already shown, that the actual size of the image it deals with is also extremely small. This minuteness becomes apparent when we consider what occurs when we look at some well-known object, such as the full moon. Taking the angular diameter of the moon as 30 minutes of arc, and the focal length of the eye at six-tenths of an inch, we find the linear diameter of the picture of the full moon on the retina is about 1/200 of an inch, and assuming that our number of the points in the retina is correct, it follows that the moon is subject to the scrutiny of 2800 of these points, each capable of dealing with the portion of the picture that falls upon it.

That is to say, the picture, as the retina deals with it, is made up to this number of separate parts, and is incapable of further division just as if it were a mosaic. I think this is really the case, and as such a supposition permits us to explain not only what occurs when we assist the eye by means of a telescope, but also what occurs when we use the telescope for photographing celestial objects, we will follow it up.

In the case of the eye we suppose the image of the moon to be made up through the agency of these 2800 points, each one capable of noting a variation in the light falling upon it. In order to make this rather important point plainer, I have had a diagrammatic drawing made on this plan. Taking a circle to represent the full moon I have divided it into this number of spaces, and into each space

I have put a black dot, large or small, according to the intensity of the light falling on that part of the image as determined by looking at a photograph of the moon. You will see by the picture of this moon the effect produced. It represents to those who are at a sufficient distance the moon much as it is really seen in the sky.

We can now with a lens of the same focal length as the eye obtain a picture of the full moon exactly of the size of the actual picture on the retina, and if we take a proper photographic process we can get particles of silver approximately of the same sizes as the dots we have used in making our diagram of the moon; the grouping is not exactly the same, but we may take it as precisely so for our purpose. I have not any photographs of the full moon of this size, but I have some here of the moon about five, seven and eight days old, which give a good idea of what I mean by the arrangements of the particles of silver being like our diagram.

It is now quite apparent that if we can by any means increase the size of the picture of the moon on the retina or make it larger on the photographic plate, we shall be able to employ more of our points in the retina of the eye or of our particles of silver in the photographic film, and so be able to see more clearly just in proportion as we increase the size of the picture in relation to the size of the separate parts that make it.

Now the telescope enables us to do this for the eye, and a lens of longer focal length will give us a larger photographic picture.

Let us assume that by means of the telescope we have increased the power of the eye one hundred times. The picture of the moon on the retina would now be one-half inch diameter, and instead of employing 2800 points to determine its shape, and the various markings upon it, we should be employing 28,000,000 of these points; and similarly with the photograph, by increasing the size of our lens we shall obtain a picture made up of this enormous number of particles of silver. But we can go further in the magnification of the picture on the retina—we can also use a still longer focus photographic lens.

A power of magnification of one thousand is quite possible under favourable circumstances; this means that the picture of one two-hundredth of an inch would be now of five inches in diameter, so we must deal with only a portion of it. Let us take a circle of one-tenth of this, equalling one-hundredth of our original picture, which in the eye, unaided by the telescope, would have a diameter of one two-thousandth of an inch, or an area of less than one five-millionth of a square inch. This means that with this magnification, we have increased the power so enormously that we are now employing for the photographic picture two thousand eight hundred million particles of silver, and in the eye the same degree of increase in the number of points of the retina employed in scrutinising the picture piece by piece as successive portions are brought into the central part.

Photography enables me to show that the result I have given of the wonderful effect of increasing the optical power is perfectly correct as far as it is concerned. We will deal with a part only of the moon, representing, as I have just said, about one-tenth of its diameter, or one-hundredth of its visible surface. Two such portions of the moon are marked, as you see, on the diagram. I have selected these portions as I am able to show you them just as taken on a large scale by photography so that you can make the comparison in the most certain manner; but let us first analyse our diagrammatic moon—let us magnify it about ten times, and see what it looks like.

I now show you a picture of this part of the diagram, inclosing the portions I wish to speak about, magnified ten times, so that you can see that about twenty-eight of our points, and by supposition twenty-eight of our particles of silver on the photographic plate, make up the picture. You will see that these dots vary in size; the difference is due to the amount of light falling within what we may call the sphere of action of each point, and should represent it exactly. The result can hardly be called a picture, as it conveys no impression of continuity of form to the mind. We have got down to the structure or separate parts, and to the limit of the powers of the eye and the photographic plate, of course on the assumption we have made as to the size of the points in the one case and the particles of silver in the other. I will now show you the same parts of the moon as represented by the circles on our diagram exactly as delineated by photography. You now see a beautiful picture giving mountains, valleys, craters, peaks, and plains, and all that makes up a picture of lunar scenery. We have thus seen how the power of the eye is increased by the enlargement of the picture on the retina by the telescope, and also how, by increasing the size of the photograph, we also get more and more detail in the picture.

We know we cannot alter the number of those separate points on the retina which determine the limit of our powers of vision in one direction, but we may be able to increase enormously the number of particles of silver in our photographic picture by processes that will give finer deposits, and so, in conjunction with more perfect and larger photographic lenses, we may reasonably look for a great improvement in our sense of vision—it may be even beyond that given by the telescope alone; although it always will be something in favour of the telescope that the magnification obtained in the eye is about fifteen times greater than that obtained by photography when the image on the retina is pitted against the photograph of the same size, unless we use a lens to magnify the photograph of the same focal length as the eye, in which case it is equal. But we *may* go much further in our magnification of the photographic image. In other ways there is great promise when we consider the difference in the action of the eye and the chemical action in the sensitive film under the action of light. As I pointed out in the discourse I gave about four years ago in this theatre, the eye cannot perceive objects that are not sufficiently illuminated, though this same amount of illumination will, by its cumulative effect, make a photographic picture, so that there are ways in which the photographic method of seeing celestial bodies can be possibly made superior to the direct method of looking with a telescope.

With some celestial objects this has been already done: stars too faint to be seen have been photographed, and nebulæ that cannot be seen have also been photographed; but much more than this is possible: we may be able to obtain photographs of the surface of the moon similar to those I have shown, but on a very much larger scale, and we may obtain pictures of the planets that will far surpass the pictures we would see by the telescope alone.

I have mentioned that the distance at which the normal eye can best see things is about nine inches, as that gives the greatest angular size to the object while retaining a sharp picture on the retina; but, as many of us know, eyes differ in this power: two of the common infirmities of the eyes are long or short-sightedness, due to the pictures being formed behind the retina, in the first case, and in front of it in the other. Towards the end of the thirteenth century it was found that convex lenses would cure the first infirmity, and, soon afterwards, that concave lenses would cure the second, as can be easily seen from what I have said about the action of these lenses; so that during the fifteenth and sixteenth centuries the materials for the making of a telescope existed; in fact, in the sixteenth century, Porta invented the camera obscura, which is in one sense a telescope. It seems very strange that the properties of a convex and concave lens when properly arranged were not known much earlier than 1608. Most probably, if we may judge from the references made by some earlier writers, this knowledge existed, but was not properly appreciated by them. Undoubtedly, after the first telescopes were made in Holland in 1608, the value of this unique instrument was fully appreciated, and the news spread rapidly, for we find that in the next year "Galileo had been appointed lecturer at Padua for life, on account of a perspective like the one which was sent from Flanders to Cardinal Borghese." As far as can be ascertained, Galileo heard of the telescope as an instrument by which distant objects appeared nearer and larger, and that he, with this knowledge only, reinvented it. The Galilean telescope is practically, though not theoretically, the simplest form. It is made of a convex lens in combination with a concave lens to intercept the cone of rays before they come to a focus, and render them parallel so that they can be utilised by the eye. It presents objects as they appear, and the picture has less colour in this form than in the other where a convex eye-glass is used. It is used as one form of opera-glass at the present time. Made of one piece of glass in the shape of a cone, the base of which is ground convex, and the apex slightly truncated and ground concave, it becomes a single-lens telescope that can be looked upon just as an enlargement of the outer lens of the eye.

Galileo was undoubtedly the first to make an astronomical

discovery with the telescope: his name is, and always will be, associated with the telescope on this account alone.

Very soon after the introduction of the Galilean telescope, the difficulties that arise from the coloured image produced by a single lens turned attention to the possibility of making a telescope by using the reflecting surface of a concave mirror instead of a lens. Newton, who had imperfectly investigated the decomposition of light produced by its refraction through a prism, was of opinion that the reflecting principle gave the greatest possibilities of increase of power. He invented, and was the first to make, a reflecting telescope on the system that is in use to the present day; thus the two forms of telescope—the refracting and reflecting—came into use within about 60 years of each other. It will be perhaps most convenient in briefly running through the history of the telescope, that I should give what was done in each century.

Commencing, then, with the first application of the telescope to the investigation of the heavenly bodies by Galileo in 1609, we find that the largest telescope he could make gave only a magnifying power of about 30.

The first improvement made in the telescope, as left by Galileo, was due to a suggestion—by some attributed to Kepler, but certainly used by Gascoigne—to replace the concave eye-lens that Galileo used by a convex one. Simple as this change looks, it makes an important, indeed vital improvement. The telescope could now be used, by placing a system of lines or a scale in the common focus of the two lenses, to measure the size of the image produced by the large lens; the axis or line of collimation could be found, and so the telescope could be used on graduated instruments to measure the angular distance of various objects; in fact, we have now in every essential principle the true astronomical telescope. It is useless as an ordinary telescope, as it inverts the objects looked at, while the Galilean retains them in their natural position. The addition, however, of another lens or pair of lenses reinverts the image, and we then have the ordinary telescope. It was soon found that the single lens surrounds all bright objects with a fringe of colour, always of a width of about one-fiftieth of the diameter of the object-glass, as we must now call the large lens; and as this width of fringe was the same whatever the focal length of the object-glass, the advantage of increasing this focal length and so getting a larger image without increasing the size of the coloured fringe became apparent, and the telescope therefore was made longer and longer, till a length of over one hundred feet was reached; in fact, they were made so long that they could not be used. A picture of one of these is shown, from which it can be easily imagined the difficulties of using it must have been very great, yet some most important measurements have been made with these long telescopes. Beyond the suggestions of Gregory and Cassegrain for improvements in the reflecting telescope, little was done with this instrument.

During the eighteenth century immense advances were made in both kinds of telescopes. With the invention of the achromatic telescope by Hall and Dollond, the long-focus telescopes disappeared.

Newton had turned to the reflecting telescopes believing from his investigations that the dispersion and refraction were constant for all substances; this was found not to be so, and hence a means was possible to render the coloured fringe that surrounds bright objects when a single lens is used less prominent, by using two kinds of glass for the lens, one giving more refraction with somewhat similar dispersion, so that while the dispersion of one lens is almost corrected or neutralised by the other, there is still a refraction that enables the combination to be used as a lens giving an image almost free from colour.

In 1733, Hall had made telescopes having double object-glasses on this plan, but never published the fact. Dollond who had worked independently at the subject, came to the conclusion that the thing could be done, and succeeded in doing it; the invention of the achromatic telescope is with justice, therefore, connected with his name.

Although this invention was a most important one, full advantage could not be taken of it owing to the difficulty of getting disks of glass large enough to make into the compound object-glass, disks of about four inches being the largest diameter it was possible to obtain. With the reflecting telescope, unhampered as it always has been by any except mechanical difficulties, advance was possible, and astronomers turned to it as the only means of getting larger instruments. Many most excellent instruments were made on the Newtonian plan. The plan proposed by Gregory was largely used, as in this instrument objects are seen in their natural position, so that the telescope could be employed for ordinary purposes.

Many were also made on the plan proposed by Cassegrain. The diagrams on the wall enable you to at once see the essential points of these different forms of reflectors.

About 1776 Herschel commenced his astronomical work; beginning with reflecting telescopes of six or seven inches, he ultimately succeeded in making one of four feet aperture. With these instruments, as everyone knows, most brilliant discoveries were effected, and the first real survey of the heavens made.

Herschel's larger telescopes were mounted by swinging them in a surrounding framed scaffolding that could itself be rotated. The smaller ones were mostly mounted on the plan of the one now before us, which the Council of the Royal Astronomical Society have kindly allowed me to bring here. The plan nearly always used by Sir William Herschel was the Newtonian, though for the larger instruments he used the plan proposed years before by Le Maire, but better known as the Herschelian, when the observer looks directly at the large mirror, which is slightly tilted, so that his body does not hinder the light reaching the telescope. In all cases the substance used for the mirrors was what is called speculum metal.

During the present century the aperture of the refracting telescope has increased enormously; the manufacture of the glass disks has been brought to a high state of perfection, particularly in France, where more attention is given to this manufacture than in any other country. Early in the century the great difficulty was in making the disk of flint glass. M. Guinand, a Swiss, beginning in 1784, succeeded in 1805 in getting disks of glass larger and finer than had been made before, and refractors grew larger and larger as the glass was made. In 1823 we have the Dorpat glass of 9·6 inches, the first large equatorial mounted with clock-work; in 1837 the 12-inch Munich glass; in 1839 the 15-inch at Harvard, and in 1847 another at Pulkowa; in 1863 Cooke finished the 25-inch refractor which Mr. Newall gave, shortly before his death last year, to the Cambridge University.

This telescope the University has accepted, and it is about to be removed to the Observatory at Cambridge, where it will be in charge of the Director, Dr. Adams. In accordance with the expressed wish of the late Mr. Newall, it will be devoted to a study of stellar and astronomical physics. There is every prospect that this will be properly done, as Mr. Frank Newall, one of the sons of the late Mr. Newall, has offered his personal services for five years in carrying on this work. Succeeding this we have the 26-inch telescope at Washington, the 26-inch at the University of Virginia, the 30-inch at Pulkowa, and the 36-inch lately erected at Mount Hamilton, California—all these latter by Alvan Clark and his sons. By Sir Howard Grubb we have many telescopes, including the 28-inch at Vienna. Most of these telescopes have been produced during the last twenty years, as well as quite a host of others of smaller sizes, including nearly a score of telescopes of about 13 inches diameter by various makers, to be employed in the construction of the photographic chart of the heavens, which it has been decided to do by international co-operation.

The first of these photographic instruments was made by the Brothers Henry, of the Paris Observatory, who have also made many very fine object glasses and specula, and more important than all, have shown that plane mirrors of perfect flatness can be made of almost any size; the success of M. Lœwy's new telescope, the equatorial *coudé*, is entirely due to the marvellous perfection of the plane mirrors made by them.

The reflecting telescope has quite kept pace with its elder brother.

Lassell in 1820 began the grinding of mirrors, he like Sir William Herschel working through various sizes, finally completing one of 4 feet aperture, which was mounted equatorially Lord Rosse also took up this work in 1840; he made two 3-foot specula, and in 1845 finished what yet remains the largest telescope, one of 6 feet aperture. All these were of speculum metal, and all on the Newtonian form. In 1870, Grubb completed for the Melbourne Observatory a telescope of 4 feet aperture, on the Cassegrain plan, the only large example. The mirror of this is of speculum metal. In 1856 it was proposed by Steinheil, and in 1857 by Foucault, to use glass as the material for the concave mirror, covering the surface with a fine deposit of metallic silver in the manner that had then just been perfected. In 1858 Draper in America, completed one on this plan of 15 inches aperture, soon after making another of 28 inches. In France several large ones have been made, including one of 4 feet at the Paris Observatory: in England this form of telescope is largely used, and mirrors up to 5 feet in diameter have been made and mounted equatorially.

Optically the astronomical telescope, particularly the refractor, has arrived at a splendid state of excellence; the purity of the glass disks and the perfection of the surfaces is proved at once by the performances of the various large telescopes. No limit has yet been set to the increase of size by the impossibility of getting disks of glass or working them, nor is it probable that the limit will be set by either of these considerations. We must rather look for our limiting conditions to the immense cost of mounting large glasses, and the absorption of light by the glass of which the lenses are made, coming injuriously into play to reduce the light-gathering power, though it will be probably a long time before this latter evil will be much felt.

With the reflecting telescope the greater attention given to the working and testing of the optical surface has enabled the concave mirror to be made with a certainty that the earlier workers never dreamed of. The examination of the surface can be made optically at the centre of curvature of the mirror in the manner that was used by Hadley in the beginning of the last century, and revived some years ago by Foucault who brought this method of testing specula to a high degree of perfection; in fact, with the addition of certain methods of measuring the longitudinal aberrations we have now a means of readily testing mirrors with a degree of accuracy that far exceeds the skill of the worker. It enables every change that is made in the surface during the progress of the figuring, as the parabolisation of the surface is called, to be watched and recorded, and the exact departure of any part from the theoretical form measured and corrected; mirrors can be made of very much greater ratio of aperture to focal length. I have one here where the focal length is only $2\frac{1}{4}$ times the aperture: such a mirror in the days of speculum metal mirrors with the methods then in use would have necessarily had a focal length of about 20 feet. The difference in curvature between the centre and edge of this mirror is so great that it can be easily measured by an ordinary spherometer, amounting as it does with one of 6 inches diameter to 3/10,000 of an inch, an amount sufficient to make the focus of the outer portion about 1 inch longer than the inner when it is tested at the centre of curvature. The diagram on the wall, copied roughly from one of the records, I keep of the progress of the work on a mirror during the figuring, shows how this system of measurements enables one to follow closely the whole operation.

The use of silver on glass as the reflecting surface is as important an improvement in the astronomical telescope as the invention of the achromatic telescope. It gives a permanency to a good figure once obtained that did not exist with the mirrors of speculum metal. To restore the surface of silver to the glass speculum is only a small matter now. How readily this is done may be seen by the practical illustration of the method I will give. I have here two liquids—one a solution of the oxide of silver, and another a reducing agent, the chief material in solution being sugar. I pour the two together in this vessel, the surface of which has been cleaned and kept wet by distilled water, which I shall partly empty, leaving the rest to mix with the two solutions; you will see in the course of about 5 minutes the silver begin to form, eventually covering the whole surface with a brilliant coating that can be polished on the outer surface as bright as that you will see through the glass.

Reflecting telescopes have advantages over the refracting telescopes in many ways, but in some respects they are not so good. They give images that are absolutely achromatic, while the other form always has some uncorrected colour. They can be made shorter, and as the light-grasping power is not reduced by the absorption of the glass of which the lenses are made, it is in direct proportion to the surface or area of the mirror. They have not had in many cases the same care bestowed upon either their manufacture or upon their mounting as has been given in nearly every case to the refracting telescope. Speaking generally, the mounting of the reflecting telescope has nearly always been of a very imperfect kind—a matter of great consequence, for upon the mounting of the astronomical telescope so much depends. To direct the tube to any object is not difficult, but to keep it steadily moving so that the object remains on the field of view requires that the tube should be carried by an equatorial mounting of an efficient character. The first essential of such a mounting is an axis parallel to the axis of rotation of the earth. The tube, being supported on this, will follow any celestial object, such as a star, by simply turning the polar axis in a contrary motion to that of the earth, and at the same rate as the earth rotates on its axis. If we make the telescope to swing in a plane parallel to the polar axis, we can then direct the telescope to any part of the sky, and we have the complete equatorial movement. There are several ways in which this is practically done: we can have a long open-work polar axis supported at top and bottom, and swing the telescope in this, or we can have short strong axes. As examples of the first, I will show you pictures of the mountings designed for Cambridge and Greenwich Observatories some forty years ago by Sir G. Airy, lately and for so long our eminent Astronomer-Royal; and as examples of the other form, amongst others, the large telescope lately erected at Nice, and also the larger one at Mount Hamilton, California, now under the direction of Prof. Holden.

The plan of bringing all the various handles and wheels that control the movement of the telescope and the various accessories down to the eye end, so as to be within reach of the observer, is carried to the highest possible degree of perfection here, as we can see by an inspection of the picture of the eye end of this telescope. The observer with the reflecting telescope is, with moderate-size instruments, never very far from the floor, but in the case of the Lick telescope he might have to ascend some thirty feet for objects low down in the sky. Thanks to the ingenuity of Sir Howard Grubb, to whom the idea is due, the whole of the floor of the Observatory is made to rise and fall by hydraulic machinery at the will of the observer—a charming but expensive way of solving the difficulty, as far as safety goes, but not meeting the constant need of a change in position as the telescope swings round in keeping up with the motion of the object to which it is directed. The great length of the tube of large refractors is well seen in this picture of the Lick telescope: it suggests flexure as the change is made in the direction in which it points, and the consequent change of stress in the different parts of the tube.

The mounting of the reflector has been treated, if not so successfully, with more variety than in the case of the refractor as we shall see from the pictures I will show you, especially where the Newtonian form is used. The 4-foot reflector at Melbourne is mounted on the German plan, in a similar way to a refractor, and an almost identical plan has been followed by the makers of the 4-foot at the Paris Observatory. Lassell, who was the first to mount a large reflector equatorially, used a mounting that may be called the forked mounting, the polar axis being forked at its upper end, and the tube of the telescope swinging between the forks; a very excellent plan, dispensing with all counterpoising. Wishing to obtain certain conditions that I thought and think now favourable to the performance of the reflector, I devised a mounting where the whole tube was supported at one end on a bent arm; a 3-foot mirror was mounted on this plan in 1879, and worked admirably. The Newtonian form demands the presence of the observer near the high end of the telescope, and the trouble of getting him there and keeping him safely close to the eye-piece is very great. As we see from the various photographs, several means have been employed to do this, none of them quite satisfactory.

All the refracting telescopes of note in the world are covered by domes that effectually protect them from the weather; these domes are in some cases comparable in cost with the instruments they cover. It is not surprising, therefore, that efforts have been made to devise a means of getting rid of this costly dome and the long movable tube.

It was suggested many years ago that a combination of plane

mirrors could be used to direct light from any object into a fixed telescope. This idea in a modified form has often been used for special work, one plane mirror being used as we see in the picture on the screen to throw a beam of light into a telescope fixed horizontally; for certain kinds of work this does admirably, but the range is restricted as can be easily seen, and the object rotates in the field of view as the earth goes round. The next step would be to place the telescope pointing parallel to the axis of the earth and send the beam of light into it from the mirror, which could now be carried by the tube so that by simply rotating the tube on its own axis the object would be kept in the field of view. Sir Howard Grubb makes a small telescope on this plan, and some years ago proposed a somewhat similar plan. A sketch of this plan I will show you. You will see, however, that here again the range is restricted, and to use the telescope, means would be required to constantly vary the inclination of the small mirror at one-half the rate of inclination of the short tube carrying the object-glass.

By the use of two plane mirrors, however, the solution of the problem of a rotating telescope tube placed as a polar axis is solved. By having such a telescope with a plane mirror at an angle of 45° to the axis of the telescope in front of the object-glass, we can, by simply rotating the telescope, see every object lying on the equator; and by adding another similar plane mirror at an angle of 45° to the axis of the telescope, *as bent out at right angles by the first plane mirror.* and giving the mirror a rotation perpendicular to this axis, we obtain the same power of pointing the telescope as we have in the equatorial. The idea of doing this was published many years ago, but it was left to the skill and perseverance of M. Lœwy, of the Paris Observatory, to put it into practical use. He devised, and had made, a telescope on this principle, of 10½ inches aperture, which was completed in 1882. It has proved itself an unqualified success, and many other larger ones are now being made in Paris, including one of 23 inches aperture, now nearly completed, for the Paris Observatory.

A lantern copy of a drawing of this latter telescope will be thrown on the screen, in order that you may see what manifest advantages exist in this form of telescope. There is but one objection that can be urged—that is, the possible damage to the definition by the plane mirrors; but this seems, from what I have seen of the wonderful perfection of the plane mirrors made by the Brothers Henry, to be an unreasonable one—at any rate not an insurmountable one. In every other respect, except perhaps a slight loss of light, this form of telescope is so manifestly superior to the ordinary form that it must supersede it in time, not only for general work, but for such work as photography and spectroscopy.

Note on a Method of Silvering Glass Mirrors.

Solutions.—Make up 10 per cent. solutions of pure recrystallised nitrate of silver, pure caustic potash, and loaf sugar. To the sugar solution add ½ per cent. of pure nitric acid and 10 per cent. of alcohol. The sugar solution is very much improved by keeping, its action being more rapid and the film cleaner when the sugar solution has been made for a long time. Make up also a weak solution, say 1 per cent. of nitrate of silver and a 10 per cent. solution of ammonia. (90 per cent. distilled water, 10 per cent. ammonia, ·880 specific gravity). *Distilled water* must be used for all the solutions.

Cleaning the Mirror.—Thoroughly clean the mirror. To do this pour on a strong solution of caustic potash, rub well with cotton wool, rinse with ordinary water, wash again with absolute alcohol, and rinse; finally pour on strong nitric acid, and rub with a piece of cotton wool, inserted in the open end of a test tube. Rinse again thoroughly with ordinary water, and then place the mirror face downwards in distilled water in a dish sufficiently large to leave two inches margin round the edge of the mirror, and to keep the face of the mirror one inch from the bottom of the dish. The liquid should stand half an inch above the face of the mirror which should not be completely submerged, and care should be taken to exclude all air-bubbles.

For Silvering a 12-inch Mirror.—Take 400 c.c. of the nitrate of silver solution and add strong ammonia until the brown precipitate first formed is nearly dissolved, then use the diluted ammonia until the solution is just clear. Then add 200 c.c. of the caustic potash solution. A brown precipitate is again formed, which must be dissolved in ammonia exactly as before, the ammonia being added until the liquid is just clear. Now add the 1 per cent. solution of silver nitrate until the liquid becomes a light brown colour, about equal in density of colour to sherry. This colour is important, and can only be properly obtained by adding the weak solution. Dilute the liquids to 1500 c.c. with distilled water.

All being ready add 200 c.c. of the sugar solution to 500 c.c. of water. Then lift the mirror out of the dish, taking care to keep its face downwards during the time it is out of the water, pour the washing water away, add the sugar solution to the silver potash solution, taking care they are thoroughly mixed, and pour them into the dish. Place the mirror face downwards in this solution, taking care to exclude all air-bubbles.

The liquid will turn light brown, dark brown, and finally black. In four or five minutes, often sooner, a thin film of silver will commence to form on the mirror, and this will thicken until in about twenty minutes the whole liquid has acquired a yellowish brown colour, with a thin film of metallic silver floating on the surface.

Lift the mirror out, thoroughly wash with distilled water, and stand the mirror on its edge, or rest it in an inclined position until it is dry; if time can be allowed, the silvered mirror may be left to soak in distilled water over night. Leave it to dry until next day, then the slight yellowish "bloom" can be polished off by rubbing softly with a pad of chamois leather and cotton wool. Carefully polish afterwards with a little dry well-washed rouge on the leather pad. The film should be opaque and brilliant, and with careful handling will be very little changed with long use.

Dishes.—Use porcelain, glass, or earthenware dishes whenever possible; but, if these are not available, a zinc dish, coated inside with paraffin or best beeswax.

For small mirrors (up to 12 inches) the easiest method of supporting them during silvering is to attach them to a wooden rod by pitch, and arrange the dish thus

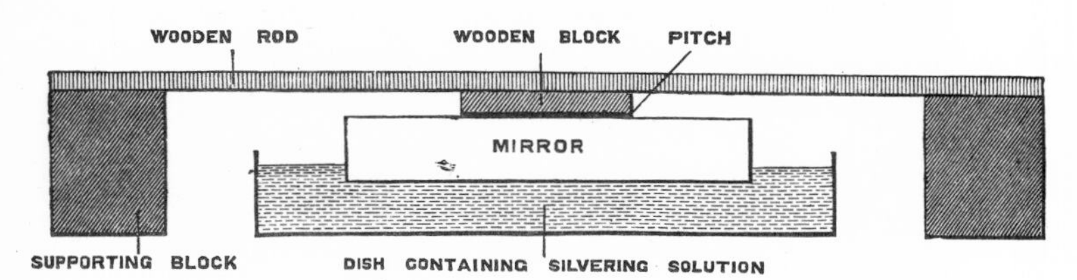

Temperature and Time.—Half an hour is the usual time taken in silvering, but this is shortened by using warmer liquids. About 65° F. is best for silvering. In colder weather longer time must be allowed for the film to be deposited. In very hot weather a smaller quantity of sugar can be used, say 150 c.c. For a 12-inch mirror it is a safe rule to allow four times the time required to get the first indications all over the mirror as the total time for the mirror to be in the bath.

In cases when it is necessary to silver face upwards, a band may be put round the mirror, and the solutions poured on. It is necessary in this case to *leave out the potash solution*, and allow a longer time for the silver to deposit; as much as two hours being sometimes necessary.

If a very thick film is required, two silvering baths can be used, the mirror being left in the first for 15 minutes, then lifted out, rinsed with distilled water, and at once immersed in the second bath, which should be ready in a second dish. The film must not be allowed to dry during the operation of changing from one bath to the other.

Friday, February 13, 1891.

WILLIAM HUGGINS, Esq. D.C.L. LL.D. F.R.S. Vice-President, in the Chair.

PROFESSOR ARTHUR SCHUSTER, Ph.D. F.R.S.

Recent Total Solar Eclipses.

SUCCESSFUL observations of solar eclipses began with the invention of the spectroscope. Though valuable results had been personally obtained, especially by De la Rue in 1860, the bulk of the observations made before 1868 are of a kind which at present can be carried on without the help of an eclipse. The steady progress of eclipse work of late years is not, however, altogether due to the spectroscope, but in great part to the science of photography, which of late years has advanced with rapid strides. In order to judge of the amount of information actually obtained, it is well to bear in mind the short time which has been at our disposal; the aggregate time during which eclipse observations have been carried on since the construction of the spectroscope hardly exceeds half an hour.

The primary object of an eclipse expedition is to investigate the regions of space which are in the vicinity of the sun, and we may hope thereby ultimately to obtain important information concerning the constitution of interstellar space.

The lecturer explained the peculiar difficulties of eclipse observations, the preparations for which often have to be conducted under great disadvantages, especially when, as happened in the West Indian eclipse of 1886, the weather is of such an unsettled character that up to the last minute of the eclipse it was uncertain whether anything could be seen at all.

Photographs of the solar corona as observed in different eclipses were thrown on the screen, and attention was drawn to the great differences in its general outline and character. Two types of corona may be distinguished; one of them principally appearing at a time of sunspot maximum, while the other is chiefly seen when there are few spots on the luminary.

The corona of sunspot minimum is characterised by long streamers spreading chiefly in directions which are not much inclined to the solar equator. In addition to these extensions, curved lines are noticed which seem to converge to two points on the solar surface which are not far removed from the solar poles. If a line is drawn through the two points of convergence, the corona is roughly symmetrical with respect to it. No such symmetry can be noticed in the corona which appears at times of maximum sunspots. The streamers extend irregularly all round the body of the sun, but do not reach so far as the long extension of the first type.

With respect to the nature of the corona, there are four alternatives. It consists of matter either (1), forming a regular atmosphere round the sun; or (2), matter projected from the sun; (3), matter falling into the sun; or finally (4), matter circulating round the sun with planetary velocity.

The choice between these observations must be made by careful analysis of the light received by us from the corona. We possess well-known methods to distinguish between light which is sent out from bodies which are self-luminous, and light which is reflected; and we may even detect violent motions of luminous matter.

The lecturer then explained in detail the photographs of the spectrum of the corona and prominences as photographed in the West Indian eclipse. The principal results may be summarised as follows:—

(1) The greater part of the light sent out by the solar corona is due to matter which is self-luminous, and probably in a solid or liquid condition; the maximum of luminous intensity is displaced towards the red end of the spectrum as compared with sunlight, showing that the temperature of the luminous matter is lower than that of the solar surface.

(2) A comparatively small part of the light is reflected sunlight. The relative importance of reflected and independent light seems to differ in different eclipses, a point which will no doubt receive careful attention on future occasions.

(3) Hydrogen and calcium, which are the main constituents of solar prominences, do not form part of the normal spectrum of the corona. The hydrogen lines are visible only in the parts overlying strong prominences. The violet calcium lines known as H and K, though visible everywhere, are stronger on that side of the corona which has many prominences at its base.

If this result is confirmed on future occasions it would prove that the matter of the corona is partly formed by substances thrown out from the body of the sun.

(4) The gaseous constituents of the corona, which seem rich in spectroscopic lines, cannot at present be identified with any terrestrial elements.

(5) The matter of the corona does not revolve with planetary velocity round the sun.

Photometric measurements have been made during the last eclipse by Professor Thorpe with instruments designed by Captain Abney, and one of their results allows us to compare the corona in this respect with the one observed by Langley in 1878. Thorpe and Abney find the luminous intensity between eight and nine minutes of arc away from the sun's limb to be about the twentieth part of the intensity of moonlight, while Langley found that at a distance of three minutes the corona radiated with one-tenth the intrinsic brightness of the moon.

Considering (1), that the brightness of the corona diminishes with the distance from the sun's limb; (2), that Langley observed at the top of Pike's Peak, at an elevation of 14,000 feet in a very dry atmosphere, while Thorpe's observations were taken at sea-level under unfavourable circumstances; also (3), that the type of corona was different on the two occasions; the results agree to a remarkable degree, and show that the eye estimates which have suggested enormous differences in the brilliancy of the corona in different eclipses are not to be trusted.

Returning to the four alternatives respecting the constitution of the corona, we may at once reject the first and fourth; for it may be proved that the sun could have no regular atmosphere to the extent indicated by the outlines of the corona, and spectroscopic results exclude the hypothesis that the bulk of its matter revolves with planetary velocity; though probably there is some meteoric material which does revolve round the sun.

Dr. Huggins, in a lecture delivered in the Royal Institution,* has suggested a theory of the corona, according to which its luminosity is due to electrical discharges, the matter conveying the discharge being projected from the sun by electrical repulsion. The author agrees with Dr. Huggins in the idea that electrical discharges are probably the cause of the streamers which form the most prominent feature of the solar corona. But before we can form any definite ideas as to the precise way in which these discharges are brought about, we must first settle the very important question whether the planetary space contains sufficient matter to be a conductor of electricity. Our present knowledge regarding electrical discharges entitles us to say that a body which is at the high temperature of the sun, surrounded by gaseous matter, cannot keep any appreciable charge of electricity, and we have some evidence for saying that once a discharge is set up in interplanetary space there is sufficient matter present to convey the discharge, so that the lecturer feels bound to believe in a direct electric connection between the sun and the planets. If then, as is probable, electric discharges take place near the sun, there must be some cause which keeps up the difference in electrical potential between the sun and outside space. The form of the corona suggests a further hypothesis, which, extravagant as it may appear at present, may yet prove to be true. Is the sun a magnet? We know that a body at such a high temperature cannot be magnetisable, but may not a revolving body act like a magnet, and may not the earth's magnetism be similarly due to the earth's revolution about its axis? It can be shown that although a revolving body may act like a magnet sufficiently to account for

* 'Proceedings,' Royal Institution, 1885.

terrestrial magnetism, our instrumental appliances would yet be quite insufficient to allow us to detect the phenomenon in bodies set into rotation artificially on the surface of the earth, so that there is no *a priori* reason against the hypothesis. Owing to the large mass of the sun the magnetic forces at his surface would be much stronger than those at the surface of the earth, and we should expect the outline of the corona to show the influence of these magnetic forces if the streamers of the corona are caused by electric discharges. The form of the corona at a time of minimum sunspot is as a matter of fact very similar to what we should expect if the sun was a magnet, discharging negative electricity near its poles. The author has shown, in his Bakerian lecture of 1884, that a magnet introduced into a hollow negative electrode drives the discharges away from the poles of the magnet, concentrating at places where the field is weakest. It is also known that the negative discharge near a magnet tends to take up the shape of the lines of force, and Bigelow has recently drawn attention to the similarity between the polar rays of the corona and the lines of force due to a magnetised sphere.

All these questions are at present in a purely speculative state, but it is only by keeping an ambitious programme that we may hope to make good use of the eclipses which are yet to come. The relation between matter and the luminiferous ether is the great question of the day, and Cosmical Physics is likely to contribute largely to its solution.

[A. S.]

Friday, May 29, 1891.

WILLIAM HUGGINS, Esq. D.C.L. LL.D. F.R.S. Vice-President, in the Chair.

DAVID GILL, Esq. LL.D. F.R.S. Her Majesty's Astronomer at the Cape of Good Hope.

An Astronomer's Work in a Modern Observatory.

THE work of Astronomical Observatories has been divided into two classes, viz. Astrometry and Astrophysics. The first of these relates to Astronomy of precision, that is to the determination of the positions of celestial objects; the second relates to the study of their physical features and chemical constitution.

Some years ago the aims and objects of these two classes of observatories might have been considered perfectly distinct, and, in fact, were so considered. But I hope to show that in more recent years their objects and their processes have become so interlaced that they cannot with advantage be divided, and a fully equipped modern observatory must be understood to include the work both of *Astrometry* and *Astrophysics.*

In any such observatory the principal and the fundamental instrument is the transit circle. It is upon the position in the heavens of celestial objects, as determined with this instrument or with kindred instruments, that the whole fair superstructure of exact astronomy rests; that is to say, all that we find of information and prediction in our nautical almanacs, all that we know of the past and can predict of the future motions of the celestial bodies.

Here is a very small and imperfect model, but it will serve to render intelligible the photograph of the actual instrument which will be subsequently projected on the screen. [Here the lecturer described the adjustments and mode of using a transit circle.]

We are now in a position to understand photographs of the instrument itself. But first of all as to the house in which it dwells. Here, now on the screen, is the outside of the main building of the Royal Observatory, Cape of Good Hope. I select it simply because being the observatory which it is my privilege to direct, it is the one of which I can most easily procure a series of photographs. It was built during the years 1824–28, and like all the observatories built about that time, and like too many built since, it is a very fair type of most of the things which an observatory should not be. It is, as you see, an admirably solid and substantial structure, innocent of any architectural charm, and so far as it affords an excellent dwelling-place, good library accommodation, and good rooms for computers, no fault can be found with it. But these very qualities render it undesirable as an observatory. An essential matter for a perfect observatory should be the possibility to equalise the internal and the external temperature. The site of an instrument should also be free from the immediate surroundings of chimneys or other origin of ascending currents of heated air. Both these conditions are incompatible with thick walls of masonry and the chimneys of attached dwelling houses, and therefore, as far as possible, I have removed the instruments to small detached houses of their own. But the transit circle still remains in the main building, for, as will be evident to you, it is no easy matter to transport such an instrument.

The two first photographs show the instrument, in one case pointed nearly horizontally to the north, the other pointed nearly vertical. Neither can show all parts of the instrument, but you can see the massive stone piers, weighing many tons each, which, resting on the solid blocks 10 feet below, support the pivots. Here are the counterweights which remove a great part of the weight of the instrument from the pivots, leaving only a residual pressure sufficient to enable the pivots to preserve the motion of the instrument in its proper plane. Here are the microscopes by which the circle is read. Here the opening through which the instrument views the meridian sky. The observer's chair is shown in this diagram. His work appears to be very simple, and so it is, but it requires special natural gifts—patience and devotion, and a high sense of the importance of his work—to make a first-rate meridian observer. Nothing apparently more monotonous can be well imagined if a man is "not to the manner born."

Having directed this instrument by means of the setting circle to the required altitude, he clamps it there and waits for the star which he is about to observe to enter the field. This is what he sees. [Artificial transit of a star by lantern.]

As the star enters the field it passes wire after wire, and as it passes each wire he presses the key of his chronograph and records the instant automatically. As the star passes the middle wire he bisects it with the horizontal web, and again similarly records on his chronograph the transit of the star over the remaining webs. Then he reads off the microscopes by which the circle is read, and also the barometer and thermometer, in order afterwards to be able to calculate accurately the effect of atmospheric refraction on the observed altitude of the star; and then his observation is finished. Thus the work of the meridian observer goes on, star after star, hour after hour, and night after night; and, as you see, it differs very widely from the popular notion of an astronomer's occupation. It presents no dreamy contemplation, no watching for new stars, no unexpected or startling phenomena. On the contrary, there is beside him the carefully prepared observing-list for the night, the previously calculated circle setting for each star, allowing just sufficient time for the new setting for the next star after the readings of the circle for the previous observation.

After four or five hours of this work the observers have had enough of it; they have, perhaps, observed fifty or sixty stars, they determine certain instrumental errors, and betake themselves to bed, tired, but (if they are of the right stuff) happy and contented men. At the Cape we employ two observers, one to read the circle and one to record the transit. Four observers are employed, and they are thus on duty each alternate night. Such is the work that an outsider would see were he to enter a working meridian observatory at night, but he would find out if he came next morning that the work was by no means over. By far the largest part has yet to follow. An observation that requires only two or three minutes to make at night, requires at least half an hour for its reduction by day. Each observation is affected by a number of errors, and these have to be determined and allowed for. Although solidly founded on massive piers resting on the solid rock, the constancy of the instrument's position cannot be relied upon. It goes through small periodic changes in Level in Collimation and in Azimuth, which have to be determined by proper means, and the corresponding corrections have to be computed and applied; and also there are other corrections for refraction, &c., which involve computation and have to be applied. But these matters would fall more properly under the head of a special lecture upon the transit instrument. I mention them now merely to explain why so great a part of an astronomer's work comes in the daytime, and to dispel the notion that his work belongs only to the night.

One might very well occupy a special lecture in an account of the peculiarities of what is called personal equation—that is to say, the different time which elapses for different observers between the time when the observer believes the star to be upon the wire and the time when the finger responds to the message which the eye has conveyed to the brain. Some observers always press the key too soon, some always too late. Some years ago I discovered, from observations to which I will subsequently refer, that *all* observers press the chronograph key either too soon for bright stars or too late for faint ones.

Other errors may, and I am sure do, arise both at Greenwich and the Cape from the impossibility of securing uniformity of outside and inside temperature in a building of strong masonry. The ideal observatory should be solid as possible as to its foundations, but light as possible as to its roof and walls—say, a light framework of iron covered with canvas. But it would be undesirable to cover a valuable and permanent instrument in this way.

But here is a form of observatory which realises all that is required, and which is eminently suited for permanent use. The walls are of sheet iron, which readily acquire the temperature of the outer air. The iron walls are protected from direct sunshine by wooden louvres, and small doors in the iron walls admit a free circulation of air. The revolving roof is a light framework of iron covered with well-painted *papier maché*.

The photograph now on the screen shows the interior of the observatory, and this brings me to the description of observations of an entirely different class. In this observatory the roof turns round on wheels, so that any part of the sky can be viewed from the telescope. This is so because the instrument in this observatory is intended for purposes which are entirely different from those of a transit circle. The transit circle, as we have seen, is used to determine the *absolute* positions of the heavenly bodies; the heliometer to determine with greater precision than is possible by the absolute method the *relative* positions of celestial objects.

To explain my meaning as to absolute and relative positions:– It would, for example, be a matter of very little importance if the absolute latitude of a point on the Royal Exchange or the Bank of England were one-tenth of a second of arc (or ten feet) wrong in the maps of the Ordnance Survey of England—that would constitute a small *absolute* error common to all the buildings on the same map of a part of the city, and common to all the adjoining maps also. Such an error, regarded as an *absolute* error, would evidently be of no importance if every point on the map had the same absolute error. There is no one who can say at the present moment whether the absolute latitude of the Royal Exchange—nay, even of the Royal Observatory, Greenwich—is known to ten feet. But it would be a very serious thing indeed if the relative positions on the same map were ten feet wrong *here* and *there*. For example, if of two points marking a frontage boundary on Cornhill one were correct, the other ten feet in error—what a nice fuss there would be! what food for lawyers! what a bad time for the Ordnance Survey Office! Well, it is just the same in astronomy.

We do not know, we probably never shall know with certainty, the absolute places of even the principal stars to $\frac{1}{10}$th of a second of arc. But $\frac{1}{10}$th of a second of arc in the measure of some relative position would be fatal. For example, in the measurement of the sun's parallax an error of $\frac{1}{10}$th of a second of arc means an error of 1,000,000 miles, in round numbers, in the sun's distance; and it is only when we can be quite certain of our measures of much smaller quantities than $\frac{1}{10}$th of a second of arc, that we are in a position to begin seriously the determination of such a problem as that of the distances of the fixed stars. For these problems we must use differential measures, that is measures of the relative positions of two objects. The most perfect instrument for such purposes is the heliometer.

Lord McLaren has kindly sent from Edinburgh, for the purposes of this lecture, the parts of his heliometer which are necessary to illustrate the principles of the instrument.

This instrument is the same which I used on Lord Crawford's expedition to Mauritius, in 1874. It was also kindly lent to me by Lord Crawford for an expedition to the Island of Ascension to observe the opposition of Mars, in 1877. In 1879, when I went to the Cape, I acquired the instrument from Lord Crawford, and carried out certain researches with it on the distances of the fixed stars.

In 1887, when the Admiralty provided the new heliometer for the Cape Observatory, this instrument again changed hands. It became the property of Lord McLaren. I felt rather disloyal in parting with so old a friend. We had spent so many happy hours together, we had shared a good many anxieties together, *and we knew each other's weaknesses so well.* But my old friend has fallen into good hands, and has found another sphere of work.

The principle of the instrument is as follows. [The instrument was here explained.]

There is now on the screen a picture of the new heliometer of the Cape Observatory, which was mounted in 1887, and has been in constant use ever since. It is an instrument of the most refined modern construction, and is probably the finest apparatus for refined measurement of celestial angles in the world.

[Here were explained the various parts of the instrument in relation to the model, and the actual processes of observation were illustrated by the images of artificial stars projected on a screen.]

Here, again, there is little that conforms to the popular idea of an astronomer's work; there is no searching for objects, no contemplative watching, nothing sensational of any kind. On the contrary, every detail of his work has been previously arranged and calculated beforehand, and the prospect that lies before him in his night's work is simply more or less of a struggle with the difficulties which are created by the agitation of the star images, caused by irregularities in the atmospheric refraction. It is not upon one night in a hundred that the images of stars are perfectly tranquil. You have the same effect in an exaggerated way when looking across a bog on a hot day. Thus, generally, as the images are approached, they appear to cross and recross each other, and the observer must either seize a moment of comparative tranquillity to make his definitive bisection, or he may arrive at it by gradual approximations till he finds that the vibrating images of the two stars seem to pass each other as often to one side as to the other. So soon as such a bisection has been made the time is recorded on the chronograph, then the scales are pointed on and printed off, and so the work goes on, varied only by reversals of the segments and of the position circle. Generally, I now arrange for 32 such bisections, and these occupy about an hour and a half. By that time one has had about enough of it, the nerves are somewhat tired, so are the muscles of the back of the neck, and, if the observer is wise, and wishes to do his best work, he goes to bed early and gets up again at two or three o'clock in the morning, and goes through a similar piece of work. In fact this must be his regular routine night after night, whenever the weather is clear, if he is engaged, as I have been, on a large programme of work on the parallaxes of the fixed stars, or on observations to determine the distance of the sun by observations of minor planets.

I will not speak now of these researches, because they are still in progress of execution or of reduction. I would rather, in the first place, endeavour to complete the picture of a night's work in a modern observatory.

We pass on to celestial photography, where astrometry and astrophysics join hands. Here on the screen is the interior of one of the new photographic observatories, that at Paris. [Brief description.]

Here is the exterior of our new photographic observatory at the Cape. Here is the interior of it, and the instrument. [Brief description.]

The observer's work during the exposure is simply to direct the telescope to the required part of the sky, and then the clockwork *nearly* does the rest—but not quite so. The observer holds in his hand a little electrical switch with two keys; by pressing one key he can accelerate the velocity of the driving screw by about 1 per cent., and by pressing the other he can retard it 1 per cent. In this way he keeps one of the stars in the field always perfectly bisected by the cross wires of his guiding telescope, and thus corrects the small errors produced partly by changes of refraction, partly by small unavoidable errors in cutting the teeth of the arc into which the screw of the driving shaft of the clockwork gears.

The work is monotonous rather than fatiguing, and the companionship of a pipe or cigar is very helpful during long exposures. A man can go on for a watch of four or five hours very well, taking plate after plate, exposing each, it may be, forty minutes or an hour. If the night is fine a second observer follows the first, and so the work goes on the greater part of the night. Next day he develops his plate and gets something like this. [Star cluster.]

Working just in this way, but with the more humble apparatus which you see imperfectly in the picture now on the screen, we have with a rapid rectilinear lens by Dallmeyer of 6 inches aperture photographed at the Cape during the past six years the whole of the southern hemisphere from 20° of south declination to the south pole.

The plates are being measured by Professor Kapteyn, of Groningen, and I expect that in the course of a year the whole work containing all the stars to 9½ magnitude (between 200,000 and 300,000 stars) in that region will be ready for publication. This work is essential as a preliminary step for the execution in the southern hemisphere of the great work inaugurated by the Astrophotographic Congress at Paris in 1887, the last details of which were settled at our meeting at Paris in April last. What we shall do with the new apparatus perhaps I may have the honour to describe to you some years hence, after the work has been done.

We now come to an important class of astronomical work more purely astrophysical, for the illustration of which I can no longer appeal to the Cape, because I regret to say that we are not yet provided with the means for its prosecution. I refer to the use of the spectroscope in astronomy, and especially to the latest developments of its use for the accurate measurement of the velocity of the motions of stars in the line of sight.*

It is beyond the province of this lecture to enter into history, but it is impossible not to refer to the fact that the chief impulse to astronomical work in this direction was given by Dr. Huggins, our chairman to-night—nay, more, except for the early contributions of Fraunhofer to the subject, Dr. Huggins certainly is the father of sidereal spectroscopy, and that not in one but in every branch of it. He has devised the means, pointed the way, and, whilst in many branches of the work he still continues to lead the way, he has of necessity left the development of other branches to other hands.

From an astrometer's point of view the most important advance that has been made in spectroscopy of recent years is the sudden development of precision in the measures of star motion in the line of sight. The method remained for fifteen or sixteen years quite undeveloped from the condition in which it left the hands of Dr. Huggins, and certainly no progress in the accuracy attained by Dr. Huggins was made till the matter was taken up by Dr. Vogel at Potsdam. At a single step Dr. Vogel has raised the precision of the work from that of observations in the days of Ptolemy to that of the days of Bradley—from the days of the old sights and pinnules to the days of telescopes. Therefore I take a Potsdam observation as the best type of a modern spectroscopic observation for description, especially as I have recently visited Dr. Vogel at Potsdam, and he has kindly given me a photograph of his spectroscope, as well as of some of the work done with it.

A photograph of the Potsdam spectroscope attached to the equatorial is now on the screen. [Description.]

The method of observation consists simply in inserting a small photographic plate in the dark slide, directing the telescope to the star, and keeping the image of the star continuously on the slit during an exposure of about an hour ; and this is what is obtained on development of the picture.

If the star remained perfectly at rest between the jaws of the slit the spectrum would be represented by a single thread of light, and of course no lines would be visible upon such a thread ; but the observer intentionally causes the star image to travel a little along the slit during the time of exposure, and so a spectrum of sensible width is obtained. (Fig. 1.)

You will remark how beautifully sharp are the faint lines in this spectrum. Those who have tried to observe the spectrum of Sirius in the ordinary way know that many of these fine lines cannot be seen or measured with certainty. The reason is that on account of irregularities in atmospheric refraction, the image of a star in the

* The older methods enabled us to measure motions at right angles to the line of sight, but till the spectroscope came we could not measure motions in the line of sight.

FIG. 1.

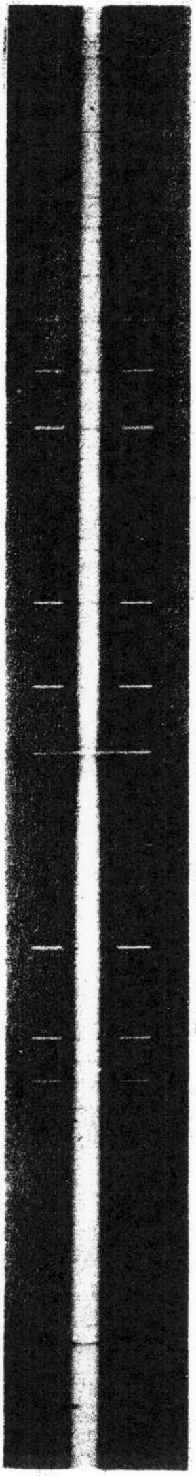

Hγ

Spectrum of Sirius with comparison Spectrum of Iron.

FIG. 2.

α Aurigæ.

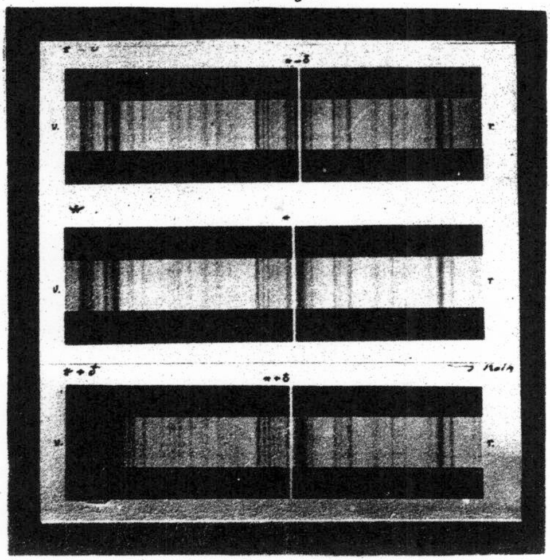

October.

December.

March.

FIG. 3.

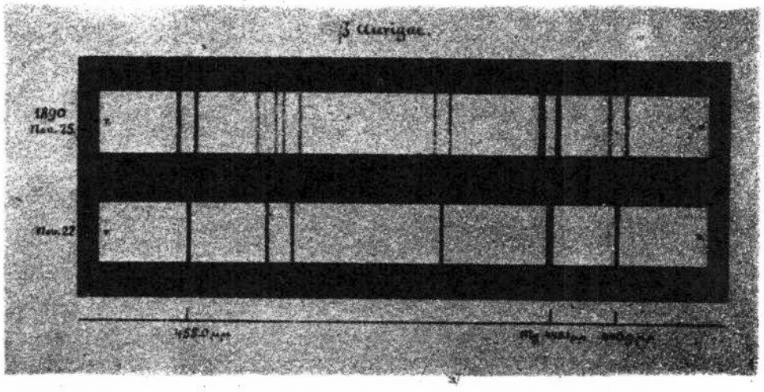

telescope is rarely tranquil, sometimes it shines brightly in the centre of the slit, sometimes barely in the slit at all, and the eye becomes puzzled and confused. But the photographic eye is not in the least disturbed; when the star image is in the slit, the plate goes on recording what it sees, and when the star is not in the slit the plate does nothing, and it is of no consequence whatever how rapidly these alternate appearances and disappearances recur. The only difference is that when the air is very steady and the star's image, therefore, always in the slit, the exposure takes less time than when the star is unsteady.

That is one reason why the Potsdam results are so accurate. And there are many other reasons besides, into which I cannot now enter. What, however, it is very important to note is this, that we have here a method which is to a great extent independent of the atmospheric disturbances which in all other departments of astronomical observation have imposed a limit to their precision. Accurate astro-spectroscopy, therefore, may be pushed to a degree of perfection which is limited only by the optical aid at our disposal and by the sensibility of our photographic plates.

And now I think we have sufficiently considered the ordinary processes of astronomical observation to illustrate the character of the work of an astronomer at night; the picture should be completed by an account of his work by day. But to go into that matter in detail would certainly not be within the limits of this lecture. It is better that I should in conclusion touch upon some recent remarkable results of these day and night labours. It is these after all that most appeal to you, it is for these that the astronomer labours, it is the prospect of them that lightens the long watches of the night and gives life to the otherwise dead bones of mechanical routine.

Let us take first some spectroscopic results. To explain their meaning let me remind you for a moment of the familiar analogy between light and sound.

The pitch of a musical note depends on the rapidity of the vibrations communicated to the air by the reed or string of the musical instrument that produces the note, a low note being given by slow vibrations and a high one by quick vibrations.

Just in the same way red light depends on relatively slow vibrations of ether, and blue or violet light on relatively quick vibrations. Well, if there is a railway train rapidly approaching one, and the engine sounds its whistle, more waves of sound from that whistle will reach the ear in a second of time, than would reach the ear were the train at rest. On the other hand, if the train is travelling at the same rate *away* from the observer, fewer waves of sound will reach his ears in a second of time. Therefore an observer beside the line should observe a distinct change of pitch in the note of the engine whistle as the train passes him, and as a matter of fact such a change of pitch can be and has been observed.

Just in the same way, if a source of light could be moved rapidly enough towards an observer it would become bluer, or if away from him it would become more red in colour. Only it would require a change of velocity in the moving light of some thousands of miles per second in order to render the difference of colour sensible to the eye. The experiment is, therefore, not likely to be frequently shown at this lecture table!

But the spectroscope enables such changes of colour to be measured with extreme precision. Here on the screen is the most splendid illustration of this that exists at present, viz. copies of three negatives of the spectrum of α Aurigæ, taken at Potsdam in October and December of 1888, and in March 1889. (Fig. 2.)

The white line (the picture being a positive) represents the bright line $H\gamma$ given by the artificial light of hydrogen, the strong black line in the picture of the star spectrum corresponds to the black absorption line which is due to hydrogen in the atmosphere of the star.

Why is it that the artificial hydrogen line does not correspond with the stellar line in these three pictures? The answer is, either the star is moving towards or from the earth in the line of sight, or the earth is moving from or towards the star. But in December the earth in its motion round the sun is moving at right angles to the direction of α Aurigæ, why then does not the stellar hydrogen line agree in position with the terrestrial hydrogen line; the simple explanation is that α Aurigæ is moving with respect to the sun.

In what way is it moving? Well, that also is clear: the stellar line is displaced towards the red end of the spectrum, that is to say the star light is redder than it should be in consequence of a motion of recession; this proves that the star is moving away from us, and measures of the photograph show the rate of this motion to be $15\frac{1}{2}$ miles per second. We also know that in October the earth in its motion round the sun is moving towards α Aurigæ nearly at the same rate as we have just seen that α Aurigæ is running away from the sun. Consequently, at that time, their relative motions are nearly insensible, because both are going at the same rate in the same direction, and we find accordingly in October, that the positions of the stellar and artificial hydrogen lines perfectly correspond. Finally, in March, the earth in its motion round the sun is moving away from α Aurigæ, and as α Aurigæ is also running away from the sun the star-light becomes so much redder than normal that the stellar hydrogen line is shifted completely to one side of the hydrogen and artificial line.

The accuracy of these results may be proved as follows:—

If we measure all the photographs of α Aurigæ which Dr. Vogel has obtained we can derive from each a determination of the relative velocity of the motion of the star with respect to our earth.

Of course these velocities are made up of the velocity of motion of α Aurigæ with respect to the sun (which we may reasonably assume to be a uniform velocity) and the velocity of the earth due to α Aurigæ—Potsdam.

Date.	Observed Relative Motion of Earth and Star. Miles per sec.	Motion of Earth.	Concluded Motion, Star Relative to the Sun.
1888.			
October 22nd	+ 2·5	− 13·0	+ 15·5
„ 24th	+ 3·1	− 12·4	+ 15·5
„ 25th	+ 3·1	− 12·4	+ 15·5
„ 28th	+ 2·5	− 11·8	+ 14·3
November 9th	+ 6·8	− 8·7	+ 15·5
December 1st	+ 11·8	− 3·1	+ 14·9
„ 13th	+ 14·9	+ 0·6	+ 14·3
1889.			
January 2nd	+ 20·5	+ 6·8	+ 13·7
February 5th	+ 32·9	+ 14·3	+ 18·6
March 6th	+ 34·2	+ 16·8	+ 17·4

α Aurigæ—Greenwich.

Date.	Observed Relative Motion of Earth and Star. Miles per sec.	Motion of Earth.	Concluded Motion, Star Relative to the Sun.
1887.			
January 26th	+ 16·4	+ 12·6	+ 3·8
February 16th	+ 34·4	+ 15·9	+ 18·5
October 22nd	+ 39·8	− 13·5	+ 52·3
„ 25th	+ 25·4	− 13·0	+ 38·4
„ 29th	+ 40·6	− 12·1	+ 52·7
1888.			
December 7th	+ 29·0	− 1·2	+ 36·2
1889.			
February 15th	+ 23·8	+ 16·0	+ 7·8
March 5th	+ 20·3	+ 17·1	+ 3·2
September 17th	+ 18·6	− 13·3	+ 33·3
„ 19th	+ 21·8	− 16·7	+ 38·5
„ 25th	+ 24·8	− 16·5	+ 41·3
November 25th	+ 24·5	− 4·9	+ 29·4

its motion round the sun. But the velocity of the earth's motion in its orbit is known with an accuracy of about one five-hundredth part of its amount, and therefore, within that accuracy, we can allow precisely for its effect on the relative velocity of the earth and α Aurigæ. When we have done so we get the annexed results for the velocity of the motion of α Aurigæ with respect to the sun. You see by the annexed table how beautifully they agree in the Potsdam results, and how comparatively rough and unreliable are the results obtained by the older method at Greenwich.

I believe that in a few years, at least in a period of time that one may hope to see, we shall not be content merely to correct our results for the motion of the earth in its orbit only, and so test our observations of motion in the line of sight, but that we shall have arrived at a certainty and precision of working which will *permit the process to be reversed*, and that we shall be employing the spectroscope to determine the velocity of the earth's motion in its orbit, or in other words to determine the fundamental unit of astronomy, the distance of the sun from the earth.

I will take as another example one recent remarkable spectroscopic discovery.

Miss Maury, in examining a number of photographs of steller spectra taken at Harvard College, discovered that in the spectrum of β Aurigæ certain lines doubled themselves every two days, becoming single in the intermediate days. Accurate Potsdam observations confirmed the conclusion.

The picture on the screen (Fig. 3) shows the spectrum of β Aurigæ photographed on November 22 and 25 of last year. In the first the lines are single, in the other every line is doubled. Measures and discussion of a number of these photographs have shown that the doubling of the lines is perfectly accounted for by the supposition of two suns revolving round each other in a period of four days, each moving at a velocity of about 70 miles a second in its orbit.

When one star is approaching us and the other receding, the lines in the spectrum formed by the light of the first star will be moved towards the blue end of the spectrum, those in the spectrum of the second star towards the red end of the spectrum. Then, as the two stars come into the same line with us, their motions become at right angles to the line of sight, and their two spectra, not being affected by motion, will perfectly coincide; but then, after the stars cross, their spectra again separate in the opposite direction, and so they go on.

Thus by means of their spectra we are in a position to watch and to measure the relative motions of two objects that we can never see apart; nay more, we can determine not only their period of revolution but also the velocity of their motions in their orbits. Now, if we know the time that a body takes to complete its revolution, and the velocity at which it moves, clearly we know the dimensions of its orbit, and if we know the dimensions of an orbit we know what attractive force is necessary to compel the body to keep in that orbit, and thus we are able to weigh these bodies. The components of β Aurigæ are two suns, which revolve about each other in four days; they are only between 7 and 8 millions of miles (or one-twelfth of our distance from the sun) apart, and if they are of equal weight they each weigh rather over double the weight of our sun.

I have little doubt that these facts do not represent a permanent condition, but simply a stage of evolution in the life-history of the system, an earlier stage of which may have been a nebular one.

Other similar double stars have been discovered both at Potsdam and at Cambridge, U.S., stars that we shall never see separately with the eye aided by the most powerful telescope; but time does not permit me to enter into any account of them.

I pass now to another recent result that is of great cosmical interest.

The Cape photographic star charting of the southern hemisphere has been already referred to. In comparing the existing eye estimates of magnitude by Dr. Gould with the photographic determinations of these magnitudes, both Professor Kapteyn and myself have been greatly struck with a very considerable systematic discordance between the two. In the rich parts of the sky, that is in the Milky Way, the stars are systematically photographically brighter by comparison with the eye observations than they are in the poorer part of the sky, and that not by any doubtful amount but by half or three-fourths of a magnitude. One of two things was certain, either that the eye observations were wrong, or that the stars of the Milky Way are bluer or whiter than other stars. But Professor Pickering, of Cambridge, America, has lately been making a complete photographic review of the heavens, and by placing a prism in front of the telescope he has made pictures of the whole sky like this. [Here two examples of the plates of Pickering's spectroscopic Durchmusterung were exhibited on the screen.] He has discussed the various types of the spectra of the brighter stars, as thus revealed, according to their distribution in the sky. He finds thus that the stars of the Sirius type occur chiefly in the Milky Way, whilst stars of other types are fairly divided over the sky.

Now stars of the Sirius type are very white stars, very rich relative to other stars in the rays which act most strongly on a photographic plate. Here then is the explanation of the results of our photographic star-charting, and of the discordance between the photographic and visual magnitudes in the Milky Way.

The results of the Cape charting further show that it is not alone to the brighter stars that this discordance extends, but it extends also, though in a rather less degree, to the fainter stars of the Milky Way. Therefore we may come to the very remarkable conclusion that the Milky Way is a thing apart, and that it has been developed perhaps in a different manner, or more probably at a different and probably later epoch from the rest of the sidereal universe.

Here is another interesting cosmical revelation which we owe to photography.

You all know the beautiful constellation Orion, and many in this theatre have before seen the photograph of the nebula which is now on the screen, taken by Mr. Roberts.

Here is another photograph of the same object taken with a much longer exposure. You see how over-exposed, in fact, burnt out, the brightest part of the picture is, and yet what a wonderful development of faint additional nebulous matter is revealed.

But I do not think that many persons in this room have seen *this* picture, and probably very few have any idea what it represents. It is from the original negative taken by Professor Pickering, with a small photographic lens of short focus, after six hours' exposure in the clear air of the Andes, 10,000 feet above sea-level.

The field embraces the three well-known stars in the belt of Orion on the one hand, and β Orionis (Rigel) on the other. You can hardly recognise these great white patches as stars; their ill-defined character is simply the result of excessive over-exposure. But mark the wonders which this long exposure with a lens of high intrinsic brilliancy of image has revealed. Here is the great nebula, of course terribly over-exposed, but note its wonderful fainter ramifications. See how the whole area is more or less nebulous, and surrounded as it were with a ring fence of nebulous matter. This nebulosity shows a special concentration about β Orionis.

Well, when Professor Pickering got this wonderful picture, knowing that I was occupied with investigations on the distances of the fixed stars, he wrote to ask whether I had made any observations to determine the distance of β Orionis, as it would be of great interest to know from independent evidence whether this very bright star was really near to us or not. It so happens that the observations were made, and their definitive reduction has shown that β Orionis is really at the same distance from us as are the faint comparison stars. β Orionis is, therefore, probably part and parcel of an enormous system in an advanced but incomplete state of stellar evolution, and that what we have seen in this wonderful picture is all a part of that system.

I should explain what I mean by an elementary or by an advanced state of stellar evolution. There is but one theory of celestial evolution which has so far survived the test of time and comparison with observed facts, viz. the nebular hypothesis of Laplace. Laplace supposed that the sun was originally a huge gaseous or nebulous mass of a diameter far greater than the orbit of Neptune. I say *originally*, do not misunderstand me. We have finite minds; we can imagine a condition of things which might be supposed to occur at any particular instant of time however remote, and at any particular distance of space however great, and we may frame a theory beginning at another time still more remote, and so on. But we can never imagine a theory beginning at an infinite distance of time or at an infinitely distant point in space. Thus, in any theory which man with his finite mind can devise, when we talk of *originally* we simply mean at or during the time considered in our theory.

Now, Laplace's theory begins at a time, millions on millions of years ago, when the sun had so far disentangled itself from chaos, and its component gaseous particles had by mutual attraction so far coalesced as to form an enormous gaseous ball, far greater in diameter than the orbit of the remotest planet of our present system. The central part of this ball was certainly much more condensed than the rest, and the whole ball revolved. There is nothing improbable in this hypothesis. If gaseous matter came together from different parts of space such coalition would unquestionably occur, and as in the meeting of opposite streams of water or of opposite currents of wind, vortices would be created and revolution about an axis set up such as we are familiar with in the case of whirlpools or cyclones. The resultant would be rotation of the whole globular gaseous mass about an axis.

Now this gaseous globe begins to cool, and as it cools it necessarily contracts. Then follows a necessary result of contraction, viz. the rotation becomes more rapid. This is a well-known fact in dynamics, about which there is no doubt. Thus, the cooling and the contracting go on, and simultaneously the velocity of rotation becomes greater and greater. At last the time arrives when, for the outside particles, the velocity of rotation becomes such that the centrifugal force is greater than the attractive force, and so the outside particles break off and form a ring. Then, as the process of cooling and contraction proceed still further, another ring is formed, and so on, till we have finally a succession of rings and a condensed central ball. If from any cause the cooling of any of these rings does not go on uniformly, or if some of the gaseous matter of the ring is more easily liquefied than others, then probably a single nucleus of liquid matter will be formed in that ring, and this nucleus will finally by attraction absorb the whole of the matter of which the ring is composed—at first as a gaseous ball with a condensed nucleus, and this will finally solidify into a planet. Or, meanwhile, this yet unformed planet may repeat the history of its parent sun. By contraction, and consequent acceleration of its rotation, it may throw off one or more rings, which in like manner condense into satellites like our moon, or those of Jupiter, Saturn, Uranus, or Neptune. Such, very briefly outlined, is the celebrated nebular hypothesis of Laplace. No one can positively say that the hypothesis is true, still less can any one say that it is untrue. Time does not permit me to enter into the very strong proofs which Laplace urged in favour of its acceptance.

But I beg you for one moment to cast your imaginations back to a period of time long antecedent to that when our sun had begun to disentangle itself from chaos, and when the fleecy clouds of cosmic stuff had but commenced to rush together. What should we see in such a case were there a true basis for the theory of Laplace?

Certainly, in the first place, we should have a huge whirlpool or cyclone of cosmic gaseous stuff, the formation of rings, and the condensation of these rings into gaseous globes.

Remembering this, look now on this wonderful photograph of the nebula in Andromeda, made by Mr. Roberts. In the largest telescopes this nebula appears simply as an oval patch of nearly uniform light, with a few dark canals through it, but no idea of its true form can be obtained, no trace can be found of the significant story which this photograph tells. It is a picture that no human eye unaided by photography has ever seen. It is a true picture drawn without the intervention of the hand of fallible man, and uninfluenced by his bias or imagination. Have we not here, so at least it seems to me, a picture of a very early stage in the evolution of a star cluster or sun-system—a phase in the history of another star-system similar to that which once occurred in our own—millions and millions of years ago—when our earth, nay, even our sun itself, "was without form and void," and "darkness was on the face of the deep."

During this lecture I have been able to trace but very imperfectly the bare outlines of an astronomer's work in a modern observatory, and to give you a very few of its latest results—results which do not come by chance, but by hard labour, and to men who have patience to face dull daily routine for the love of science—to men who realise the imperfections of their methods and are constantly on the alert to improve them.

The mills of the astronomer grind slowly, and he must be infinitely careful and watchful if he would have them like the mills of God, to grind exceeding small.

I think he may well take for his motto these beautiful lines:—

> "Like the star
> Which shines afar,
> Without haste,
> Without rest,
> Let each man wheel
> With steady sway,
> Round the task
> Which rules the day,
> And do his best."

[D. G.]

Friday, May 13, 1892.

Sir James Crichton-Browne, M.D. LL.D. F.R.S. Treasurer and Vice-President, in the Chair.

William Huggins, Esq. D.C.L. LL.D. Ph.D. F.R.S. *M.R.I.*

The New Star in Auriga.

We depend so absolutely at every moment, and in every action, upon the uniformity of Nature, that any event which even appears to break in upon that uniformity cannot fail to interest us. Especially is this the case if a strange star appears among those ancient heavenly bodies by the motions of which our time and the daily routine of life are regulated, and which through all ages have been to man the most august symbols of the unchanging. For, notwithstanding small alterations due to the accumulated effects of changes of invisible slowness which are everywhere in progress, the heavens, in their broad features, remain as they were of old. If Hipparchus could return to life, however changed the customs and the kingdoms of the earth might appear to him, in the heavens and the hosts thereof, he would find himself at home.

Only some nineteen times in about as many centuries have we any record that the eternal sameness of the midnight sky has been broken in upon by even the temporary presence of an unknown star, though there is no doubt that in the future, through the closer watch kept upon the sky by photography, a larger number of similar phenomena will be discovered.

According to Pliny it was the sudden outburst into splendour of a new star in 130 B.C. which inspired Hipparchus to construct his catalogue of stars. Passing at once to more modern times we come to the famous new star of 1572 discovered by Tycho Brahe in the Constellation of Cassiopeia which outshone Venus, and could even be seen as a bright object upon the sky by day. But its brilliancy, like that of the new stars before and since, was transitory; within a few weeks its great glory had departed from it, and it then continued to wane until at last it had fallen back to its original low estate, as a star invisible to the naked eye.

The star of 1886 which, on May 2nd of that year, burst forth as a star of the second magnitude in the Northern Crown, is memorable as the first of those objects which was subjected to the searching power of the spectroscope. Two temporary stars have appeared since, one of the third magnitude in 1876 in Cygnus, and a small star in the Great Nebula of Andromeda in 1885.

It may be asked whether these temporary stars are in reality new stars, the creations of a day, or but the transient outbursts into splendour of small stars usually invisible; and, indeed, whether they may be but extreme cases of the large class of variable stars which wax and wane in periods more or less regular.

In the case of the more modern temporary stars the evidence is forthcoming that they did exist before and do exist still. The star of 1866 may be seen as one of about the ninth magnitude, with nothing to distinguish it from its fellows. So the star of 1876 in Cygnus, which rose to the third magnitude, is still there as a star of about the fourteenth magnitude. To these may be added, perhaps, Tycho's star.

The new star which makes the present year memorable is, indeed, so far as our charts go, without descent. But there is no improbability in assuming that in its usual low estate, to which it has now returned, it is of smaller magnitude than would bring it within our catalogues and charts.

Of great value in similar cases, in the future, will be the plates of the International Star Chart, which begins its existence this year. Such a photographic record, like the partial ones already made at the Cape Observatory and at the Harvard Observatory, will enable us to put back at will the dial of time, and to re-observe the heavens as they appeared when the plates were taken.

The absence of any previous record of the new star of the present year is not necessarily to be regarded as a proof that it did not exist as a star emitting light. Visibility and invisibility in our largest instruments are but expressions in terms of the power of the eye. The photographic plate, untiring in its power of accumulation, has brought to our knowledge multitudes of stars which shine, but not for us. The energy of their radiation is too small to set up the changes in the retina upon which vision depends.

A striking illustration is presented by plates taken of the neighbourhood of η Argus by Mr. Russel at Sydney, and later by Dr. Gill at the Cape. In these photographs a crowd of stars reveal themselves for the first time, which have hitherto shone in vain for the dull eye of man.

It is not improbable that the new star in Auriga did exist as a very faint star; but what were the conditions under which it woke up into sudden splendour? Such information as is forthcoming has been gained chiefly from that particular application of the spectroscope by which we can measure motion in the line of sight. It is not too much to say that this method of observation has opened for us in the heavens a door through which we can look upon the internal motions of binary and multiple systems of stars, which otherwise must have remained for ever concealed from us.

With every increase of telescopic aperture more stars are resolved into double or multiple systems, but no conceivable progress in instrument-making could have put it in our power, as the spectroscope does, to discover within the point-like image of a star, in many cases, a complex system of whirling suns, gigantic in size, and revolving with enormous speed, close about each other. An object-glass, as large in diameter as this theatre, if it could be constructed, would fail to show close systems of stars which the prism easily lays open to our view.

It is as many as twenty-three years ago since I had the honour of describing in this place the first successful application of this mode of using the spectroscope to the heavenly bodies. The method is now too well known for me to say more than that the change of wave-length or pitch of the light shows itself by a shift of the lines in the spectrum; towards the blue for an approach, towards the red for a recession between the light-source and the observer. It is obvious that the prism can take note only of the motions which are precisely in the line of sight. The stars, as seen from the earth, are moving in all directions; the spectroscope selects out of the star's motion, whatever it may be, that part only which is in the line of sight. It is of this component only of the complete motion that we can gain information directly by the spectroscope.

My original observations of the motion of Sirius were made in 1868, and of other stars in the following years, but the advance since then, and especially in recent years, in the improvements of instruments and in the use of the sensitive gelatine plate, has made a much higher degree of accuracy in the determination of motions attainable now, than was then possible. To Prof. Vogel is due the working out of a photographic method by which he has now determined the motions in the line of sight of more than fifty stars.*

This method is applicable not only to the drift of star-systems, but what is of more immediate interest in connection with the new star, to the internal motions within those systems. The simplest case of such systems is where one body only is bright enough to produce a spectrum. Unless the plane of the orbit is across the line of sight the star will have alternate periods of approach and of recession, and the lines in its spectrum will be seen to swing backwards and forwards relatively to a terrestrial line of the same substance in times corresponding to the star's orbital period. A grand example of this state of things was revealed by the discovery at Potsdam of the orbital motion of the bright star of Algol showing that the variation of its light is caused by its being partially eclipsed at intervals by a dusky companion star, the existence and motions of which were thus brought to light.

---

* Photographs of the spectrum of Sirius compared with that of iron, and photographs of the spectra of other stars, showing motions in the line of sight taken at Potsdam were thrown upon the screen.

I wish to express my great obligations to Professor Vogel, Professor Pickering, Professor Holden, M. Deslandres, MM. Henry, Dr. Belopolsky, Dr. Roberts, and Father Sidgreaves, for photographs of star-motions and of the New Star, and its spectrum, many of which were specially prepared for this lecture. The photographs not suitable for throwing upon the screen were exhibited in the Library.

If the plane of the star-system is inclined to the line of sight, the dark body might pass above or below the bright one as seen from the earth, and not eclipse it. Vogel had the good fortune to discover such a system in Spica, which he showed to consist of a pair of great suns, one bright and the other dark, or nearly so, whirling round their common centre of gravity in about four days.

If, however, in a binary system both stars are bright, the minute stellar point formed in the telescope will contain the light of both stars; and its spectrum will be a compound one, the spectrum of one bright star being superposed upon that of the other. If the spectra are identical, all the lines will be really double, though apparently single when the stars have no relative motion; and will open and close in periods depending upon the stars' motions.

Such a system was first made known to us spectroscopically by Prof. Pickering from his photographs of Mizar, which consists of a pair of gigantic blazing suns, equal together to forty times the sun's mass and whirling round their common centre of gravity with the speed of about 50 miles a second. Then followed at Harvard the discovery in β Auriga of an order of close binary stars hitherto unknown. In Fig. 2 of Plate I. are reproduced the original photographs showing the duplication every second day of the lines in the spectra of this double star; the doubling is well seen in K, which is very narrow in this star.

Now it is to this method of spectroscopic observation that we are indebted for the revelation of the remarkable state of things existing in the new star. I may remark, in passing, that it is not a little surprising that a new star as bright as the fifth magnitude should have burst out almost directly overhead in the heavens, and yet have remained undiscovered for nearly seven weeks. Europe and the United States bristle every clear night with telescopes pointed from open observatories, which are served by an army of astronomers; and yet the honour of the discovery of the new star is due to an amateur, Mr. Anderson, possessed only of a small pocket-telescope and a star-chart. Happily the days are not over when discoveries can be made without an armoury of instruments.

As soon as the news reached Cambridge, U.S., Prof. Pickering, by means of photographs which had been taken there, was able to cause the part of the sky where the new star appeared, to pass again under his examination, precisely as it had appeared at successive intervals during the last six years; but the new star's place had remained unoccupied all that time by any star so bright as of the eleventh magnitude.

For about a year a still closer watch has been kept upon the sky at Cambridge by means of a photographic transit instrument driven by clockwork, which automatically patrols the sky every clear night, and registers upon one plate all stars as bright as of the sixth magnitude, within a great zone 60° in breadth, and three hours of Right Ascension in length. On December 1 the Nova did not appear upon

Fig. 1.

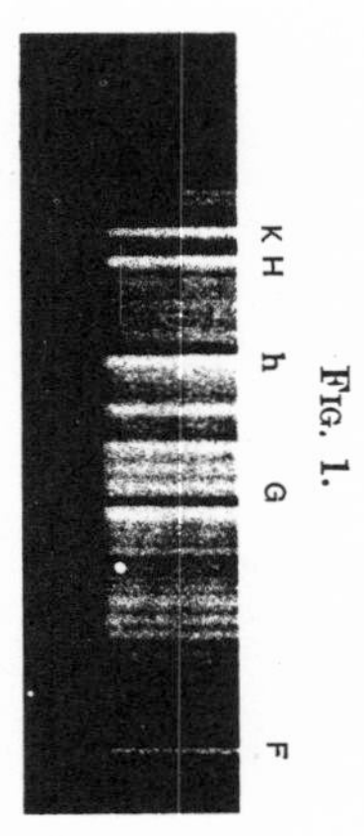

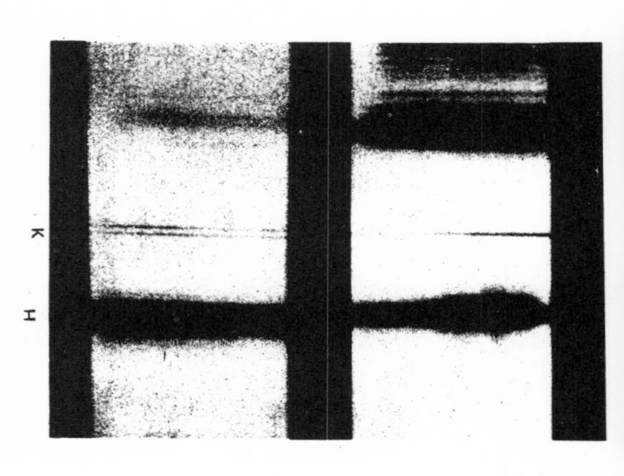

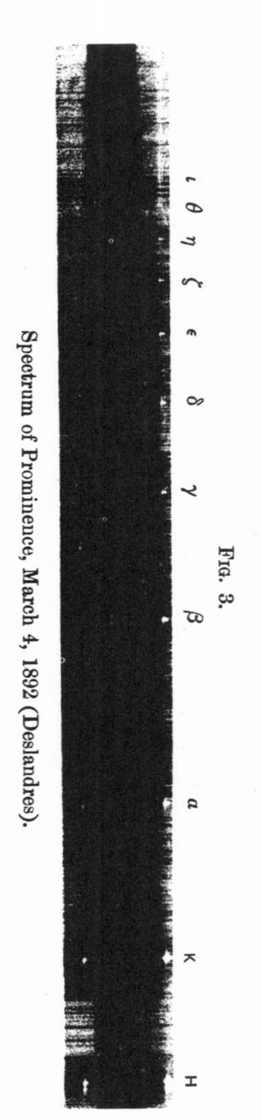

Fig. 3.

Spectrum of Prominence, March 4, 1892 (Deslandres).

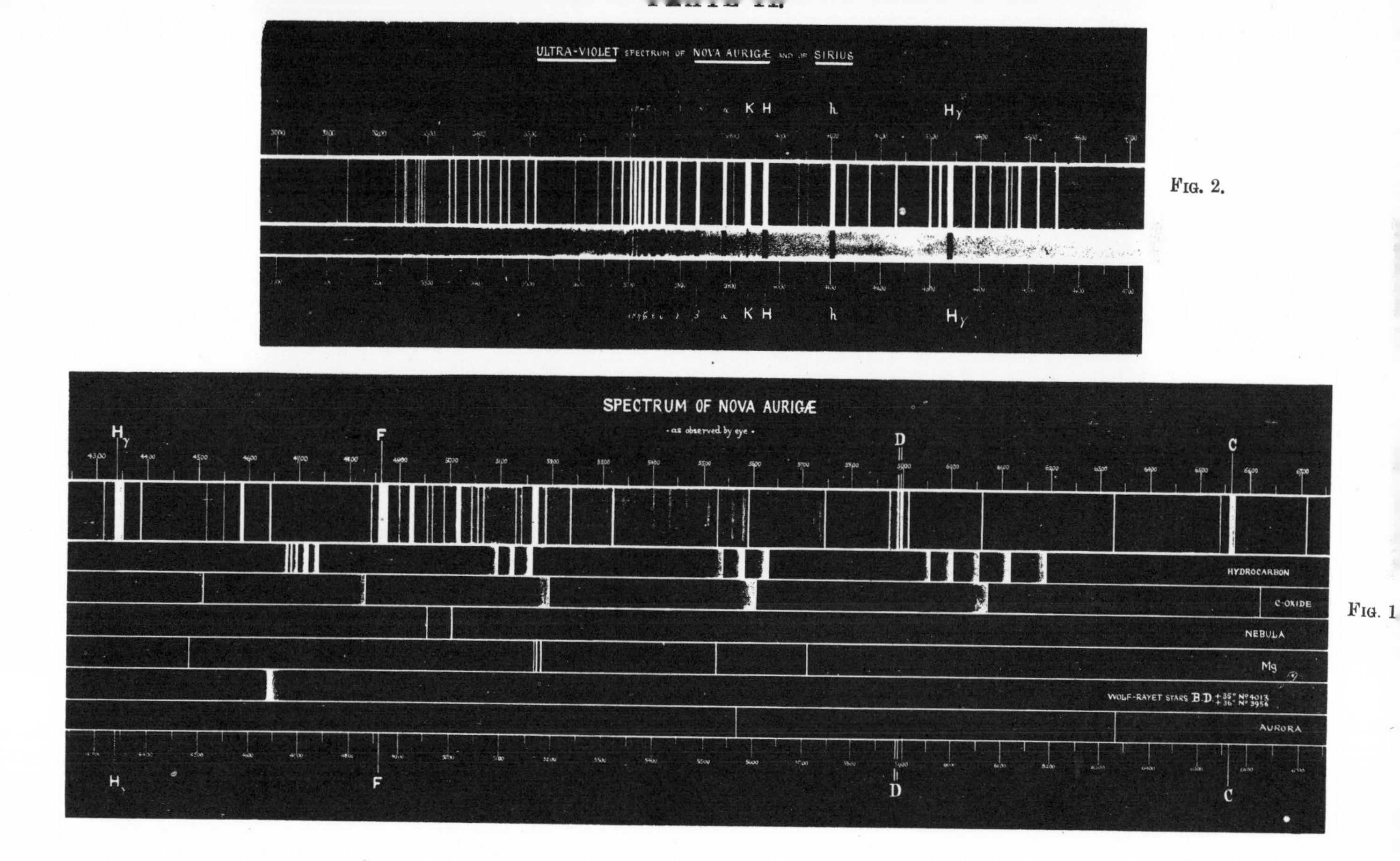
ULTRA-VIOLET SPECTRUM OF NOVA AURIGÆ AND OF SIRIUS
K H
h
Hγ
SPECTRUM OF NOVA AURIGÆ
· as observed by eye ·
Hγ
F
D
C
HYDROCARBON
C-OXIDE
NEBULA
Mg
WOLF-RAYET STARS B.D. +35° No 4013 +36° No 3956
AURORA

Fig. 2.

Fig. 1

the plate, but the next night that was clear, December 10th, the Nova is recorded as of the fifth magnitude. Most fortunately, on December 8, Dr. Max Wolf photographed this part of the constellation of Auriga, but no star so bright as of the ninth magnitude was to be found where the Nova afterwards appeared.

The new star must therefore have sprung up to the brightness of the fifth magnitude between the 8th and the 10th of December last.

At that early time the Nova was not nebulous on Prof. Pickering's plates. The question has been raised since whether the star was surrounded by a faint nebula. To us, in our observations, it appeared like an ordinary star; but the point may be considered set at rest by photographs taken by Mr. Roberts, which by his kindness I am able to throw upon the screen. With an exposure of over three hours, there is only the usual small fringe of nebulosity due to our atmosphere, and which is present also about the other stars on the plate. How searching a test for faint nebulous matter is so long an exposure, is strikingly shown by contrasting a short exposure plate of the Pleiades, where nebula do exist, with a photograph on which the light action has been prolonged for nearly four hours. The Nova is free from nebulosity in photographs which have been sent to me by the Brothers Henry, and by Prof. Holden of the Lick Observatory.

The changes of magnitude of the Nova as shown by photographs taken at Greenwich, from eye-observations at Prof. Pritchard's observatory, and by Mr. Stone and by Mr. Knott, are recorded in the diagram on the wall. These observations show that notwithstanding continual fluctuations a slow but steady decline had set in, carrying the light of the star from nearly the fourth and one-half magnitude down to the sixth magnitude by the early days of March; but after March 7th, these swayings to and fro of its light, set up probably by commotions attendant on the causes of the star's outburst, calmed down, and the light fell rapidly and with regularity to about the eleventh magnitude by March 24th, and then down to 14·5th magnitude by April 1st. On April 26th, however, it was still visible at Harvard observatory as a star of the 14·5th magnitude.

We commenced our observations on February 2nd. The spectrum of the star in the visible region is represented in Fig. 1 of Plate II. Below the star's spectrum are placed the terrestrial spectra with which it was directly compared. The spectrum showed a brilliant array of bright lines, among which four in the green were very conspicuous.

The brightest of these we recognised as the second line of hydrogen, and passing the eye to the red, we saw blazing the first line of hydrogen at C; the blue line near G was also visible. But a remarkable phenomenon presented itself in that each bright line seemed to cast a shadow, for on the blue side of each was a narrow space of intense blackness. When we threw into the spectroscope for comparison the bright lines of hydrogen, the secret of this unusual appearance was revealed. The hydrogen line did not fall upon the middle of the F line, but upon one side. We had before us a magnificent example, on a grand scale, of motions in the line of sight—two mighty streams of hydrogen fleeing from each other; the hotter one, emitting the bright lines, going from us; the cooler producing the dark shadows by absorption, coming towards us, indicating a relative velocity of about 550 miles a second.

Direct comparisons of the bright line near the position of the chief nebular line, with lines of nitrogen and lead, showed that the stellar line was less refrangible than the principal nebular line. The second nebular line was not present in the star. The spectrum of the Nova showed, therefore, no relationship with the well-known spectrum of the bright-lined nebulæ.

A similar want of relationship of the spectrum of the new star with the usual hydro-carbon spectrum of comets was shown by direct comparison with the Bunsen flame. The bright line near *b* differs in position and in character from the beginning of the brightest band of the Bunsen flame spectrum, and no bright lines were found in the star at the positions of the other bright bands of this spectrum.

This bright line in the star falls very near the magnesium triplet at *b*, but a careful comparison of the spark spectrum of magnesium leaves little doubt that it does not owe its origin to this substance.

The sodium line at D is bright in the spectrum of the star, in which appears also a thin bright line at about the position of D_3. The continuous spectrum extended, when the star was brightest, below C in the red, and as far into the blue beyond G as the eye could follow it. The spectrum in Fig. 2 of Plate II. is from a photograph of the spectrum of the Nova which we took on February 22nd, using a mirror of speculum metal and a spectroscope with a prism of Iceland spar and lenses of quartz, so that the extreme violet part of the star's light was not cut off by passing through glass. The brilliant lines followed by absorptions, and the fainter continuous spectrum were found to extend upon the plate nearly as far as the light of Sirius, and not far short of the place where our atmosphere cuts off all celestial light. A photograph of the spectrum of Sirius showing the group of lines near the end of the spectrum has been added for comparison. In the star the whole range of the hydrogen lines, including the ultra-violet series and the calcium lines H and K, were bright, each accompanied on the blue side by a dark absorption band. In this respect, as well as in the positions of the principal bright lines in the visible region, the Nova suggested a state of things not unlike what we find in the erupted matter at the solar surface.

M. Deslandres permits me to reproduce in Fig. 3, Plate I., the photograph of a remarkable prominence taken on March 4th, 1892, in which are reversed not only H and K, and the known hydrogen series, but three additional members are to be seen at the more refrangible end, the positions of which M. Deslandres informs me fall into Balmer's formula for the hydrogen series.*

The resemblance of the spectrum of the Nova to that of the erupted solar surface is further shown in a remarkable feature of great significance in the character of the hydrogen lines both bright and dark. On February 2nd we noticed that the F line was not of uniform brightness throughout its breadth. We soon came to the conclusion that it was divided, not quite symmetrically, by a very narrow dark line. The more refrangible component was brighter, and rather broader than the other. Later on in February, we were sure that small alterations were taking place in this line, and that the component on the blue side no longer maintained its superiority. We suspected, indeed, at times that the line was triple, and towards the end of February and in the beginning of March we had no longer any doubt that it was occasionally divided into three bright lines by the incoming of two very narrow dark lines.

Similar alterations, giving a more or less apparent multiple character to the lines, are to be seen not only in the bright lines, but also in those of absorption in contemporary photographs taken of the spectrum of the star. I may mention those taken at Potsdam, Stonyhurst, and the Lick Observatory. These changes were specially watched and measured by M. Bélopolsky at Pulkova.

Prof. Pickering informs me that on a photograph taken at Cambridge, U.S., on February 27th, H, K, and α are triple, and that Miss Maury recorded, "the dark hydrogen lines rendered double, and sometimes triple, by the appearance of fine bright threads superposed upon the dark bands."

Now, when on the sun's surface, or in the laboratory, portions of the same gas at different temperatures come in before each other, the cooler gas may cause a narrow absorption line to form upon a broader bright line, and thus impart to it the appearance of a double line; or in the case of hotter gas, a narrow bright line upon a dark line. Prof. Liveing and your distinguished Professor of Chemistry, Prof. Dewar—whose researches with the electric arc-crucible have made them specially familiar with the ever-changing guises and disguises of this protean phenomenon of reversal—have recorded cases not only of double reversals, giving apparent triplicity to single bands, but also of threefold reversals. The phenomenon of the unsymmetrical division of the bright and dark lines which was occasionally seen in the Nova frequently presents itself in the laboratory from the unequal expansion on the two sides of the line on which the reversed line falls; and at the solar surface from the relative motions in the line of sight of the hotter and cooler portions of the gas taking part in the phenomenon. Unless we accept this obvious interpretation of the apparent multiple character of the stellar lines, we should have

---

* M. Deslandres has detected since two more lines, thus adding five new lines to the hydrogen series.

to assume a system of at least six bodies, all moving with different velocities.

In Fig. 1, Plate I., is reproduced a photograph of the blue part of the spectrum taken at Harvard Observatory with a prism placed over the object-glass. The dark absorption lines on the blue side of the bright lines are well shown.

It is of great importance to state that the waning of the star was not accompanied by any material change of its spectrum, but only of such apparent changes as might well come in when parts of an object differ greatly in brightness. On March 24th, when the star's light had fallen so low as to nearly the eleventh magnitude, we could still glimpse the faint continuous spectrum, upon which the remarkable quartet of bright lines still shone out without any great change of relative intensity. Prof. Pickering informs me that on his plates the principal lines in the photographic part of the spectrum "faded in the order K, H, α, F, h, G, the latter becoming brighter as the star was faint." Omitting the calcium lines H and K, which varied, the order of disappearance agrees with that of the sensitiveness of the plate for these parts of the spectrum, and is in accordance with the view that the star's spectrum remained without material alteration through this great range of magnitude.

How are we to account for the appearance and doings of this new star, or rather stars? For, as we have seen, the great shifts in the spectrum of the bright and dark lines, the bright to the red, and the dark to the blue, appear to show two bodies having relative motion in the line of sight of about 550 miles a second. Now, during the whole time, some seven weeks, that the star was under observation, this relative velocity was maintained without any great alteration, though it is probable that small changes, beyond the reach of our instruments, took place.

A reasonable explanation of these phenomena may perhaps be found if we venture to assume, though with considerable hesitation, as the subject is obscure, two gaseous bodies, or bodies with gaseous atmospheres, moving away from each other after a near approach, in parabolic or hyperbolic orbits, with our sun nearly in the axis of the orbits; the components of the motions of the two bodies in the line of sight, after they had swung round, might well be as rapid as those observed in the new star, and might continue for as long a time without any great change of relative velocity. Unfortunately, information as to the motions of the bodies at the critical time is wanting, for the event through which the star became suddenly bright had been over for some forty days before any observations were made with the spectroscope.

Analogy from the variable stars of long period would suggest the view that the near approach of the two bodies may have been of the nature of a periodical disturbance, arising at long intervals in a complex system of bodies. Chandler has recently shown in the case of Algol that the minor irregularities in the variation of its light are probably caused by the presence of one or more bodies in the system, besides the bright star and the dusky one which partially eclipses it. To a similar cause are probably due the minor irregularities which form so prominent a feature in the waxing and waning of the variable stars as a class. We know that the stellar orbits are usually very eccentric. In the case of γ Virginis the eccentricity is as great as 0·9, and Auwers has recently found the very considerable eccentricity of 0·63 for Sirius.

The great relative velocity of the component stars of the Nova, however, seems to force us to look rather to the casual near approach of bodies possessing previously considerable motion, unless we are willing to concede to them a mass very great as compared with that of our sun. Such a near approach of two bodies of great size is very greatly less improbable than would be their actual collision. The phenomena of the new star scarcely permit us to suppose even a partial collision; though if the bodies were very diffuse, or the approach close enough, there may have been possibly some mutual interpenetration and mingling of the rarer gases near their boundaries.

A more reasonable explanation of the phenomena, however, may be found in a view put forward many years ago by Klinkerfues, and recently developed by Wilsing, that under such circumstances of near approach enormous disturbances of a tidal nature would be set up, amounting it may well be to partial deformation in the case of gaseous bodies, and producing sufficiently great changes of pressure in the interior of the bodies to give rise to enormous eruptions of the hotter matter from within, immensely greater, but similar in kind, to solar eruptions; and accompanied probably by large electrical disturbances.

In such a state of things we should have conditions so favourable for the production of reversals undergoing continual change, similar to those exhibited by the bright and dark lines of the Nova, that we could not suppose them to be absent; while the integration of the light from all parts of the disturbed surfaces of the bodies would give breadth to the lines, and might account for the varying inequalities of brightness at the two sides of the lines.

The source of the light of the continuous spectrum upon which were seen the dark lines of absorption shifted towards the blue, must have remained, as seen by us, behind the cooler absorbing gas, so as to form a background to it; indeed, must have formed with it the body which was approaching us, unless we assume that both bodies were moving exactly in the line of sight, or that the absorbing gas was of enormous extent.

The circumstance that the receding body emitted bright lines, while the one approaching us gave a continuous spectrum with broad absorption lines similar to a white star, may, perhaps, be accounted for by the two bodies being in different evolutionary stages, and consequently differing in diffuseness and in temperature. Indeed, in the variable star β Lyræ, we have probably a binary system, of which one component gives bright lines, and the other dark lines of absorption. We must, however, assume a similar chemical nature for both bodies, and that they existed under conditions sufficiently similar for equivalent dark and bright lines to appear in their respective spectra.

We have no knowledge of the distance of the Nova, but the assumption is not an improbable one that its distance may be of the same order of greatness as that of the Nova of 1876, for which Sir Robert Ball failed to detect any parallax. In this case, the light-emission suddenly set up, certainly within two days and possibly within a few hours, was probably much greater than that of our sun; yet within some fifty days after it had been discovered, at the end of January, its light fell to about 1/300th part, and in some three months to nearly the 1/10,000th part. As long as its spectrum could be observed the chief lines remained without material alteration of relative brightness. Under what conditions could we suppose the sun to cool down sufficiently for its light to decrease to a similar extent in so short a time, and unaccompanied with the incoming of very material changes in its spectrum. It is scarcely conceivable that we can have to do with the conversion of gravitational energy into light and heat. On the theory we have ventured to suggest, the rapid calming down, after some swayings to and fro of the tidal disturbances, and the closing in again of the outer and cooler gases, together with the want of transparency which might come in under such circumstances, as the bodies separated; might account reasonably for the very rapid and at first curiously fluctuating waning of the Nova, and also for the observed absence of change in its spectrum.

I may, perhaps, be permitted to remark that the view suggested by Dr. William Allen Miller and myself, in the case of the Nova of 1866, was essentially similar, in so far as we ascribed it to erupted gases. The great suddenness of the outburst of that star, within a few hours probably, and the rapid waning from the 3·6 magnitude to the 8·1 magnitude in nine days, induced us to throw out the additional suggestion that possibly chemical actions between the erupted gases and the outer atmosphere of the star may have contributed to its sudden and transient splendour, a view which, though not impossible, I should not now, with our present knowledge of the light changes of stars, be disposed to suggest.

The subject is necessarily obscure, but we must not on this account feebly relinquish the hope of conquest. The words of a great Seer may well be taken as the watchword of the Astronomer:—

. . . "Fervent love,
And lively hope, with violence assail
The kingdom of the heav'ns, and overcome"

* * *

[W. H.]

Friday, June 9, 1893.

Hugo Müller, Esq. Ph.D. F.R.S. Vice-President, in the Chair.

Professor T. E. Thorpe, D.Sc. F.R.S. *M.R.I.*

*The Recent Solar Eclipse.**

Most people who take any interest in those larger problems with which men of science are nowadays concerned are aware that there are certain questions relating to the chemistry and physics of the sun which, at present, can only be solved by observations made during the fleeting moments of the total phase of a solar eclipse. Thanks to the action of the *Nautical Almanac* office in this country, and of similar institutions in other countries, we have not only ample warning of the advent of an eclipse of the sun, but we are furnished with such details concerning the time of its occurrence, the direction of the path of the moon's shadow on the earth, and the duration of the various phases, that we are enabled to decide whether it is expedient to attempt to seize the precious seconds during which the sun is obscured by the moon, in order to get further light on those questions which, as has been said, can only be at present solved, or at least studied, at such times.

During the eclipse of last April the moon's shadow swept over a considerable expanse of land. It touched the coast of Chili in latitude 29° S. at about 8.15 A.M. of local time, passed over the highlands of that country, across the borders of Argentina and Paraguay, and over the vast plains and forests of Central Brazil, emerging, at about noon of local time, at a short distance to the north-west of Ceara on the North Atlantic seaboard. Crossing the Atlantic, at about its narrowest part, it struck the coast of Africa north of the river Gambia, and finally disappeared somewhere in the Sahara. It would seem, therefore, there was ample choice in the selection of stations. But all situations were not equally good or equally available. There were, indeed, special reasons why every effort should be put forth to observe this eclipse as completely as possible. To begin with, it had an unusually long totality—upwards of four minutes at places at or near the central line of the shadow. Next, it occurred at about a period of maximum of solar energy, and hence we had an

* A full report of the discourse is given in the 'Fortnightly Review,' July 1893.

opportunity of solving certain questions as to the connection between the character of the corona and the solar cycle. Further, it was hoped that by multiplying the stations along the path of the eclipse, and therefore by making observations at considerable intervals of time, the photographic records might decide upon the possibility of changes in the form and internal disposition of the corona—a question of the greatest importance in regard to the physical nature of this solar appendage.

After careful consideration of sites, and of the various suggestions which were made as to the nature of the work to be undertaken, a committee, representing the Royal Society, the Royal Astronomical Society, and the Solar Physics Committee of the Science and Art Department, decided to send two observers to Para Curu, in the province of Ceara in Brazil; and four observers to some station in Senegal, preferably Fundium, on the Salum river. Substantially the same scheme of work was arranged for the two parties. Spectroscopic observations with the Prismatic Camera and a series of photographs with what is now known as the Duplex Coronagraph were to be taken at each station. In the case of the African station, it was further decided that photometric measurements of the coronal light should be made by the method adopted by Captain Abney and myself on the occasion of the West Indian eclipse of 1886.

It will be understood that the work of both parties was entirely confined to the study of the corona. In the first place, photographic records of its form, its extension and internal structure were to be made according to a uniform plan at both stations. The apparatus to be used consisted of a sort of double camera, in one compartment of which was placed a 4-inch lens of 60 inches focus, belonging to Captain Abney, which has already seen much service in eclipse photography. It was employed in Egypt in 1882, in the Caroline Islands in 1883, in the West Indies in 1886, and in the Salut Isles, in French Guiana, where that veteran eclipse observer, Father Perry, lost his life, in 1889. One special reason for using this lens was that the continuity of the series of photographs which have been obtained by it might be maintained. It gives pictures on the scale of about half an inch to the moon's diameter. In the other compartment was a 4-inch Dallmeyer photo-heliograph lens mounted in combination with a 2½-inch Dallmeyer negative lens of 8-inch negative focus, giving with the total length of 68 inches pictures on the scale of over 1½ inch to the moon's diameter. This double camera was fitted with special plate-carriers, enabling two plates to be exposed at the same time, one to each lens, so that by one operation of changing and exposing, two pictures of the eclipsed sun could be simultaneously obtained. The times of exposure were so arranged that the longest exposed picture with the enlarging combination should receive the same photographic effect as the shortest exposed picture with the Abney lens. The whole arrangement was equatorially mounted, so that the plates were kept in a constant position with respect to the sun during the times of exposure.

Three different methods were employed to obtain photographic records of the spectrum of the corona. In the first, which was suggested by Professor Norman Lockyer more than twenty years ago, the eclipsed sun was to be photographed through a prism attached to a telescope of 6-inch aperture. In this manner an image of the corona would be obtained corresponding with each kind of light emitted by it. Thus, if the corona consisted entirely of glowing hydrogen, there would be an image in the position occupied by each of the lines in the hydrogen spectrum. If, as may be expected, the materials composing the corona are different in different regions, the images obtained will not exactly resemble each other, but the form of each image will depend upon the distribution of that particular spectral line through the corona. The complete spectrum of every part of the corona which is bright enough to be photographed will, therefore, be obtained with a single exposure.

The other method of studying the spectrum of the corona is by means of the ordinary slit spectroscopes. The arrangement, employed by Captain Hills, consisted of two spectroscopes, each provided with a condensing lens and camera, mounted on an equatorial stand. The spectroscopes were of different dispersive power, one having two prisms, and the other one. The slits were placed parallel to each other, and were so arranged as to cut across opposite limbs of the sun at right angles to the sun's equator. An image of the sun is thrown on the slit by the condensing lens, and the slit is long enough to cover the whole width of the corona. The resulting photographs ought then to show at least three different spectra: a continuous spectrum over the dark body of the moon, on either side of which will appear the prominence spectrum, and outside of which again will be the true corona spectrum, which may or may not be broken up into bands by the occurrence of rifts or dark spaces in the corona. This method has the great advantage of discriminating between the different spectra of every portion of the corona along the line of the slit; the main difficulty connected with it is the want of light, which makes it almost impossible to give a sufficient exposure unless the slit is opened rather wide. It was decided, therefore, to make only one exposure with each spectroscope; this was to last as nearly as possible the whole time of totality, and the most rapid photographic plates procurable were to be used.

The measurement of the visual brightness of the coronal light was to be effected by the following arrangement. An image of the corona is accurately focused on a white screen by means of an equatorial of 6-inch aperture and 78-inch focal length, and the intensity of the light from different portions of the corona at definite distances from the limb is compared with that of a standard glow-lamp by means of an arrangement constructed on the principle of the Bunsen photometer, the light from the standard glow-lamp being varied by introducing a variable amount of resistance into the current and measuring the current strength at the moment of comparison.

In order to ascertain the total intensity of the coronal light Mr. Forbes employed a similar contrivance, his screen, however, having only one large translucent spot or disc, as in the ordinary Bunsen photometer. Concurrently with these observations it was arranged that the photographic intensity of the coronal light, as distinct from the visual intensity, should also be measured by a method devised by Captain Abney, which consists in impressing standard intensity scales along the edges of the photographic plates to be exposed in the coronagraph, these being developed at the same time as the coronal pictures. The photographic plates to be used in the split spectroscope were also provided, in like manner, with standard scales, with a view of measuring the comparative luminosity of different portions of the coronal spectrum, a point which has an important bearing on the question of the possibility of photographing the corona in ordinary sunlight.

On the day preceding that of the eclipse the French gunboat *Brandon* came up the river, bringing with her the Governor of Senegal. His Excellency M. de la Mothe, together with the administrator of the district, M. Allys, to whom the expedition is indebted for many courtesies, paid a visit to the English camp at Fundium and witnessed the final rehearsal of our operations. They arranged for a guard to protect the enclosure during the time of the eclipse, and gave orders that all chanting, screaming, or beating of tom-toms in the village was to be forbidden.

On Sunday, the 16th, the day of the eclipse, although the morning was bright and clear, the effects of the comparatively moist winds from the sea were to be seen in the changed colour of the sky and the prevalence of thin haze. Still the sky was almost cloudless, save for a few thin wispy cirri, which floated almost motionless near the horizon. A gentle air from the west made scarce a ripple on the yellow waters of the Salum. As the day advanced, the sky became even lighter in colour, and there was a perceptible haze in the neighbourhood of the sun; the wind almost died away, and everything betokened that we should have to face—as indeed we fervently hoped might be the case—the pitiless glare of that fiercest of all suns—the African sun at noon. At 12.30 our party went ashore, the huts were uncovered, the equatorials adjusted, clocks wound, and the instruments set running on the sun. Shortly before 2 P.M. the officers and men from the *Alecto*, bringing their lanterns, came to the camp and took up their several positions. As the light waned there was a distinct feeling of chilliness in the air, and the wind suddenly rose in short sharp gusts. The few natives who had congregated round the stockade began to show signs of trepidation, but no sound of distress or fear was heard save the plaintive cry of a tethered goat near the administrator's house. There was a great hush as the last gleam of sunlight died away. The corona seemed almost to flash into existence, so suddenly did its light grow in intensity. Faint indications of its appearance could, indeed, be perceived on the photometer screen some seconds before the last trace of the yellow crescent disappeared. The phenomenon known as "Baily's beads" was plainly visible. The lower corona was wonderfully bright, and a whole row of prominences started into view. The panaches, sheafs, and other evidences of "structure" were distinctly marked on the white screen. The general sky illumination was so great that only some five or six stars were visible. The gloom, indeed, was nothing like so intense as I had seen in previous eclipses, and there would have been little difficulty in reading the second-hand of the chronometer or the scales of the ammeters without the aid of the lighted lanterns. And now the oft-repeated programme was being gone through for the last time, with a quickened sense and a concentrated earnestness springing from the consciousness that the veritable four minutes—the 240 and odd seconds—on which our thoughts for months past had been dwelling, were now speeding away, and that, with the first rush of sunlight on the other side of the black disc of the moon, our opportunities would be gone for ever. The silence was most impressive; it was broken by the stentorian voice of the quartermaster as he told us at intervals, by the aid of his log-glass, the number of seconds that still remained to us. Now and again, too, one heard from the adjoining huts the command to expose, and the sharp click of the carriers as slide after slide was inserted and withdrawn. Thanks to the repeated drills, everything went with the smoothness and regularity of clockwork. There was no hitch or stoppage, and no undue haste on the part of anybody. Sergeant Kearney secured ten out of the twelve corona pictures that he had been instructed to make. Mr. Fowler, in all, made thirty exposures in the prismatic camera, including a number taken during the five minutes before and after totality; and Captain Hills obtained both his slit-spectroscope photographs. Mr. Gray and I made twenty photometric measurements of the light from different parts of the corona, and Mr. Forbes obtained eleven concordant observations of its total intensity. The full measure of our success was not yet known to us, but every man had the certain knowledge that he had secured enough to make the eclipse of April 16, 1893, take its place as one of the best observed eclipses of recent times, and that his work, done at the sacrifice of much personal comfort, and under the trying circumstances of a fierce temperature and an unhealthy climate, would contribute towards the solution of one of the most profoundly interesting of all physical problems.

After a short rest the command, "Down huts," was given, and in a few hours the *Alecto*, with all our cases once more packed and safely stowed, was groping her way amongst the shallows and banks of the Salum down to the sea. The memory of our green-canvassed structures and of the strange instruments of brass and iron with which we English sought to shoot the moon for trying to eat up the sun has now doubtless become one of the traditions of the Wolofs and Sereres of Fundium.

M. Deslandres, I am happy to say, was not less successful. In a communication which he has just made to the French Academy he gives a brief account of some of the main results he has gathered from the photographs which he was able to take. His instrumental equipment enabled him to obtain photographs of the corona, to study its spectrum, to examine the coronal light in the most refrangible part of the ultra-violet, and to measure the rotation of the corona by the method of displacement of the lines in its spectrum. His coronal photographs showed luminous jets of a length equal to twice the diameter of the sun, while the general form was similar to that usually observed at times of maximum sun-spot frequency. The spectrum photographs have revealed the existence of at least fifteen new coronal and chromospheric lines. But the most novel of M. Deslandres's observations relate to the rotation of the corona. His negatives showed the spectra of two points on exactly opposite sides of the corona, situated in the equatorial plane of the sun, at a distance equal to two-thirds of his diameter. The lines in the spectra indicated large displacements, and from the measurements, M. Deslandres concludes that the corona must travel nearly with the disc in its motion, and thus be subject to its periodical rotational movement.

M. Bigourdan, who had been stationed at Joal, on the coast of Senegal, since December last, for the purpose of observing southern nebulæ and making pendulum observations, was commissioned by the Bureau des Longitudes to search during the eclipse for the inter-mercurial planet which Leverrier assumed to exist, and which he named Vulcan. M. Bigourdan was also requested to make careful determinations of all the four contacts, with a view of obtaining additional data for correcting the tables of the motion of the sun and moon.

As regards Vulcan, M. Bigourdan was not more successful than his predecessors, but he determined with great accuracy the time of the total phase at Joal, which he found to be 4 min. 1 sec. My own observations at Fundium, which is about as much to the south of the probable central line as Joal is to the north, gave 4 min. 3 sec. as the time of totality, which is in very fair accord with M. Bigourdan's determination. M. Coculesco, a young Roumanian astronomer, who volunteered to accompany M. Deslandres to Fundium, found 4 min. 11 sec.

As yet we have only meagre information of the results obtained by other observers. In spite of the many chances against them, Mr. Taylor and Mr. Shackleton were successful at Para Curu.

Although large portions of the sky were covered with cumuli, the sun was not clouded over at the period of totality; the atmosphere, of course, was nearly saturated with aqueous vapour, but no haze or precipitation of moisture seems to have occurred, and in consequence of the remarkable transparency of the air the photographs are certain to be of exceptional interest.

The Americans, who were mainly stationed in Chili, were equally fortunate. At Minas Aris, the Harvard College station, the atmospheric conditions are said to have been all that could have been wished for; there was no passing cloud or haze to mar the observations. The corona is reported by Professor Pickering to resemble that of 1857, as portrayed by Liais, and that of 1871, as observed by Captain Tupman. There were four streamers, two of which had a length exceeding the sun's radius, or stretching out more than 435,000 miles. Several dark rifts were visible, extending outwards from the moon's limb to the utmost limit of the corona. No rapid movement was observed within the streamers. The moon appeared of almost inky blackness, while from behind it, streamed out on all sides radiant filaments, beams, and sheets of pearly light. The inner corona was of dazzling brightness, but still more dazzling were the eruptive prominences which blazed through it, to use the words of Professor Young, like carbuncles. Generally, the inner corona had a uniform altitude, forming a ring of four minutes of arc in width, but separated with more or less definiteness from the outer corona, which projected to a far greater distance, and was much more irregular in shape. The outer corona seems to have been much larger than in 1879 or 1889, as, indeed, might have been expected at a period of maximum solar energy. The party seems to have been successful in photographing for the first time the "reversing layer" of the solar atmosphere.

Professor Schaeberle, from the Lick Observatory, who observed at Mina Bronces, in the Desert of Atacama, reports that the corona was similar to that of 1883. He obtained in all fifty photographs, eight of which are ten by twenty inches in size, and one of which shows an image of the sun four inches in diameter, the corona covering a plate eighteen by twenty-two inches—a truly "record" result. The photographs are said to afford strong presumptive evidence of the truth of the mechanical theory of the corona which is associated with Professor Schaeberle's name.

I cannot close without some reference to the debt of gratitude we are under to Captain Lang and his officers, for the readiness, zeal, and intelligence with which they co-operated in our work. Indeed, the whole crew of the gunboat did all in their power, often under circumstances of no little personal hardship, to minister to our success, and to contribute to our comfort. The best-laid schemes of astronomers, as of other men, "gang aft a-gley." There is a spanner to make, or a bit of soldering to be done, or a piece of woodwork to be altered. Assistance of this kind was always most cheerfully and promptly rendered. Lastly, it remains to be said, the recollection of the hospitality of H.M.S. *Alecto* and of H.M.S. *Blonde*, which took us away from the fever-stricken coast, will ever remain one of the pleasantest of the associations connected with the successful expedition of the African eclipse party.

[T. E. T.]

Friday, May 25, 1894.

William Huggins, Esq. D.C.L. LL.D. F.R.S. Vice-President, in the Chair.

Sir Howard Grubb, F.R.S. F.R.A.S.

The Development of the Astronomical Telescope.

Since I last had the honour of lecturing in this theatre, astronomical research has had opened up to it a totally new field of work, one which appears almost without limit in its scope.

When Dr. Gill, of the Cape of Good Hope Observatory, made arrangements to photograph the comet of 1882 with a long exposure, using only an ordinary photographic lens, even his fertile and sanguine spirit did not, I think, anticipate the possibilities opened to astronomical research since photography has been called in to its aid.

I do not propose in this lecture to discuss any of the interesting astronomical and physical problems that have been opened up by the adoption of the "New Astronomy," as it is aptly called. I leave these subjects to be dealt with by those who have made such their special study, and are better qualified to speak of them.

I propose to-night to discuss the differing conditions which astronomical instruments are required to fulfil under the new system as compared with the old, and to point out the possibilities that appear to exist for improvement and future development, touching only on the astronomical part of the work so much as may be necessary to explain the instrumental equipment required.

It would be well, perhaps, that I should first call to your recollection some of the chief lines of work which have been opened by the introduction of the new photographic method.

The International Photographic Survey of the Heavens has been undertaken by sixteen of the principal observatories of the world, which have agreed to co-operate in producing two series of photographic pictures of the entire heavens; the first series including all stars down to the 11th magnitude, a catalogue of which is to be formed from the photographic plates; while the other series will include all stars to the 14th magnitude, and will be photographically reproduced as a chart.

The enormous advantages of photography for this work, as compared with the old system, have been brought so often and so prominently before you by Dr. Gill and other lecturers that it is unnecessary for me to dwell upon them here. As a supplement to this international work we have the independent surveys of Dr. Gill and Professor Pickering, taken photographically with short focus lenses on a small scale.

We have also the recording by the aid of photography of specially interesting objects, star clusters and groups, comets, &c., and more particularly very faint objects such as nebulæ, which require long exposure. The results in this field of work demonstrate perhaps more than any other the powerful agent that photography becomes in the hands of the astronomer. It is not necessary to go into any detail in considering this work, the results of which are tolerably familiar to you. It is only necessary to mention the names of Draper, Common, Gill, the Brothers Henry, Roberts, Gothard, Barnard Russell and Dr. Max Wolf, to indicate how much we owe to long exposure photography on these objects.

Of new minor planets, 33 were discovered by photography in 1893, and several lost planets were rediscovered.

We have also the study of the parallax of fixed stars and of nebulæ by this means, as carried out by Rutherford and Jacoby in America, Professor Pritchard at Oxford, Sir Robert Ball and Professor Rambaut at Dunsink and Dr. Wilsing at Potsdam.

Turning to spectroscopy, we find again the enormous importance of the photographic method. The development of Fraunhofer's original idea of a slitless spectroscope has given us the objective prism of to-day, and with this instrument we are able to simultaneously photograph the spectra of several hundreds of stars on a single plate, these spectra being available for study at leisure, so that they can be classified and selected for future work and more crucial examination and investigation. The results from a single plate are more accurate than could possibly be given by months of very close observation by the older method. The Draper Catalogue, which we owe to Professor E. C. Pickering, of Cambridge, Mass., gives the spectra of over 10,000 northern stars obtained in this manner. In addition to this he has in preparation a similar catalogue of the southern stars. The beautiful photographs obtained by Professor Lockyer at South Kensington, in which the detail is so fine that the spectra can be enlarged to a length of five feet, show the great value of this method for complete study of the spectra after the first rough cataloguing. With the slit spectroscope equally important work has been done in the "New Astronomy." Not only can we get the spectra of celestial objects with comparison spectra of terrestrial substances on the same plate, and thus investigate the chemical and physical constitution of these bodies, but by the adaptation of the beautiful discovery of Dr. Huggins we can detect and measure the motion in the line of sight, the photographic method giving far more accurate results than can be obtained by visual means. In the hands of Dr. Vogel, of Potsdam, this spectrographic method has been used for nearly fifty stars, and he is only waiting for a larger instrument to further extend this work. We have also in this connection the discovery of spectroscopic double stars, stars so close that we cannot hope to see them double by any possible optical means, and yet of which we know sufficient by the aid of photography to calculate their masses, distances, periods and rates of motion with considerable accuracy. In this work Professor Pickering, at Cambridge, Mass., and Dr. Vogel, at Potsdam, led the way, but Father Sidgreaves, at Stonyhurst, has extended the work by his beautiful analysis of the variations in the spectrum of β Lyræ.

Lastly may be mentioned the work of Professor Hale at Chicago, on the photography of solar prominences and faculæ with the spectro-heliograph. In this apparatus, instead of using the full light of the sun, only light of one wave-length is allowed to act on the plate. There are several methods of accomplishing this, but the latest form is that of an ordinary spectroscope with a metal plate to receive the spectrum, a narrow slit being arranged in this plate to select the particular wave-length in which it is desired to work. The photographic plate is behind a second slit and in the image formed of the spectrum. The whole spectroscope is given a motion such that the front slit passes over the image of the sun formed by a photographic object glass, while the selecting slit moves at the same rate in front of the photographic plate. A complete picture of the prominences, chromosphere and faculæ of the sun is thus obtained, and by an ingenious adaptation of clock-work Professor Hale has been able to make his apparatus automatic, and to set it to take 36 plates of the sun at any desired interval of time between each, without any superintendence whatever from the observer in charge. Photography has also been extended to the study of solar spots, lunar and planetary detail, and many other departments of astronomical research too numerous to mention, but not having any special interest for us at present in their bearing upon the instrumental arrangements.

Every one of these branches of work has already been not only suggested but put into actual practice with more or less success, but, as usual in the inception of such work, there are many lessons to be learned from the first few years' experience. The most evident fact, and one easily learned from any one of the various branches mentioned, is that the utmost perfection is necessary in the apparatus which enables the telescope to follow the object to be photographed. Before enumerating the various points necessary to be attended to to ensure this accuracy, perhaps it would be well to explain why this increased amount of accuracy is necessary when using the photographic method of observation. In the older methods it sufficed if the star remained on the wires of the micrometer during the actual observation, which rarely lasted many seconds, and even if the star did move off the wire the observer could see that it did so, and would move up his wire again to the star, repeat the observation, and would not record it unless he was satisfied that all was right at the moment of bi-section.

In the photographic method, however, the record of the observation is not that of any one moment, it is the aggregate of all the impressions made every second and every part of a second during the exposure. The photographic plate, unlike the eye, takes note of, and records every position of the star image, and not the one selected position as the eye does; hence you can easily see the great necessity of having the utmost possible perfection in the clock driving arrangements.

This condition of perfection is popularly supposed to be satisfied by having a perfect clock, but there is hardly a portion of the instrument that can be mentioned, the perfection of which does not contribute in some way or other to the accuracy of the motion.

1. The instrument must have a stability far beyond what is necessary for ordinary work, otherwise the very handling of the slow motions will sensibly affect the positions of the images and injure the results.

2. It is evident that the axis on which the instrument revolves must be of extreme accuracy, otherwise the instrument will not move truly.

3. The anti-friction arrangements must be of the most efficient nature in order to give the clock a fair chance of doing its work.

4. The slow motions must be extremely perfect, as otherwise it will be impossible to bring the guiding star on the cross wires of the "guider" with that accuracy necessary for the best results.

5. The arrangements, even for the adjustment of the instrument, so far at least as the placing of the polar axis parallel to the pole of the earth, must be such as to enable the observer to make this adjustment with an accuracy not at all necessary for visual observations.

Professor Rambaut, Royal Astronomer of Ireland, has recently shown that whereas for ordinary visual work it suffices if the polar axis of a telescope be adjusted in altitude and azimuth with an accuracy of 1 minute of arc, errors of a few seconds of arc only are allowable with a photographic telescope, and that this great degree of accuracy is best obtained from measurements of the photographic plate itself.

An instrument, therefore, that is required to give the best results used photographically, should be made with a view to such work in all its details from the very beginning, for an instrument that may be excellent for ordinary observations will most probably break down under the stringent conditions necessary for the more modern work.

In the usual form of mounting it is necessary to reverse the instrument when observations have been made on a star as it passes from the east to the west of the meridian. This is a great disadvantage in photographic work, and in designing the new 26-inch photographic telescope, which Sir Henry Thompson has undertaken to present to Greenwich, I have arranged to allow complete circumpolar motion so that this reversing on the meridian will not be necessary, and the telescope will follow any star through the whole of its path in the heavens so long as it is above our horizon.

But no matter how perfect the instrument may be in all the details above spoken of, it is not possible to attain the necessary perfection of motion without a good clock, and I thought it would be interesting for you to see the working of such a clock, and have here one which is identical with those used in the standard instruments of the International Photographic Survey. This clock is the combination of a good frictional governor, supplemented by a system of control from an independent pendulum. Perhaps you will allow me to explain why this control is necessary. A clock such as this will go well and smoothly and keep good time from second to second, but no uniform motion clock that I have ever met with can be depended on for long periods. This one, I find, can be depended on to about 1 second in 600, but as it is necessary, or at least desirable, to be able to depend on the clock for longer periods than this, while no error of more than $\frac{1}{20}$th part of a second can be permitted, it is evidently necessary to supplement this by control from an independent pendulum which can be relied upon to the required amount of accuracy.

There is another very important reason why an independent control is necessary. When an error occurs in the clock driving, owing perhaps to some morsel of adventitious matter in the bearings of the polar axis, or some little extra stiffness due to want of perfect balance, &c., the tendency of all these governors is to bring the *rate* of the clock which has been disturbed back again to the normal. This answers perfectly well for visual work because, if such an accidental error does take place occasionally, it merely means that the star slightly shifts in the field and the wires can be again brought up to the star and a satisfactory measurement taken, provided that the image does not again shift during the few seconds required for taking the observation.

But in the case of the photographic telescope such an error would be fatal, because the star has already impressed the photo plate at one certain point. When the error occurs the image shifts, and even if the rate of the clock continues perfectly right for the whole of the remainder of the exposure, the result will be of course a double or distorted image.

In photographic work we require some arrangement by which any error which is accidentally introduced will be effectually and as quickly as possible wiped out, the star image brought back again to its original position on the plate and then the clock to resume its normal rate, and this is a condition which no uncontrolled clock can fulfil. The only solution which has yet been suggested to fulfil these conditions is to have some means by which the clock of the equatorial (which, as I said, goes well and smoothly for short intervals) is checked and controlled every second from an independent pendulum.

The lecturer here exhibited in action an equatorial clock controlled from an independent pendulum as above described, with an arrangement of bells added by which the audience were enabled to judge of the perfect synchronism of the controlled clock and the controlling pendulum. The lecturer purposely introduced errors into the clock train to illustrate the power of the controlling apparatus to erase these errors.

Suppose now we have our clock as perfect as is possible, it is further necessary to see that that perfectly uniform motion is transmitted to the instrument; in other words, that any gearing between the controlled clock and the polar axis be as far as possible without error. This gearing consists mainly of the endless screw, called the right ascension screw, and the toothed sector. The precautions taken for the ensuring of this perfection have been elsewhere described, and are of too technical a character to deal with here, but one observation only I would desire to make.

In a recent paper by Professor Pickering, commenting on instruments he saw during a recent visit over here, he is kind enough to make complimentary allusions to some of these arrangements, but he takes exception to the use by us in this country of long radius sectors for driving the polar axis instead of entire circles. Perhaps it may be well, therefore, if I take this opportunity of saying why we prefer the sector.

Bear in mind that the greatest possible perfection of clock driving is what we are aiming at, and you will easily see our reason for adopting the sector. When a sector or portion of a circle only is used it is possible to get a radius much greater than in the case of an entire circle.

No mechanism ever made is absolutely free from error. Call the residual error of this screw anything you like, one 10-thousandth or one 20-thousandth of an inch; whatsoever that error be its effect on the accuracy of the driving of the telescope will be exactly in the inverse ratio of the radius at which it acts; therefore any given error will only have one-third the effect on the driving of the telescope if working (as it may in a sector) at three times the radius. One 10-thousandth of an inch at say 10-inch radius will produce an angular error of about 2 seconds of arc; at 30 inches radius it will only produce two-thirds of a second error. This may seem a small advantage, but the nearer we approach to perfection the more difficult it is to obtain any given increment.

It does not take much coal to increase the speed of a locomotive from 10 to 11 miles an hour, but it is very different if we want to increase it the same 10 per cent. from say 60 to 66.

Another fact that has been brought to light by the experiments of the last few years is that atmospheric disturbance, the *bête noire* of the astronomer, has not so much effect on most of these photographic results as in the case of visual observations.

In the number of the 'Observatory' published in December 1889, Dr. Gill makes the following remarks in a note accompanying a specimen photograph which he sent over. "The picture is sent in corroboration of a fact I have suspected for some time, viz. that for stellar photographs, after a certain period of exposure, it is quite immaterial whether the atmospheric definition is good or bad, the photographic images of stars will be equally sharp in either case. That good measurable pictures can be taken on nights when refined eye observations of any value are impossible is a very remarkable fact, and one that *a priori* would probably be deemed unlikely. The explanation appears to be that the discs on the developed film which represents stars are very much larger than the minute circle formed by the converging cone of rays from the object glass at its intersection with the plane of the film. These discs are produced by so-called photographic irradiation; in other words, by chemical action set up in the film, having origin in the central point of light, and extending gradually and symmetrically over a wider radius from that centre. This being so, whenever the radius of the disc becomes greater than the radius of extreme oscillation of the optical image from a mean point, the resultant photographic action produced by the rapidly moving point of light becomes identical with the effect produced by a similar steady point of light occupying the same mean position."

Again, in the case of some of the spectroscopic methods of observation, more particularly when a slit is used, this peculiarity of the photographic method is perhaps still more apparent.

On this subject Dr. Gill, in a lecture delivered in this theatre just three years ago, says, "On account of irregularities in atmospheric refraction, the image of a star in the telescope is rarely tranquil, sometimes it shines brightly in the centre of the slit, sometimes barely in the slit at all, and the eye becomes puzzled and confused. But the photographic eye is not in the least disturbed; when the star image is on the slit the plate goes on recording what it sees, and when the star is not on the slit the plate does nothing, and it is of no consequence whatever how rapidly these alternate appearances and disappearances recur. The only difference is that when the star is steady and the star's image therefore always on the slit, the exposure takes less time than when the star is unsteady. That is one reason why the Potsdam results, in the determination of stellar motion in the line of sight, are so accurate. And there are many other reasons besides, into which I cannot now enter. What, however, it is important to note is this, that we have here a method which is to a great extent independent of the atmospheric disturbances which in all other departments of astronomical observation have imposed a limit to their precision."

Those who are familiar with the use of large telescopes, know only too well that the larger the aperture the fewer are the opportunities on which it can be used with advantage, and the question has often been discussed as to whether the useful limit of aperture has not already been reached, except in cases when it may be possible to transport the instrument to Arequipa or some such favoured locality. No doubt large instruments so placed ought to be, and are, capable of doing much more and better work than if placed in a less favoured spot, say in the neighbourhood of a town, but experience has shown that other influences often arise which militate against the possibility of taking full advantage out of the improved locality. The conditions of life in some such isolated stations are not the pleasantest, and though human nature may put up with inconveniences and unpleasantness as a temporary arrangement, for the sake of science, yet, as a permanency, this state of things is not found compatible with the production of the best work, and in some cases it has been found necessary to send relays of workers to these isolated stations, a plan, no doubt, which meets to some extent this difficulty, but is evidently open to other objections.

If however, as it appears, the new photographic system is to a great extent independent of atmospheric disturbance, it ought to be possible not only to use, and use with efficiency, large instruments in situations within measurable distance of the haunts of civilisation (a great gain in itself), but it will also be possible to use with advantage, even in such accessible positions, instruments of far greater power than have ever yet been built, and of whose practical value there have been well founded doubts so long as the old system of eye observations was the only one available.

When this fact forces itself upon the attention of the scientific world, as it must do before long, and the necessity of adding to the power of our telescopes becomes apparent, there is little doubt but that the means will be found to satisfy the necessity; but as the magnitude of these instruments becomes greater, the importance of studying beforehand the necessary conditions to fulfil and the mistakes to be avoided becomes all the greater, and therefore I have thought it may not be amiss to bring under your notice a few suggestions as to the possibility of obtaining increased optical power in our telescopes.

Before we discuss the conditions to be fulfilled in the case of the mounting of more powerful telescopes, perhaps it would be well to get a clear idea of what is meant by the power of a telescope as distinct from *magnifying* power. You are aware that most of the work done with our very large refractors is done with magnifying powers which are equally useable with instruments of half the aperture or less, but it must not be assumed that the power of the instrument (used in its broad sense, i.e. its capability of distinctly viewing minute objects, or the details of such) is then only the same as that of the smaller instrument used with the same magnifying power. On the contrary, Jupiter or Saturn viewed with a power of 600, with 28 inches aperture, is a very different object to what it is when viewed with the same magnifying power and an aperture of say 8 or 10 inches. This is not due to extra brilliancy from the larger amount of light collected by the bigger object-glass, for even in the case of observing the moon (say), when it is necessary to use tinted screens to moderate the brilliancy, this superiority of the larger aperture is just as evident, but it may be explained in this way:—

In a lecture by Dr. Common, delivered in this theatre in May 1890, he gave a very neat explanation of the fact that a certain amount of magnification of image is required in order to see a certain amount of detail in that image. He showed that the sensation by which the brain is excited is carried from the retina by an enormous number of fine nerves, which are excited by small bodies called technically "rods and cones," and that each of these produces one distinct sensation as relating to the particular part of the image which falls on that particular part of the retina; the image, therefore, as presented to the brain is a kind of mosaic, and it is evident that the larger the image that falls on the retina the finer will the mosaic be in proportion to the details of the image, and therefore the better will the details be appreciated.

A similar explanation may be given of the different character of the image given by large aperture telescopes and small. The image of a star, as given by a telescope's objective, is not exactly a point; it is, owing to certain physical reasons which it would be impossible to enter into in this lecture, in the form of a small disc of light which, if the object is of a sufficient brilliancy, is surrounded by diffraction rings. The diameter of this spurious disc depends amongst other conditions on the diameter of the object-glass; the larger the diameter of the object-glass the smaller the diameter of the disc; in other words, the discs, as seen in large object-glasses, are smaller than those seen in smaller object-glasses; or putting it in another way, if a certain size of object-glass be found to give a spurious disc of a certain size, *reducing* the aperture of the object-glass will *increase* the size of the spurious disc.

Every object may be considered to be made up of an infinity of points, of every one of which the object-glass gives an image in the form of a little disc. It is evident that the image that is made up of the larger spurious discs will not be as fine or as delicate, or show as much detail as that made up with the smaller discs. The image of such an object as Jupiter or Saturn as seen in the small telescope, compared with that as seen in the large telescope, will be as a drawing made in the first case with a coarse crayon or stump, to that made in the second case with a finely pointed lead pencil; or we may compare the first to a very coarse mezzotint engraving, while the second may be compared to the very finest work that can possibly be turned out.

This is the reason that the larger aperture telescope, even when used with powers corresponding only to those which can be effectively used with a smaller instrument, show objects with a clearness and distinctness that it is impossible to obtain in the smaller instrument, no matter how perfect the workmanship may be.

And now we come to the question of the probabilities of our

ω Centauri.

η Argus.
45 minutes' exposure.

Reproductions of photographs taken by Dr. D. Gill with the 13·2 inch Capetown Astrographic Telescope.

η Argus.
3 hrs. 12 min. exposure.

η Argus.
12 hrs. 12 min. exposure.

Reproductions of photographs taken by Dr. D. Gill with the 13·2 inch Astrographic Telescope.

being able to increase to any great extent the powers of our telescopes, and with this also naturally arises a question which, judging by the number of queries that reach me about it, seems highly interesting to the general astronomical world. "Will the great telescopes of the future be Refractors or Reflectors?"

Mr. Alvan Clark, whose large refractors in the United States testify to his great skill, declares emphatically for refractors, perhaps naturally so, but his reasons do not appear altogether convincing, and there are others well qualified to judge who give an opposite opinion. It is a question which only the future can decide. Of course, if we all make up our minds that the coming telescope is to be a refractor it will be so, for all our energies will be devoted to its development; but the same might be said of the reflector, which, I believe, is capable of being greatly improved if attention were directed to it.

There is one reason that I believe has been overlooked, which explains to some extent why the reflector has not been developed of late years as has the refractor. This matter is not of a scientific, but purely of an economic character, and I should, perhaps, ask pardon for introducing it into a scientific lecture; still, it is necessary for explanatory purposes.

Reflectors are, unfortunately for themselves, much less costly than refractors, and I believe that this has much to do with their comparatively neglected condition at present; this may seem curious, but it is easily explained. An object-glass of 18 inches is worth, say 1000*l.*; a mirror of 18 inches is worth, say 100*l.* No one who wanted to have good mounting would object to pay 1000*l.* to mount the 1000*l.* object-glass, but there are many who would object to pay the same 1000*l.* to mount the 100*l.* mirror; and yet why should it not be equally well mounted? and if not so, how can it be expected to give as good results as the refractor? As a matter of fact there are greater difficulties in mounting a reflector than a refractor, and these greater difficulties mean increased cost for an equally good mounting. I believe this simple economic question has much to answer for in bringing the reflector into disrepute with many. It has often been remarked that the reflectors that have been best worked have been constructed and worked by amateurs, the reason, to a great extent, being that this economic point does not then enter so largely into the question.

There are great difficulties in the mounting of reflectors, more especially when required for use as a photographic telescope, and these difficulties have never yet been satisfactorily solved, but I believe there is nothing unsolvable in them, and that it only wants attention to be drawn to them to ensure a solution.

Only within the last month Dr. Johnstone Stoney has devised a most ingenious arrangement for supporting the great mirrors of reflecting telescopes on an air support, graduating the pressure according to the angle of inclination of the telescope by an automatic contrivance. Possibly this apparatus in its present form may be capable of improvement, but it is at least a step, and a very important step, in the direction of solving one of the most troublesome and difficult problems to be met with in the attempt to obtain a really satisfactory mounting for reflecting telescopes.

That the reflecting telescope is capable of doing excellent work in the hands of those who take sufficient care and trouble with the adjustment and in the working, is sufficiently evidenced in the results obtained by Draper, De la Rue, Common and Roberts, but the fact is that to obtain good results with the reflector in its present imperfect state of development, more labour and patience is required than most observers care to bestow on the work, and there is much to be said in excuse for this, for if an astronomer's time be taken up with the necessary attention to the details of his instrument he will not be able to pay that undivided attention otherwise possible, and at all times desirable, for his more legitimate work in the obtaining of results with that instrument. The fact that the reflector brings all rays of light to a common focus, irrespective of their wave-lengths, while the refractor is at best but a compromise, tells strongly in favour of the ultimate success of the reflector over its rival.

True, it may be said that the experiments in glass-making that have been carried on for some years at the Jena glass manufactory may yet eventuate in producing qualities of glass which will remove this reproach from the refractor, and enable us to perfectly balance the chromatic error, and at the same time be of a sufficiently permanent character to justify its use in the case of large objectives. No one would be foolish enough to attempt to make a large objective of any material which was not known from previous experience to be capable of preserving its perfection of surface for at least 20 or 25 years.

Unfortunately there is no test of permanency except the lapse of years; even if therefore some such glass were in existence at the present time, no maker who had any desire that his name should live after him in his work, would care to use this untried material until actual experience proved its character for permanence. This, and the fact that it has not yet been found possible to produce perfect pieces of this Jena glass of one-tenth of the weight of those already produced of the more ordinary varieties of optical glass, cuts off any hope we might otherwise have of being able for the present to produce large objectives with perfect correction for the chromatic aberration, and so long as this is the case, the reflector, which treats rays of all refrangibilities alike, has in this respect the advantage.

When we consider that the largest optical discs ever yet produced, and which were rightly considered a perfect triumph of art, are only 40 inches in diameter, and that on the other hand Lord Rosse's reflector of 72 inches diameter is now half a century old, it is tolerably evident that for the present, at least, we must look to reflectors if we want to increase to any large extent the power of our telescopes.

With this view, and understanding that it is likely an attempt will be made to build an 8 or 10-foot reflector for the great exhibition to be held in Paris in 1900, it may be interesting to consider the conditions desirable to be fulfilled for such an instrument, and the most promising construction to satisfy those conditions. If a monster telescope, such as this, is to be mounted only in such a manner as will satisfy the ordinary conditions of star gazing, I fear the results will be disappointing, but let it be mounted in such a manner as to render it useable for the more delicate and refined work of the modern astronomy, and a grand and productive field of work is open to it.

But the problem of mounting an enormous instrument such as this, whose weight would probably amount to from 50 to 100 tons, so perfectly poised and so accurately driven by clockwork as never to vary from its true position by a quantity greater than the apparent motion of a star in one-twentieth of a second of time, is

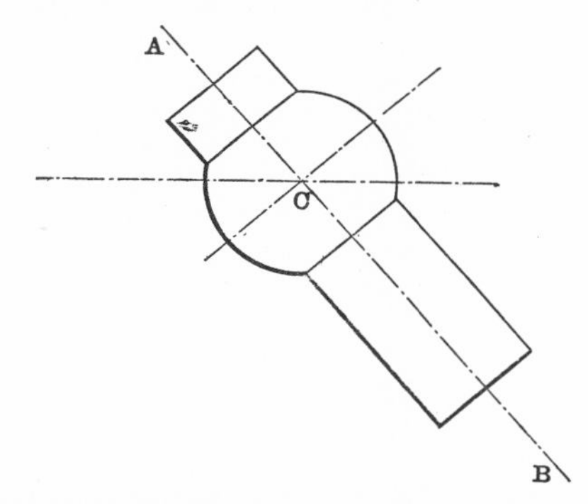

sufficiently difficult to justify almost a doubt of its possibility, and this difficulty has been appreciated by others; for Dr. Common, who, as the maker of the largest equatorially mounted reflector ever completed, must be considered as the first authority, proposed some short time since to resort to the alt-azimuth form of mounting, with which it would, of course, be impossible to satisfy the above condition.

Dr. Common himself has made a splendid advance in adopting the system of flotation of the polar axis; this principle of flotation appears to me to be capable of further development. It is perfectly possible to make a tube for a Newtonian reflecting telescope (which is necessarily closed at the lower end) of such a weight, and with its weight so distributed that it will not only float in water submerged to a certain point (preferably near the upper end), but will be in a state of equilibrium when placed at any or in every position down to a certain angle, which angle depends on the exact outside form of the tube. For instance, if *A B* (Fig. 1) be a tube closed at *B* and perfectly symmetrical round the axis *A B*, and the total weight of the tube be equal to the weight of water which is displaced when the tube is sunk to *C*, the weight of the different

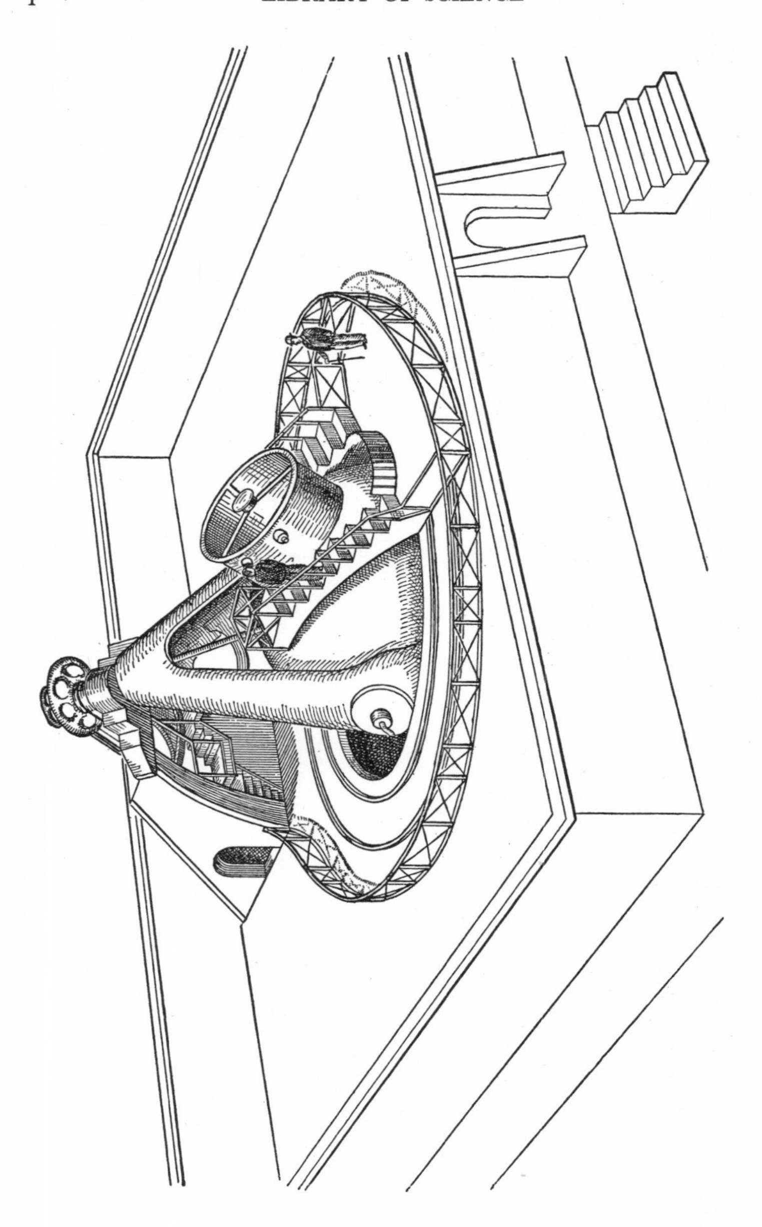

sections along the axis $A\,B$ can be so distributed that the tube will equally well remain in any other position, except it be so far turned over that the cylindrical part of the tube is lifted out of the water at one end and dipped at the other.

By making the spherical part of about the proportions of the figure, the tube can be depressed to within 25° of the horizon, and still remain in perfect equilibrium.

Now, suppose the tube to have a pair of trunnions attached at the water line, and these carried on a polar axis of, say, the English type (see Fig. 2), we have an equatorially-mounted telescope of any size, without any weight whatever on the bearings of the Dec axis, or, the tube may be lightened by an amount nearly equal to the weight of the polar axis, and there will then be practically no weight whatever on the bearings of that axis. So here we have a case of, say, an 80-ton telescope mounted and carried by an equatorial, but without throwing any weight whatever on that equatorial; and the force necessary to drive the instrument is independent of the weight of the telescope, and dependent only on the friction necessary to be overcome in carrying the tube at an exceedingly slow rate through the water.

Let us inquire into any possible disadvantages that may be urged against this form of mounting:—

1. That the temperature of the water will often be different from that of the air; and consequently that there will be a detrimental mixture, at the mouth of the tube, of air from inside the tube, which will partake of the temperature of the water, with the outside air. This I would propose to avoid by making the tube double, with a space of some 3 inches between inside and outside tubes, hermetically closed except at the lower end, where there would be apertures in the inside envelope. The space between the two tubes would be connected through the trunnions with an air pump, worked by a gas or other motor, which would continually exhaust the air from between the two tubes, and thus cause a current of the outside air to pass continually down the tube and back to the pump by the space between the two tubes. This would keep the temperature of the inside tube and the air in the tube constant with that of the outside air.

2. The limited range of the equatorial. I have stated that the instrument would be in perfect balance down to 25° from the horizon. If desired, though no longer perfectly balanced, it can be used lower by employing a chain or wire rope connected between the lower end of the tube and the upper end of the polar axis, and the amount which the instrument would be out of balance, between 25° and 20°, would be very trifling.

Again, it will not be convenient to use the instrument within some 15° of the pole. It could be planned to go somewhat closer, but when it is considered that nine-tenths of the work required to be done can be commanded by this instrument, it is clearly better to design it to do that nine-tenths well than to strain it into doing another 5° that would only be useful on very rare occasions.

3. It may be urged that the friction of the water will prevent the rapid setting of the instrument. In a telescope of this size all the motions would be effected by motors of some description, guided by the observer from a commutator-board at the eye end, and there would be no difficulty in setting the telescope quite as quickly as could be expected considering its great size.

4. It may be objected that currents will be set up in the water by the moving of the telescope, which currents will affect the steadiness. No doubt this will be the case to some extent, but these will soon subside, and the motion necessary for following the stars will be so slow that no perceptible effect of this kind will be felt from it.

As to convenience in getting at the eye end, there need be no difficulty whatever in this form. As the eye-piece is only about 15 feet from the centre of motion, the movement of the observer is never more than 3 feet per hour. By means of a platform such as that shown in Fig. 2, running on rails, and quite independent of the instrument, the eye end is readily accessible at all times. To overcome the rotation of the tube as the instrument moves in right ascension, I would pierce the tube for eye-pieces every 30° round its circumference, and mount the flat mirror and cell in a collar so as to enable it to be readily rotated through intervals of 30°. By these means the image of the celestial object to be observed could be sent through either or any of the perforations of the tube, and the observer always observe in the direction most convenient to himself.

There are various difficulties about this construction which may naturally suggest themselves, but there are none, I believe, which cannot be overcome.

This is hardly the place to discuss details, but if there be any here who are sufficiently interested in this new form of equatorial to desire further information, I would refer them to an article in the present month's issue of 'Knowledge' which deals with most of the difficulties.

Putting aside now the question of reflectors *versus* refractors, there are some directions, applicable equally to reflectors and refractors, in which it is evidently possible to improve our designs for large equatorial instruments. It is not the first time that I have urged similar developments, but the advantages of what I recommended were not so apparent then as they are now in the present advanced state of astronomical research.

It may be remembered that when I lectured here in the year 1886 I strongly urged the desirability of employing some form of motive power for carrying out the various motions required in manipulating the large equatorials, domes, &c., and exhibited a model design I had made for the Lick Observatory, which illustrated the manner in which these various motors could be controlled by the observer, as well as the then newly devised lifting floor arrangement. The latter, that is the lifting floor, was, as you are aware, adopted by the trustees of the Lick Observatory, and I suppose we may assume that it was considered a success, as it has been copied in the case of the two large observatories built since in the United States, viz. the Washington and the Chicago Observatories. The Lick Trustees, however, rejected the proposals for the employment of motors for the equatorial movements and dome. That experience gained since the construction of this telescope has confirmed the correctness of my views as respects the desirability of adopting this system is evident, as, in their latest and most perfect instrument, that of the Yerkes equatorial at Chicago, practically all the suggested improvements in that model exhibited before you in 1886 have been adopted, and I hope will contribute in no small degree to the quantity as well as the quality of the output of work we may expect from that splendid instrument; but to show that this by no means represents all that can be done in this direction, I have here a rough and unfinished model of an observatory in which the principle is carried still further.

It will be observed that in the Lick design the astronomer is relieved of all physical exertion, but still his attention is required during all the process of setting of the instrument. In our latest design we are able to relieve the astronomer of even the mental strain in this way. In some of our modern equatorials the setting of the circles can be placed (by a new arrangement) at the eye end of the telescope. Suppose this to be so arranged in the large telescope and that an arrangement be added something similar in principle to the steam steering gear of our large steamers, with which every one is now familiar. In this machine the construction is such that when a small light wheel is turned any quantity to port or starboard the motors are automatically set to work and force the helm over in the right direction, and do not stop until the position of the helm itself exactly corresponds to that of the light guiding wheel. A very simple arrangement of electrical contacts suffices to effect this in the case of the telescope, and the working of the instrument is effected thus :—

The astronomer decides what AR and what DEC he desires the telescope to be set at, he walks over to the eye end of the telescope, which, as will be seen further, *can only be* at a convenient height from the floor, and sets a pair of pointers at the eye end to the particular readings he wishes, and then presses a button to start the motors and "awaits developments."

Without any more attention from the astronomer the instrument now sets itself in exactly the position he requires, the motors continuing to revolve the telescope on its axes till that position is attained, and *then they stop.* Meanwhile as certain contacts are arranged at the upper end of the tube connected with the motor which drives the dome, and at the lower end with motors which elevate and depress the floor, the dome revolves if necessary and keeps the opening opposite the upper end of the telescope, while the floor rises and falls as may be necessary, always keeping at a convenient distance below the eye end; and so, as I said, without any attention or physical or mental strain on the part of the observer, he finds after a few minutes his telescope set correctly in AR and Decn—the dome opening is opposite the object-glass and the floor at a height most convenient for observation.

The idea may seem almost Utopian, but there are no particular difficulties in carrying it out, nor would it add sensibly to the cost of a large instrument.

If I have succeeded, even very imperfectly, in rendering a necessarily technical, and therefore somewhat dry, subject sufficiently interesting to have enabled you to follow me, you will have no difficulty in seeing that the principal ideas I desired to convey may be summed up shortly in this way:—

That, while on the one hand, the adoption of new methods for prosecuting astronomical research have created a set of conditions under which it is possible to use, and use with advantage, instruments of greater optical power than hitherto, yet, on the other hand, the mechanical arrangements for mounting these instruments must be of a much higher standard than has been necessary for the older methods; in fact, they must be mounted as instruments of precision, in the highest sense of the term, and while the mounting of instruments even larger than we have at present and suitable to the older conditions would not present any serious engineering difficulties, the problem of mounting them as instruments of precision is one of considerable magnitude—a fact which is well recognised by those who have studied the subject.

The time has gone by when with very few inches increase of aperture some sensational discovery is expected.

The astronomer of the future will not be satisfied with mere stargazing instruments. The experiments of the last few years show that there is no royal road to great astronomical discoveries, but that patient, honest and self-recording work is necessary to enable us to add (as our ambition is) course after course to the great edifice of astronomical truth.

But the ever-advancing work of the astronomer demands an ever-increasing perfection of his instruments, and the records of the past show that few of these improvements are to be credited either entirely to the astronomer, be he ever so practical, or to the instrument-maker, however ingenious, but to both working harmoniously together, the astronomer finding out the weak points of existing instruments, and the instrument-maker continually devising new contrivances to meet the difficulties of the astronomer.

I myself would like to take this opportunity of saying that whatever measure of success I may have had in my work is due in no small degree to the helpful kindness I have invariably received from those astronomers with whom I have been in communication.

As it has been in the past, so, let us hope, it will be in the future, for in this harmonious working together of those who design and those who work the instruments lies our strongest hope of the future development of the astronomical telescope.

[H. G.]